Ultrastructural Atlas of the Inner Ear

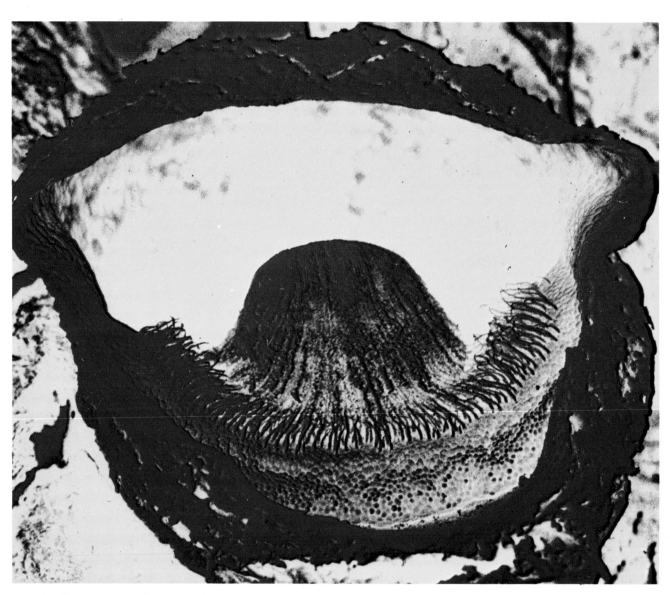

Ampulla of semicircular canal. Scanning electron micrograph, proces-sed photographically in colour (Reproduced by courtesy of Dr Ivan Hunter-Duvar)

Ultrastructural Atlas of the Inner Ear

Edited by

Imrich Friedmann
MD, DSc, FRCPath, FRCS

Emeritus Professor of Pathology, University of London
Formerly Director, Department of Pathology, The
Institute of Laryngology and Otology, University of
London
Honorary Consultant Pathologist, Royal National
Throat, Nose and Ear Hospital, and Northwick Park
Hospital and Clinical Research Centre, Harrow
Research Fellow, Imperial Cancer Research Fund,
London
Visiting Professor, University of California, San
Francisco, USA
Consultant in Electron Microscopy, House Ear
Institute, University of Southern California, Los
Angeles, USA

John Ballantyne
CBE, FRCS, Hon FRCS(I)

Consulting Ear, Nose and Throat Surgeon, Royal Free
Hospital, London
Consultant Ear, Nose and Throat Surgeon, King
Edward VII Hospital for Officers, London

Butterworths   London   Boston   Durban   Singapore   Sydney   Toronto   Wellington

First published 1984

© **Butterworth & Co. (Publishers) Ltd, 1984**

British Library Cataloguing in Publication Data

Ultrastructural atlas of the inner ear.
  1. Labyrinth (Ear)   2. Ultrastructure (Biology)
  I. Friedmann, Imrich   II. Ballantyne, John
  612'.858   QP461

  ISBN 0-407-00221-9

Library of Congress Cataloging in Publication Data
Main entry under title:

Ultrastructural atlas of the inner ear.

  Bibliography: p.
  Includes index.
  1. Labyrinth (Ear)–Anatomy–Atlases.
  2. Ultrastructure (Biology)
  I. Friedmann, Imrich   II. Ballantyne, John
  QM507.U38 1983   611'.85   83-10043
  ISBN 0-407-00221-9

Photoset by Butterworths Litho Preparation Department
Printed and Bound in Great Britain by Butler & Tanner Ltd, London & Frome

Dedication
To Hans Engström, MD

Pioneer and friend

Contributors

Matti Anniko *MD, PhD*
King Gustav V Research Institute
Karolinska Institute, Stockholm, Sweden

Dan Bagger-Sjöbäck *MD*
Department of Otorhinolaryngology,
Karolinska Hospital, Sweden
Institute of Laryngology and Otology, London

John Ballantyne *CBE, FRCS, Hon FRCS(I)*
Royal Free Hospital and King Edward VII Hospital, London

Björn Bergström, *MD*
Department of Otolaryngology,
Central Hospital, Karlstad, Sweden

Imrich Friedmann, *MD, DSc, FRCPath, FRCS*
Emeritus Professor of Pathology, University of London
Honorary Consulting Pathologist, Royal National Throat, Nose and Ear Hospital,
and Northwick Park Hospital and Clinical Research Centre, London
Consultant Electron Microscopist, House Ear Institute, Los Angeles, USA

Frank R. Galey, *PhD*
The House Ear Institute, Los Angeles, USA

Raul Hinojosa, *MD*
University of Chicago, Section of Otolaryngology,
Chicago, Illinois

Jon M. Holy, *MS*
University of Wisconsin, Madison, Wisconsin, USA

Ivan M. Hunter-Duvar, *PhD*
Department of Otolaryngology, The Hospital for Sick Children,
Toronto, Ontario, Canada

Robert S. Kimura, *PhD*
Department of Otolaryngology, Harvard Medical School,
Boston, USA
and Massachusetts Eye and Ear Infirmary, Boston, USA

David J. Lim, *MD*
Otological Research Laboratories, Department of Otolaryngology, Ohio State University College of Medicine, Columbus, Ohio, USA

Per-G. Lundquist, *MD*
Departments of Otorhinolaryngology, Linköping Regional and University Hospitals, Sweden

Helge Rask-Andersen,
Department of Otolaryngology, Uppsala University Hospital, Sweden

M. R. Romand
Laboratorie de Neurophysiologie, Université de Montpellier II,
Montpellier, France

R. Romand, *PhD*
Laboratoire de Neurophysiologie, Université de Montpellier II,
Montpellier, France

Jerzy E.Rose, *MD*
University of Wisconsin, Madison, Wisconsin, USA

Grayson L. Scott, *BSc*
University of Wisconsin, Madison, Wisconsin, USA

Hanna M. Sobkowicz, *MD, PhD*
University of Wisconsin, Madison, Wisconsin, USA

H. Spoendlin, *MD*
Universitätsklinik für Hals-, Nasen- und Ohrenkrankheiten, Innsbruck,
Austria

Jukka Ylikoski, *MD*
Department of Otolaryngology, University of Helsinki, Finland

Contents

Introduction

1
Anatomy of the Ear

John Ballantyne

The aim of this atlas is to present an up-to-date account, mainly in the form of illustrations with accompanying legends, of the cytoarchitecture of the inner ear; and the object of this introductory chapter is to show where its several parts fit into the whole, and to indicate the impressive degrees of magnification achieved by the various techniques of ultrastructural analysis.

A century has passed since Gustav Retzius, the Swedish scientist, published his classic studies on the inner ear, first in fish and amphibians (1881), and later in reptiles, birds and mammals (1884); but it was not until 1953 that three of his compatriots – Engström, Sjöstrand and Wersäll – published the first electron microscopic study of the inner ear.

For the next two decades, transmission electron microscopy was the most widely used method of ultrastructural examination and, when Salvatore Iurato, of Bari, edited an excellent collection of essays on the 'Submicroscopic structure of the inner ear', in 1967, most of his 200-odd illustrations were transmission electron micrographs (TEM).

Shortly afterwards, towards the end of the 1960s, a new type of microscope, the scanning electron microscope (SEM), became available for biomedical research. In contrast to the transmission electron microscope, which transmits an electron beam *through* the specimen, the image received by the SEM is created by secondary electrons emanating from the surface of the specimen.

The first SEM studies of the mammalian inner ear were made by Lim (1969) and by Lim and Lane (1969), and in the following year an excellent study of the organ of Corti was published by Bredberg and his colleagues (1970).

The resolving power of the transmission electron microscope (10 Ångstrom; 1 nm) is approximately 100 times that of the conventional light microscope (1000 Å; 100 nm); the resolving power of the SEM is 100 Å (10 nm). The depth of field achieved with the SEM is approximately 500 times greater than that of the light microscope, and its magnification ranges from ×20 to ×20 000, with higher degrees of magnification obtainable by photographic enlargement.

Anatomy of the Inner Ear

The inner ear, or labyrinth – literally, a 'structure of winding passages' (*Figure 1.1*) – consists of a continuous series of membranous sacs and ducts (the otic labyrinth) within a protective 'shell' of bone (the periotic labyrinth). Between the two is the perilymph fluid, similar in composition to cerebrospinal and other extracellular fluids; within the membranous labyrinth are the endolymph and sensory epithelia.

The otic labyrinth is divided into: a *pars superior*, or vestibular labyrinth; a *pars inferior*, or cochlea; and an endolymphatic system. It consists of: three membranous semicircular ducts, within their corresponding bony canals; the utricle and saccule, within the bony vestibule; the ductus cochlearis (scala media) in the bony cochlea; and the endolymphatic sac and duct, in the bony aqueduct of the vestibule.

It is with the sensory epithelia and primary neurons of the cochlear and vestibular receptor organs (*Figures 1.2 and 1.3*) that this atlas is principally concerned; and these structures are just visible to the naked eye, and to the low magnifications (×6–×16) of the operating microscope.

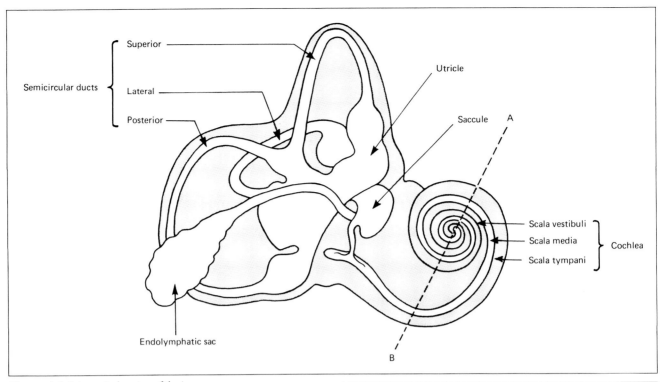

Figure 1.1 *Schematic drawing of the inner ear*

Set upon the basilar membrane, throughout the length of the membranous cochlea, is the spiral organ (of Corti). In conventional light microscopy (*Figure 1.4*), the scala media is triangular in shape, the base being formed by the basilar membrane, the outer wall by the stria vascularis, and the third side of the triangle by the vestibular (Reissner's) membrane.

The expanded (ampullary) end of each semicircular duct contains a *crista* (*Figure 1.5*); and in the utricle and saccule there are also patches of specialized epithelia, the *maculae* (*Figure 1.5*), the utricular macula being in the horizontal plane, that of the saccule in a vertical plane.

The epithelia of these receptor organs share three common features: sensory cells, with hairs on their free surfaces; supporting cells; and a gelatinous substance over the hairs. Surmounting the organ of Corti is the tectorial membrane (*Figures 1.6* and *1.7*); in the ampullary cristae are the dome-shaped cupulae (*see Figure 1.5*); and in the utricular and saccular maculae are the flattened otoconial membranes, in which are embedded numerous crystals, or statoconia (*see Figure 1.5*).

The sensory cells of the organ of Corti (*Figures 1.6 and 1.7*) are disposed in a single row of inner hair cells and three or four rows of outer hair cells, astride (respectively) the inner and outer pillars (or rods) of Corti which enclose the tunnel of Corti.

The inner hair cells (*Figure 1.8, left*) are bulbous in shape, and their hairs are arranged in two rows in the form of a double V, with their apices directed towards the outer hair cells. The outer hair cells (*Figure 1.8, right*) are columnar in shape, their hairs being arranged in three rows in the form of a wide triple W (*Figure 1.9*),

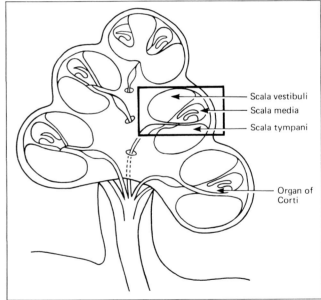

Figure 1.2 *Section through the cochlea at the point indicated by the line A–B in Figure 1.1. The organ of Corti is seen to ascend in a spiral from base to apex*

with their apices pointing in the same direction as those of V's in the inner hair cells. All the hairs are stereocilia.

The terminal fibres of the cochlear nerve end in contact with the hair cells, and there are two types of nerve fibres: type I fibres, which are sparsely granulated and probably afferent; and type II fibres, which are richly granulated and probably efferent. Most of the afferent neurons end on the inner hair cells; only a small minority are associated with the outer hair cells (Spoendlin, 1969).

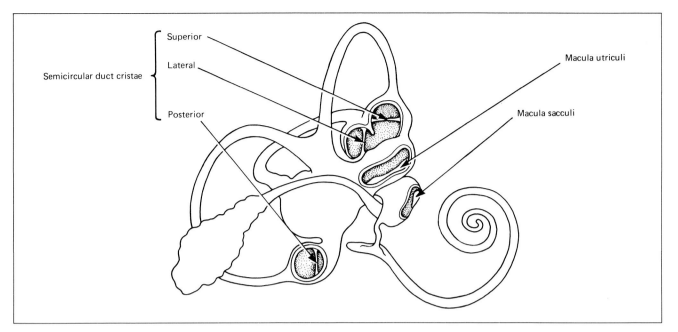

Figure 1.3 *The vestibular sensory epithelia: cristae in the semicircular ducts; maculae in the utricle and saccule*

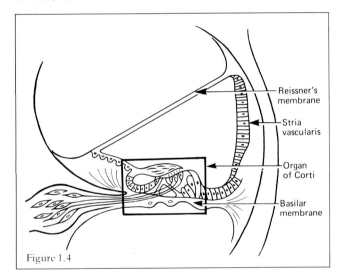

Figure 1.4

After piercing the basilar membrane through the habenula perforata (or foramina nervosa), most of these fibres pass through the osseous spiral lamina to the bipolar cells of the spiral ganglion: proximal to the habenula, the fibres are myelinated; distal to it, they are unmyelinated.

The vestibular sensory cells (*Figure 1.10*) are of two types; the type 1 cell, which is rounded and flask shaped, thus resembling the inner hair cell of the cochlea; and the type 2 cell, cylindrical in shape and thus

Figure 1.4 *Higher magnification of part of the cochlea (approximately one turn) outlined by the rectangle in Figure 1.2, as seen in the light microscope*
Figure 1.5 *Diagrammatic representation of the vestibular epithelia. The crista of each semicircular duct has a dome-shaped cupula; the utricular and saccular maculae have a flattened otoconial membrane*
Figure 1.6 *The organ of Corti with surrounding structures, in the area outlined by the rectangle in Figure 1.4, as seen by light microscopy*

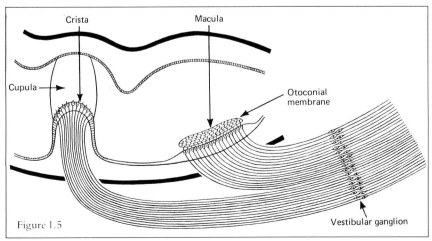

Figure 1.5

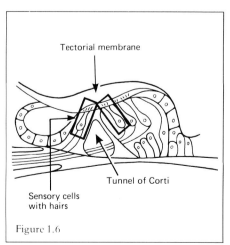

Figure 1.6

5

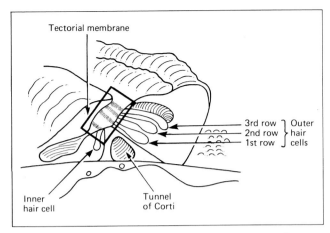

Figure 1.7 *Drawing of freeze-fractured specimen of organ of Corti (same area as in Figure 1.6), as seen by scanning electron microscopy (Original magnification approximately ×600 reduced to 66%; after G. Bredberg.)*

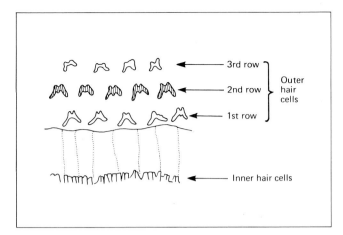

Figure 1.9 *Hairs of organ of Corti, from the rectangular area in Figure 1.7. Between the inner hair cells and the first row of outer hair cells are the tops of the supporting pillars. As seen by scanning electron microscopy*

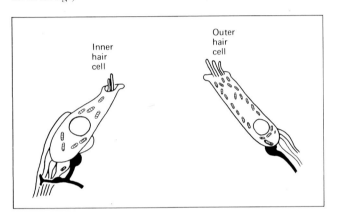

Figure 1.8 *Sensory cells of organ of Corti, as shown in the rectangle in Figure 1.6. Left: inner hair cell; right: outer hair cell. As seen by transmission electron microscopy*

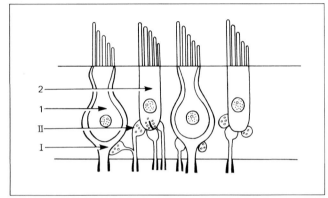

Figure 1.10 *Sensory cells of vestibular labyrinth, as seen by transmission electron microscopy*

resembling the outer hair cells of the organ of Corti. The type 1 cells are embraced by nerve chalices; the type 2 cells have none.

Sensory hairs protrude from the surface of each sensory cell. On each cell there is one kinocilium and numerous stereocilia, the kinocilium being longer than the stereocilia; the length of the stereocilia diminishes gradually as their distance from the kinocilium increases. Stimulation of the hairs in a direction away from the stereocilia towards the kinocilium produces 'depolarization' of the cell, with increase in the frequency of nerve impulses; stimulation in the opposite direction produces 'hyperpolarization', with reduction in the frequency of nerve impulses (Flock and Wersäll, 1963).

Lindeman (1969) has shown that, in the ampullary cristae, all the cells are morphologically polarized in one direction; in the horizontal semicircular duct, *towards* the utricle; in the two (superior and lateral) vertical ducts, away from it. However, in the utricle and saccule, the maculae are divided by an arbitrary line, the striola: in the utricle, the sensory cells are polarized *towards* the striola; in the saccule, away from it.

The terminal fibres of the vestibular nerve ramify around the sensory cells, and they too are of two types: type I fibres, probably afferent and including the nerve chalices around the type 1 cells; and type II fibres, richly granular and probably efferent.

Proximal to the neuroepithelium, the fibres become myelinated and they pass to the large bipolar cells of the vestibular (Scarpa's) ganglion.

The dual techniques of transmission and scanning electron microscopy have been applied to a variety of pathological conditions in the inner ear, notably to damage by drugs and noise, but essentially this atlas is concerned with the normal inner ear.

We have been exceptionally fortunate in enlisting the help of so many outstanding contributors, and to all of them we extend our sincere thanks.

It is our hope that the atlas will be of interest not only to biologists, otologists, physiologists, and students of anatomy, both normal and morbid, but also to all those who share with us a sense of wonder at the singular beauty of this most exquisite little structure.

6

References

Bredberg, G., Lindeman, H. H., Ades, H. and West, R. (1970) Scanning electron microscopy of the organ of Corti. *Science, 170,* 861–863

Engström, H., Sjöstrand, F. S. and Wersäll, J. (1953) The fine structure of the tone receptors of the guinea pig cochlea as revealed by the electron microscope. In *Proceedings of the Fifth International Congress of Oto-Rhino-Laryngology,* Eds. P. G. Gerlings and W. H. Struben, pp. 563–568. Assen: Van Gorcum.

Flock, Å. and Wersäll, J. (1963) Morphological polarization and orientation of the hair cells in the labyrinth and the lateral line organ. *Journal of Ultrastructural Research,* 8, 193–194.

Iurato, S. (1967) *Submicroscopic Structure of the Inner Ear.* Oxford: Pergamon Press.

Lim, D. J. (1969) Three dimensional observation of the inner ear with the scanning electron microscope. *Acta Otolaryngologica (Stockholm),* Supplement 255, 1–38.

Lim, D. J. and Lane, W. C. (1969) Vestibular sensory epithelia. *Archives of Otolaryngology,* 90, 283–292.

Lindeman, H. H. (1969) Regional differences in structure of the vestibular sensory regions. *Journal of Laryngology and Otology,* 83, 1–17.

Retzius, G. (1881) *Das Gehörorgan der Wirbeltiere: I. Das Gehörorgan der Fische und Amphibien.* Stockholm: Samson and Wallin.

Retzius, G. (1884) *Das Gehörorgan der Wirbeltiere: II. Das Gehörorgan der Reptilien, der Vögel und der Säugetiere.* Stockholm: Samson and Wallin.

Spoendlin, H. (1969) Innervation patterns in the organ of Corti of the cat. *Acta Otolaryngologica (Stockholm),* 67, 239–254.

2
Techniques in Ultrastructural Anatomy

Dan Bagger-Sjöbäck   Matti Anniko   Per-G. Lundquist

The techniques used in ultrastructural anatomy have been vastly improved during the last decade. Several methods and tools enable the investigator to choose various routes in the processing of his or her material. To choose the best possible method or methods, the first question one should ask oneself is what one wishes to demonstrate. Owing to the large number of animal models and processing techniques it may sometimes be difficult to select the right procedure from the beginning (Anniko and Lundquist, 1980a, b; Anniko, 1981). The purpose of this chapter is to facilitate the choice of experimental model and to describe the standard techniques available.

Various animal models have been used and studied in morphological investigations of the inner ear. As long ago as 1881 and 1884 Retzius described in detail the morphology of the inner ear in various species. Different animal models display different features which influence the choice of specimen. Of late, the ultimate choice for an inner-ear model – the human ear – has attracted considerable interest. Amphibians and reptiles are cold-blooded animals with a variable metabolism, generally slower than that of mammals. Their tissues, therefore, are less susceptible to post-mortem autolysis and degeneration. Their vestibular apparatus resembles that of more developed species, but their auditory organs are more primitive than the mammalian cochlea. It has been shown, however, that the amphibian and reptilian inner-ear tissues react to toxic substances, such as antibiotics, in a similar fashion to those of mammals (Bagger-Sjöbäck and Wersäll, 1976).

The mammalian inner ear shows a fairly uniform morphology throughout the evolutionary chain. This favours the assumption that many conditions are fairly constant and that data regarding these conditions can be transferred from one species to the other. The inner-ear tissues, however, are very fragile and are readily altered by toxic influence or mechanical disruption during handling. The most widely used model is the guinea-pig inner ear with its readily accessible cochlea and fairly thin bony vestibular capsule (*Figures 2.1–2.4*). The cat inner ear is favoured by some investigators owing to its greater resemblance to the human inner ear. Developmental studies have been performed mainly on the mouse inner ear, but fowl, guinea-pig and other inner-ear models have also been used (van de Water *et al.*, 1980). Investigations of the human inner ear present special problems inherent in difficulties with fixation, with long post-mortem delays, and with the difficult and time-consuming microdissection of the inner ear.

Several standard ultrastructural techniques have been used in the investigation of the inner ear; a combination of two or more techniques often offers a better comprehension of its complicated microanatomy. Three standard ultrastructural techniques will be discussed: transmission electron microscopy (TEM); scanning electron microscopy (SEM); and freeze-fracturing. The energy-dispersive X-ray microanalysis/microprobe technique is a new tool in inner-ear research. This method is technically difficult and time consuming and has so far not been used for routine purposes (Anniko and Wróblewski, 1983; Anniko, Lim and Wróblewski, 1983).

Transmission Electron Microscopy (TEM)

TEM maintains its position as the basic ultrastructural

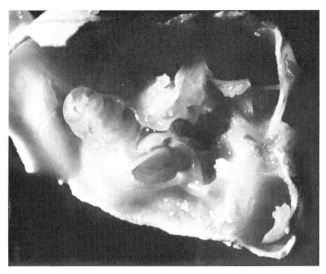

Figure 2.1 *Fixed guinea-pig temporal bone. The bulla has been removed and the medial wall of the middle ear is exposed, with the cochlea to the left and the vestibular apparatus to the right. In the centre, the middle-ear ossicles and round and oval windows are shown*

Figure 2.3 *Ampulla of a semicircular canal in the guinea-pig. The specimen has been fixed with osmium tetroxide and stains dark. In the centre of the ampulla, the crista ampullaris is readily visible. Dark-staining nerve fibres project from the base of the crista ampullaris. The specimen was embedded in Epon*

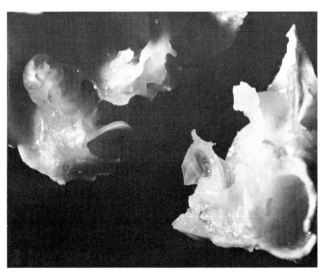

Figure 2.2 *Same specimen as in Figure 2.1. The temporal bone has been fractured parallel to the basal coil of the cochlea. The fracture line runs through the round and oval windows. A perfect sectioning of the temporal bone in this way preserves the whole cochlea, including the entire basal coil. The fracture also exposes the vestibule and the vestibular end-organs. These can be microdissected and processed further according to choice*

Figure 2.4 *Epon-embedded half-turn of guinea-pig cochlea. The specimen has been stained with osmium tetroxide revealing the lateral wall, the organ of Corti and the radiating myelinated nerve fibres. This type of processing is suitable for surface preparation studies as well as for ultrathin sectioning and transmission electron microscopy*

technique. Owing to the great variety of available techniques only a standard method used in many inner-ear laboratories will be described.

The processing of the material starts with fixation, possibly the most sensitive and difficult step in the whole procedure; if not performed correctly it can introduce serious artefacts. Fixation can be performed either by vital perfusion through a cannula placed in the ascending aorta, or through post-mortem immersion of the specimen in the fixative. A combination of these two methods generally offers the best results, but this is more tedious and time consuming than simply immersing the opened specimen into the fixative after killing the animal (*Figure 2.2*). Routinely, a combination of glutaraldehyde and paraformaldehyde in buffer is used (Karnovsky, 1965). This fixation is followed by a second fixation with osmium tetroxide (*Figure 2.12*). This method provides good contrast and preserves the membranes and other subcellular structures in a satisfactory way (*Figures 2.13* and *2.15*). Primary fixation with osmium tetroxide (Wersäll, 1956; Millonig and Marinozzi, 1968) is also satisfactory (*Figures 2.14* and *2.16*). After removal of the temporal bones, it is important to flush the fixative through the portion of the inner ear to be investigated (*see Figure 2.1*). TEM-fixatives generally penetrate poorly, thus increasing the risk of post-mortem autolysis (Wersäll, Kimura and Lundquist, 1965; Anniko and Bagger-Sjöbäck, 1977). After fixation, the specimens are carefully rinsed, dehydrated in alcohol, and embedded in epoxy resin (*see Figures 2.3–2.6*) such as Epon, Araldite, etc. (Luft, 1961). Sections are prepared on an ultramicrotome (*Figures 2.7–2.10*), mounted on formvar-coated copper grids, and stained with uranyl acetate and lead citrate (*Figure 2.11*; Reynolds, 1963).

Figure 2.5 *Epon-embedded cochlear tissue. The tissue intended for ultramicrotomy is embedded in Epon contained within a gelatine capsule. The specimen is oriented as required and the surrounding embedding material is removed with a razor blade*

Figure 2.6 *Epon-embedded scala media of the human cochlea. After removing excess embedding material, a small broad-based pyramid of Epon is left surrounding the specimen. It is desirable to make the top of the pyramid, i.e. the sectioning area, as small as possible since large areas create problems such as chatter and uneven sections. The gross outlines of the organ of Corti can be easily distinguished in the sectioning area, making it easy to trim the pyramid to perfect shape*

Figure 2.9 *Schematic diagram of the LKB ultramicrotome. The specimen moves in relation to a fixed glass or diamond knife. The whole process is viewed through a binocular (With permission from LKB Inc., Stockholm.)*

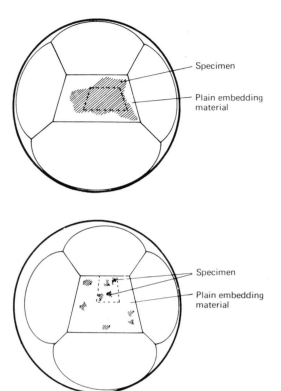

Specimen

Plain embedding material

Specimen

Plain embedding material

Figures 2.7 and 2.8 *Schematic representation of different types of specimens embedded in Epon. Figure 2.7: a solid specimen is shown. If the sectioning area were to be too large, the pyramid could be reduced to the size outlined by the dotted line. Figure 2.8: the specimen is scattered throughout the embedded material. The specimen could be reduced along the dotted line (From Glauert, 1975, reproduced by kind permission of author and publishers.)*

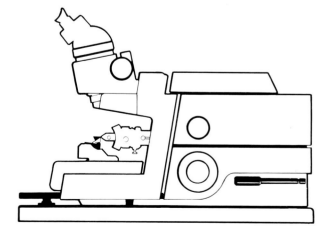

10

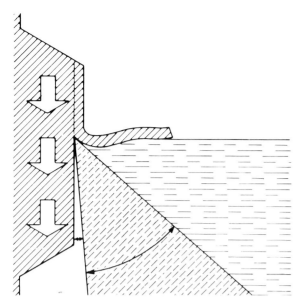

Figure 2.10 *Schematic representation of glass knife cutting the specimen according to the LKB principle. The section floats out on to the water surface to be picked up on the grid (With permission from LKB Inc., Stockholm.)*

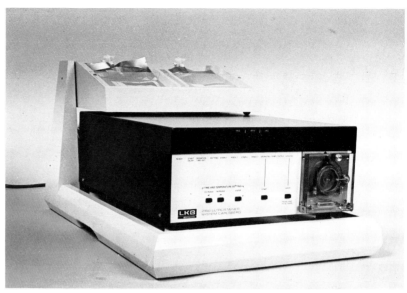

Figure 2.11 *LKB automatic staining apparatus. In order to minimize the contact with the toxic staining solutions, lead citrate and uranyl acetate, LKB has developed a new automatic staining device. Sections, mounted on formvar-coated copper grids, are fed into the apparatus and run through a staining programme chosen as required (With permission from LKB Inc., Stockholm.)*

Date:...............			Specimen:	1................
				2................
				3................
				4...............
Researcher:...............				

1.	Glutaraldehyde, Paraformaldehyde - Buffer		2 hours
2.	Rinse: Ringer´s solution or Buffer	change 1, 2, 3, 4	24 -"-
3.	Osmiumtetroxide 1 % or 2 % - Buffer		2 -"-
4.	Rinse: Ringer´s solution or Buffer	change 1, 2, 3, 4	1 -"-
5.	Dissection		
6.	Alcohol 70 %	change 1, 2	1/2-"-
7.	Alcohol 90 %	change 1, 2	1/2-"-
8.	Alcohol 95 %	change 1, 2	1/2-"-
9.	Alcohol 100 %	change 1, 2, 3, 4	1 -"-
10.	Propylenoxide	change 1, 2, 3, 4	1 -"-
11.	Propylenoxide + Epon 1:1		1 -"-
12.	Epon		12 -"-
13.	Epon	Embedding 45°C	24 -"-
		60°C	48 -"-

Figure 2.12 *A processing scheme for transmission electron microscopy*

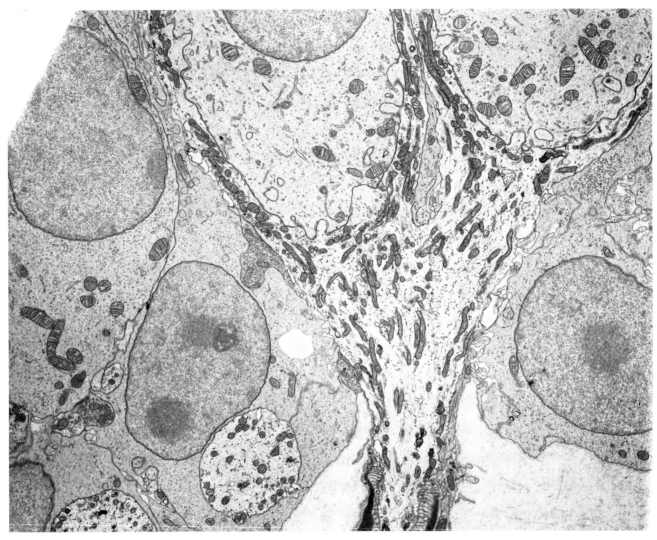

Figure 2.13 *Transmission electron micrograph of basal portion of two type I hair cells in the guinea-pig crista ampullaris. The hair cells are surrounded by one common nerve calyx which projects into a myelinated nerve fibre. Note the loss of myelinization in the region of the basal membrane. Specimen fixed with glutaraldehyde–formaldehyde and postfixed in osmium tetroxide*

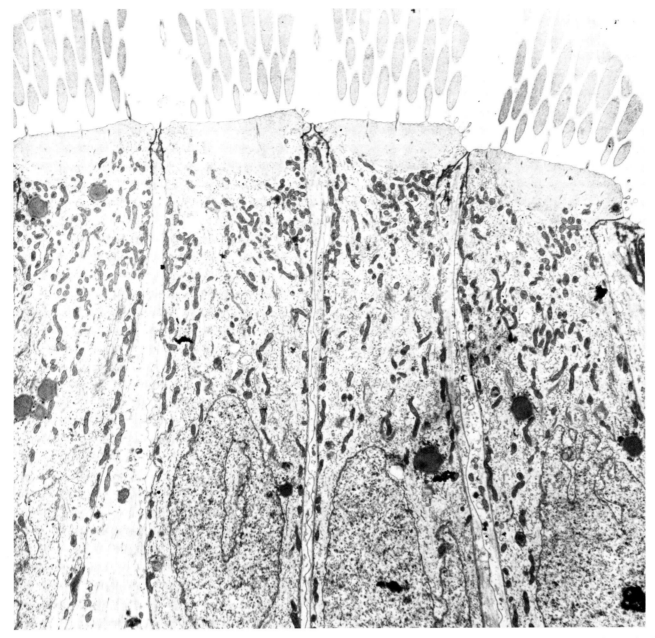

Figure 2.14 *Transmission electron micrograph of the apical portions of hair cells and supporting cells in the lizard basilar papilla (hearing organ). The contrast is good and the cytoplasmic organelles are clearly demonstrated by primary osmium tetroxide fixation*

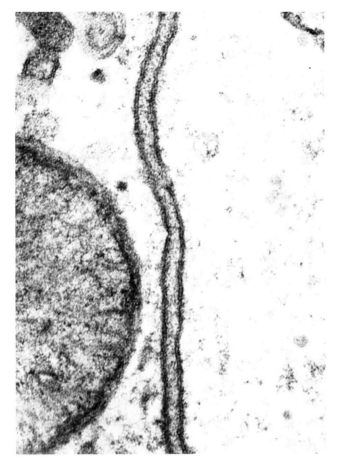

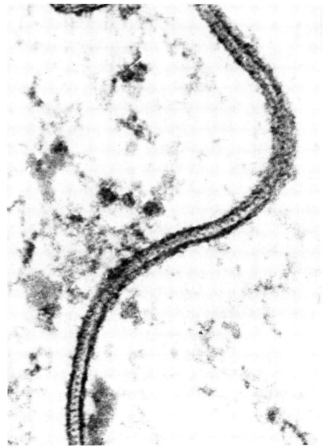

Figure 2.15 *High-resolution transmission electron micrograph of the cell border within the guinea-pig crista ampullaris. The two adjoining cells are separated by an intercellular cleft. The plasma membranes are well fixed and display both membrane layers (Primary glutaral-dehyde–formaldehyde fixation.)*

Figure 2.16 *High-resolution transmission electron micrograph of the cell border within the sensory epithelium of the basilar papilla of the lizard. The specimen was primarily fixed with osmium tetroxide which has denaturated the membranes so that the two membrane layers have fused and cannot be separated. Based on this, primary osmium tetroxide fixation cannot be recommended for high-resolution membrane studies*

Scanning Electron Microscopy (SEM)

SEM is best suited for surface preparations. The technique is extremely useful when it comes to examining whole organs such as the cochlea (*Figure 2.17*), the vestibular end-organs or the luminal surface of the endolymphatic sac. It is easy to analyse the condition of the surface morphology during normal and pathological conditions. Microdissection, for example of the cochlea, also makes it possible to view the interior of the organ of Corti (Bredberg, 1981).

The fixation procedure is similar to that used in TEM. It is equally important to be thorough in the perfusion of the specimen so as to minimize artefacts. Either aldehyde-based or osmium-tetroxide fixatives can be used. After fixing and rinsing the tissue, the specimen has to be dried completely. The method most often used at present is critical-point drying. The freeze-drying procedure has lately been favoured by some investigators in preference to the critical-point method. After microdissection of the dried specimen, it is

mounted on specimen holders. Owing to the optical properties of the scanning electron microscope, the correct mounting of the specimen on the holder is important so as to obtain the best possible view. Subsequently, the specimen is coated with an ultrathin layer of a heavy metal (mostly gold), in order to prevent charging in the microscope. This is performed either in a vacuum evaporator or in a diode sputter coater. The sputter coating method is advantageous when examining an irregular surface such as that of the cochlea or the vestibular end-organs. The stages of the method are as follows:

1. Fixation (glutaraldehyde, glutaraldehyde/para-formaldehyde, osmium tetroxide, etc.).
2. Rinse – buffer, Ringer's solution, etc.
3. Drying – critical point, freeze-drying.
4. Mounting.
5. Coating – vacuum evaporator, diode sputter coater.

The microvasculature of small specimens, such as the inner-ear end-organs or the facial nerve, can be studied

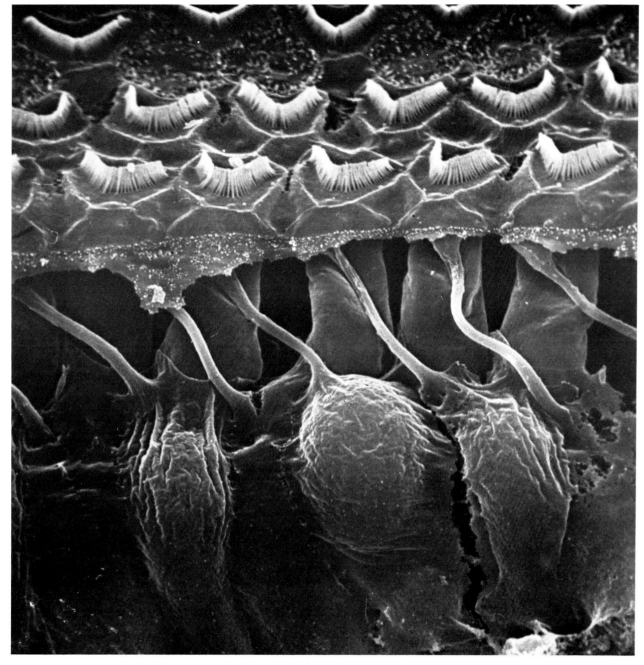

Figure 2.17 *Scanning electron micrograph of guinea-pig organ of Corti. The specimen is fractured, revealing the outer hair cells with their characteristic W-shaped hair bundles. Below the cuticular plate the hair cell bodies as well as supporting elements are apparent. In order to obtain the best possible results the specimen should be carefully oriented and mounted (By courtesy of Birgitta Björkroth, Stockholm.)*

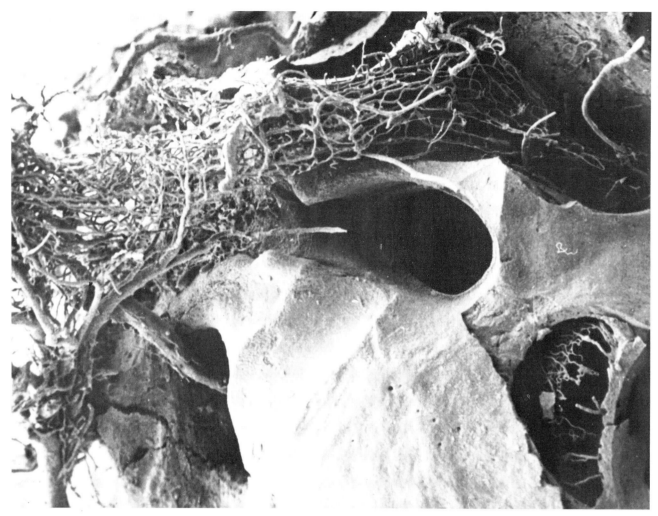

Figure 2.18 *Low-power scanning electron micrograph of injection replica of the vascular tree of the facial nerve in the mouse. The large ascending vessels, the stylomastoid vessels, ramify into numerous small branches along the length of the entire Fallopian canal. The medial wall of the middle ear with the round and oval windows is also shown. The basal coil of the cochlea is exposed with the ramifications of the strial vessels*

under the SEM using the injection replication method. The vascular system of the animal is fixed by vital perfusion with an aldehyde-based fixative. After rinsing with buffer, a quick-hardening epoxy resin is injected into the vessels (Miodonski, Hodde and Bakker, 1976). Under ideal conditions this forms a perfect cast of the whole vascular tree including the smallest capillaries (*Figures 2.18* and *2.19*). The specimen, with surrounding tissue, is dissected and kept in 20% KOH and rinsed daily. When most of the biological tissue is corroded, the remaining vascular cast is carefully dissected, mounted on a specimen holder and coated with gold according to the standard procedure. This method provides a very good three-dimensional impression of the microvascular supply of the tissue.

Freeze-fracturing

The routine methods used in freeze-fracturing have been developed during the last two decades. Freeze-fracturing is a replica method which is best suited for studies of membrane structures, even within the cell. In contrast to SEM, freeze-fracturing can be used to study membrane surfaces within almost every organ (*Figure 2.21*). In the inner ear, the interest has been focused on intercellular junctions, synaptic membranes, and endothelial cell surfaces in the microvasculature. The intercellular junctions in the inner ear are of special interest owing to the high ionic gradients between the perilymphatic and endolymphatic spaces (Anniko and Bagger-Sjöbäck, 1982; Bagger-Sjöbäck, Rask-Andersen and Lundquist, 1981).

(*continued on page 19*)

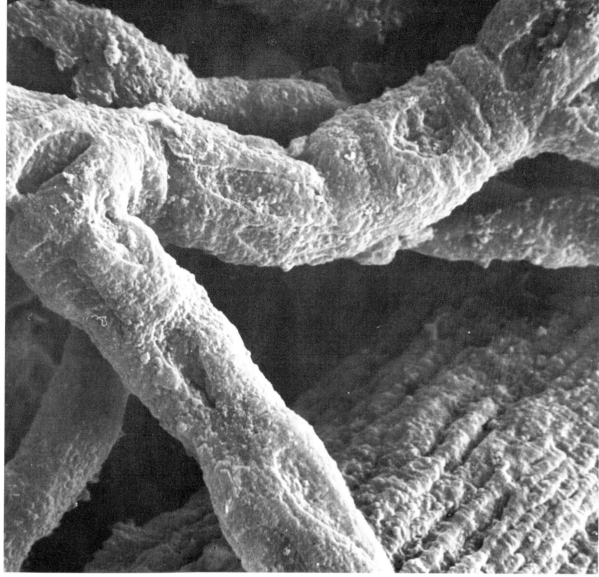

Figure 2.19 *High-power scanning electron micrograph of the same specimen as in Figure 2.18. The ultrastructure of the vessel walls is clearly displayed. The larger vessel surface has a striated appearance while in the smaller precapillaries and capillaries, the single endothelial cell surfaces are well visualized*

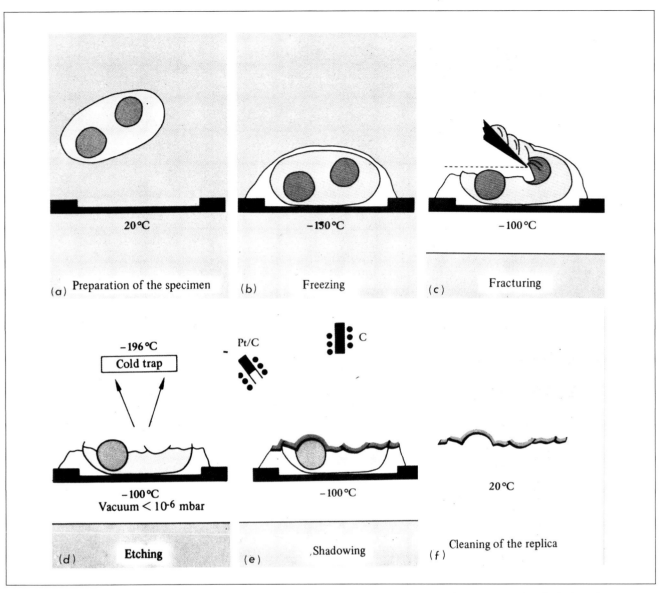

Figure 2.20 *Schematic representation of processing for freeze-fracturing replicas. (Schematic drawings by courtesy of Dr Wolfgang Niedermeyer, West Berlin, BRD and AG Balzers, Lichtenstein.) (a) The aldehyde-fixed specimen is immersed in a 30% buffered glycerol solution and mounted on a gold specimen holder at room temperature. (b) The mounted specimen is covered by some excess of buffered glycerol and rapidly frozen in liquid Freon or super-cooled nitrogen slush. (c) In a high-vacuum chamber at −100°C, the specimen is fractured, either by a knife as in the single replica method, or with the sandwich technique as in the double replica method. (d) After fracturing, the raw fracture surfaces of the specimen are exposed to etching. The degree of etching is based on the biological properties of the specimen. (e) After etching the fracture surface, platinum and carbon are evaporated at an angle of about 45 degrees to the specimen. The specimen is thus covered by a replica approximately 2 nm thick which is reinforced by subsequent linear carbon coating. The thickness of the resultant replica is approximately 22 nm. (f) After the shadowing procedure, the covered specimen is removed from the vacuum chamber and treated in a series of corrosive baths. All the biological tissue is removed in this way, leaving the isolated replica to be mounted on a copper-mesh grid*

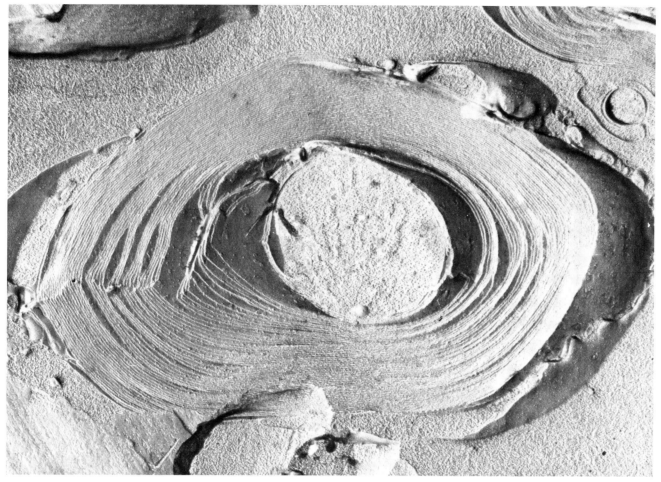

Figure 2.21 *Freeze-fracturing replica of a myelinated nerve fibre. The membrane faces of the various myelin lamellae are well visualized with this technique. Note the three-dimensional appearance of the replica*

The specimens are fixed as for TEM and SEM, with one exception. Osmium tetroxide fixatives are not suitable for this technique since they denaturate the biomembranes in such a way that freeze-cleaving becomes impossible. After fixation, the specimen is transferred to a buffered 30% glycerol solution (*Figure 2.20(a)*) and rapidly frozen in liquid Freon or super-cooled nitrogen slush (*Figure 2.20(b)*). The specimen has to be frozen in the glycerol solution to minimize the production of ice crystals which would be prominent if the specimens were frozen in water alone. The specimen, still frozen, is introduced into a high-vacuum chamber in the freeze-fracturing apparatus. Under high vacuum (approximately 2×10^{-6} torr) the specimen is fractured either by a knife or by applying the double replica method (*Figure 2.20(c)*).

The surface of the fractured specimen faces two electron guns within the vacuum chamber. If desirable, the surface of the specimen is 'etched', i.e. water is briefly evaporated from the surface (*Figure 2.20(d)*). From an angle of about 45 degrees a mixture of evaporated platinum and carbon is allowed to coat the fracture-surface with a thin layer about 2 nm thick. To reinforce this very thin replica a layer of carbon approximately 20 nm thick is evaporated on top of it (*Figure 2.20(e)*). The specimen is subsequently removed from the vacuum chamber and treated in a series of corrosive baths (hypochlorite, etc.) (*Figure 2.20(f)*). After the removal of all biological material, the resultant replica is mounted on a coppermesh grid and examined under a transmission electron microscope.

Investigators of the inner ear should be encouraged not only to master one technique, but also to learn two or maybe all the techniques described in this chapter. Only in this way is it possible to obtain a thorough knowledge of the complicated cytoarchitecture of the normal and pathological inner ear.

Acknowledgements
The authors are supported by grants from the Karolinska Institute, the Swedish Medical Research Council, No. 12X-00720 and the Foundation Tysta Skolan.

References

Anniko, M. (1981) Decalcification effects on the fine structure of the mammalian inner ear. *Micron*, **12**, 267–278.

Anniko, M. and Bagger-Sjöbäck, D. (1977) Early post-mortem change of the crista ampullaris. A light and electron microscopic study of the guinea pig. *Virchows Archiv, B, Cell Pathology*, **25**, 137–149.

Anniko, M. and Bagger-Sjöbäck, D. (1982) Maturation of junctional complexes during embryonic and early postnatal development of the inner ear secretory epithelia. *American Journal of Otolaryngology*, **3**, 242–253.

Anniko, M. and Lundquist, P-G. (1980a) Temporal bone morphology after systemic arterial perfusion or intralabyrinthine *in-situ* immersion. I. Hair cells of the vestibular organs and the cochlea. *Micron*, **11**, 73–83.

Anniko, M. and Lundquist, P-G. (1980b) Temporal bone morphology after systemic arterial perfusion or intralabyrinthine *in-situ* immersion. II. Secretory and reabsorptive epithelia in the cochlea and the vestibular organs. *Micron*, **11**, 103–114.

Anniko, M. and Wróblewski, R. (1983) X-ray microanalysis in the studies on developing and mature inner ear. Scanning Electron Microscopy/1982. AMF O'Hare, Il.: SEM. (In press).

Anniko, M., Lim, D. and Wróblewski, R. (1983) Elemental composition of individual cells and tissues in the cochlea. *Acta Otolaryngologica*. (In press).

Bagger-Sjöbäck, D. and Wersäll, J. (1976) Toxic effects of gentamicin on the basilar papilla in the lizard *Calotes versicolor*. *Acta Otolaryngologica*, **81**, 57–65.

Bagger-Sjöbäck, D., Rask-Andersen, H. and Lundquist, P-G. (1981) Intercellular junctions in the endolymphatic sac; a freeze-fracturing and TEM study on the guinea pig's labyrinth. In *Ménière's Disease*, Eds. K.H. Vosteen *et al*. Stuttgart: George Thieme Verlag.

Bredberg, G. (1981) SEM studies of the organ of Corti with special reference to its innervation. *Biomedical Research*, **2**, Supplement, 403–413.

Glauert, M. M. (1975) *Fixation, Dehydration and Embedding of Biological Specimens*. Amsterdam: North-Holland/American Elsevier.

Karnovsky, M. H. (1965) A formaldehyde–glutaraldehyde fixative of high osmolality for use in electron microscopy. *Journal of Cellular Biology*, **35**, 2–3.

Luft, H. H. (1961) Improvements in epoxy resin embedding methods. *Journal of Biophysics, Biochemistry and Cytology*, **9**, 409.

Millonig, G. and Marinozzi, B. (1968) Fixation and embedding in electron microscopy. In *Advances in Optical and Electron Microscopy*, Eds. R. Barer and V. E. Coslett, Vol. 2, p. 251. New York: Academic Press.

Miodonski, A., Hodde, K. C. and Bakker, C. (1976) Rasterelektronen Mikroskopie von Plastik-Korrosion Präparaten, Morphologische Unterschiede zwischen Arterien und Venen. Beitr. *Elektronen Mikroskop*, **9**, 435–442.

Retzius, G. (1881) *Das Gehörorgan der Wirbeltiere I*. Stockholm: Samson och Wallin.

Retzius, G. (1884) *Das Gehörorgan der Wirbeltiere II*. Stockholm: Samson och Wallin.

Reynolds, E. S. (1963) The use of lead citrate of high pH as an electron opaque stain in electron microscopy. *Journal of Cellular Biology*, **17**, 208.

Van de Water, T. R., Cheuk, W. L., Ruben, R. J. and Shea, C. A. (1980) Ontogenetic aspects of mammalian inner ear development. *Birth Defects: Original Article Series*, **XVI**, 5–45.

Wersäll, J. (1956) Studies on the structure and innervation of the cristae ampullares in the guinea pig. *Acta Otolaryngologica*, Supplement **126**, 1–85.

Wersäll, J., Kimura, R. and Lundquist, P-G. (1965) Early postmortem changes in the organ of Corti (guinea pig). *Zeitschrift Zellforschung*, **65**, 220–237.

3
Organ Culture of the Avian and Mammalian Embryo Otocyst

Imrich Friedmann

Studies on the isolated embryonic otocyst in tissue culture began with the pioneering work of Fell (1929). The original watch-glass technique of Fell and Robinson (1929) has been successfully employed in the study of differentiation of the neuroepithelial elements of the inner ear (Friedmann, 1959) and in the evaluation of various ototoxic agents (Friedmann and Bird, 1961). In all of these studies the otocyst of 3½-day-old chick embryos has been used, reaching *in vitro* complete differentiation of the neuroepithelium, extending to the ultrastructural level. The system, representing a 'model ear' (Friedmann, 1968) has lent itself well to the study of the innervation, cilia and membrane structures, and of the agents that may play a role in the causation of deafness. Notwithstanding earlier attempts by Lawrence and Merchant (1953), the first successful organ cultures of the isolated mouse embryo otocyst were achieved in 1973 by Van de Water and Ruben, who subsequently succeeded in making 'the organ culture system for the development of the mammalian inner ear into a tool with which some forms of inner-ear deafness may be studied'.

Yamashita and Vosteen (1975) have cultured isolated outer and inner hair cells of the organ of Corti of the newborn guinea-pig and they have been maintained *in vitro* for more than 20 days. The development of hair cells of the embryonic chick's basilar papilla was studied by Cohen and Fermin (1978) and the neuronal growth in the organ of Corti in culture has been elegantly investigated by Sobkowicz *et al.* (1980).

The development of the embryonic chick otic placode was studied by Meier (1978). A chemically defined medium was used by Friedmann, Hodges and Riddle (1977) in their studies of the avian and mammalian

otocyst. There was full differentiation of the isolated embryo otocyst in both the chick and the mouse. The neuroepithelial structures, ciliary apparatus and organelles have developed in a normal fashion. The sensory cells are connected by desmosomes and tight junctions (Van De Water, Heywood and Ruben, 1973).

Non-myelinated axons emerge from their non-myelinated neurons and often form large structures composed of several axons surrounded by a single Schwann cell and may be mistaken, under the light microscope, for neurons. Eventually, the myelinated axons lose their myelin sheaths and, penetrating the basal lamina, spread among the basal cells and between the supporting cells; protected and nourished by them, the nerves ultimately reach the synaptic area of the sensory hair cells. The micellar orientation of the medium and the guiding surfaces of the supporting cells *in vitro* and *in vivo* assist greatly in the spread and direction of the nerves, together with certain intricate cytochemical processes.

The growth of normal cells is controlled by the interaction of cells with various polypeptide hormones or hormone-like growth factors contained in the surrounding nutrient fluids (Holley, 1975). In organ cultures some of these may be supplied by the explanted tissue or by the addition of serum, which includes various growth-promoting factors (such as nerve growth factors) or agents contributing to the differentiation of specific cells and tissues.

The behaviour of dissociated and re-aggregated cells of the chick embryo otocyst was studied by Orr (1975), who recognized the important influence of neural tissue on sensory cell differentiation. This is supported by the evidence presented by Friedmann (1969) and by Sher

(1971) of the inductive influence that nerves exert on the undifferentiated epithelium of the otocyst.

In contrast, Van De Water (1976) noted that removal of the statoacoustic ganglion complex of murine-cultured otocysts at different critical gestational periods did not adversely affect differentiation. None the less, it can be argued that the nerves had already imparted their influence before the excision when the ciliary apparatus and other cellular features may have been forming, as suggested by Hilding (1969). Functional differentiation, however, would not be completed before synaptic linkage between hair cells and nerve had been established (Friedmann, 1969).

As regards the theories of innervation, the neuronal or 'outgrowth' theory of Granville Ross Harrison (1907), supported by Cajal (1960) has been challenged by Hensen (1864), and by Proctor and Proctor (1967). It has been suggested that the neurons of the acoustic ganglion are formed by the neuroepithelial cells and then migrate from the otocyst. Our observations have led to the conclusion that the innervation of the otocyst, ceteris paribus, of the inner ear, develops centrifugally from the neurons of the statoacoustic ganglion, itself a product of the neural crest.

Permission to include material from the following publications is gratefully acknowledged: *Acta Otolaryngologica, Stockholm; Annals of Otology, Rhinology and Laryngology, St. Louis;* Blackwell Scientific Publications Ltd., Oxford; and *Journal of Laryngology.*

Figure 3.1 *12-day-old mouse embryo otocyst (resembling a fowl embryo otocyst of 3½–4 days) used as explants in the tissue cultures*

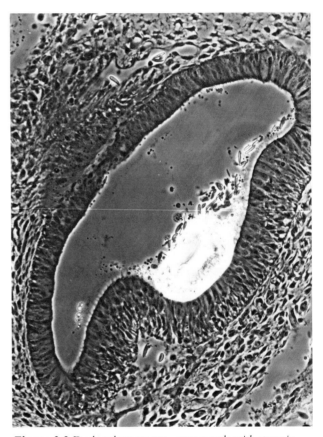

Figure 3.2 *Duck embryo otocyst – note macula with otoconia (Phase contrast microscopy.)*

Figure 3.3 *Otocyst transplanted on to the chorioallantoic membrane of a fertilized egg. Although satisfactory growth and differentiation can be achieved, the rich vascularization of the sensory areas has proved to be a disadvantage*

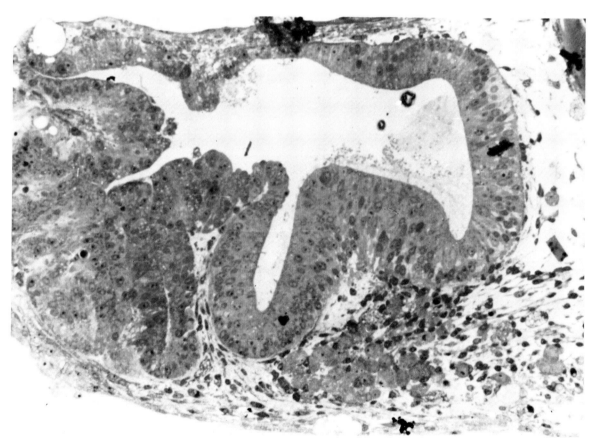

Figure 3.4 *Survey picture of a fully differentiated tissue culture of the fowl embryo otocyst (12 days in vitro). For details see Figures 3.7–3.9. Toluidine blue stained, semi-thin section*

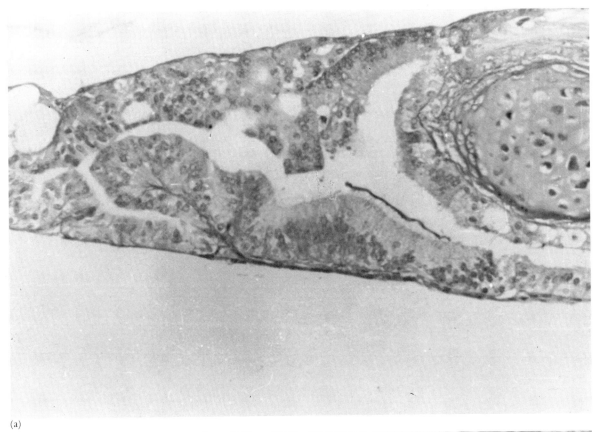

(a)

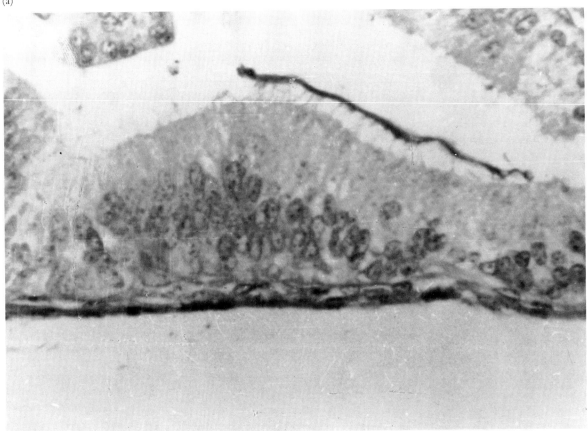

(b)

24

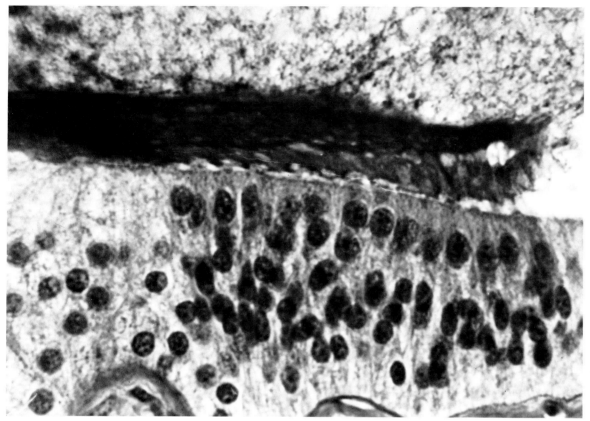

Figure 3.6 *Fowl embryo otocyst in tissue culture (13 days in vitro). Detail of 'organ of Corti' (or papilla basilaris). Note row of hair cells on the surface with bundles of the stereocilia indenting the dense and darkly stained (PAS positive) tectorial membrane. Note stratified supporting cells. ×1200*

◀ **Figure 3.5** *(a) Fowl embryo otocyst in tissue culture (12 days in vitro) showing Corti's organ-like sensory epithelium (papilla basilaris) covered by the tectorial membrane. ×225. (b) As in (a) showing hair cells with the stereocilia attached to (or touching) the tectorial membrane; a well-defined basilar membrane supports the sensory epithelium (papilla basilaris). ×600*

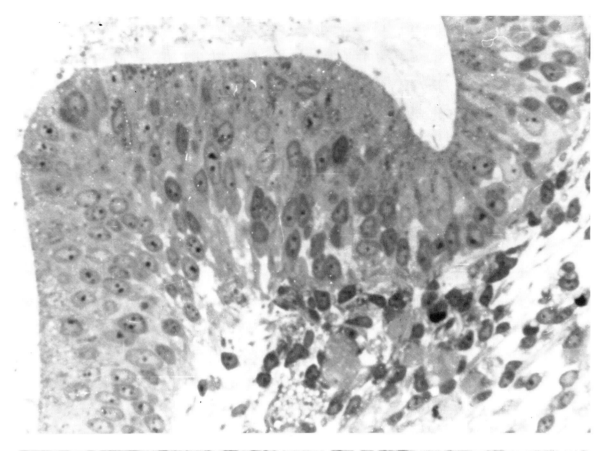

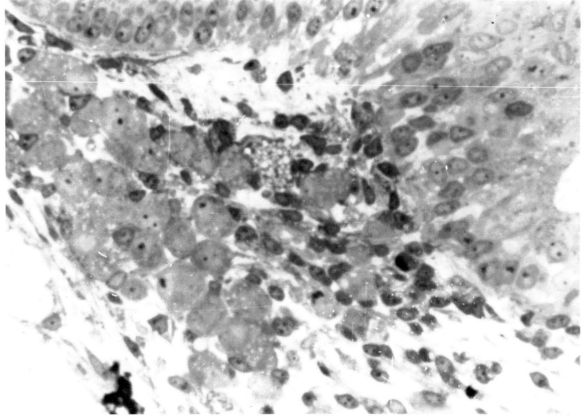

26

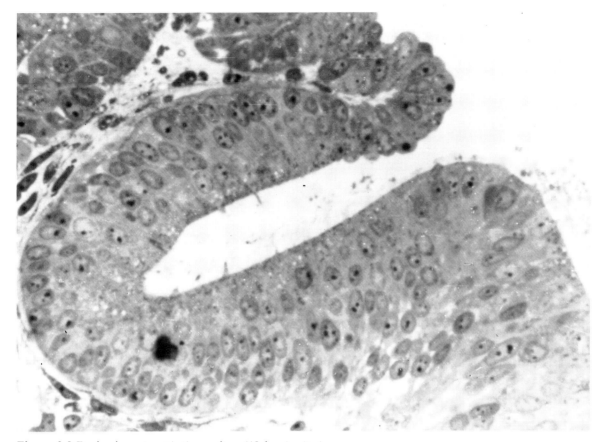

Figure 3.9 *Fowl embryo otocyst in tissue culture (12 days in vitro) showing macula-like area lined by hair cells. Toluidine blue stained, semi-thin sections (detail of Figure 3.4).* ×600

◀ **Figure 3.7** *(top) Fully developed organ culture of fowl embryo otocyst (12 days in vitro) showing a sensory epithelium resembling Corti's organ (papilla basilaris). Note hair cells covering the surface and the large group of neurons beneath the basal layer. Toluidine blue stained, semi-thin section (detail of Figure 3.4).* ×600

◀ **Figure 3.8** *(bottom) Group of neurons forming statoacoustic ganglion beneath the well-differentiated sensory area (as in Figures 3.4 and 3.7).* ×600

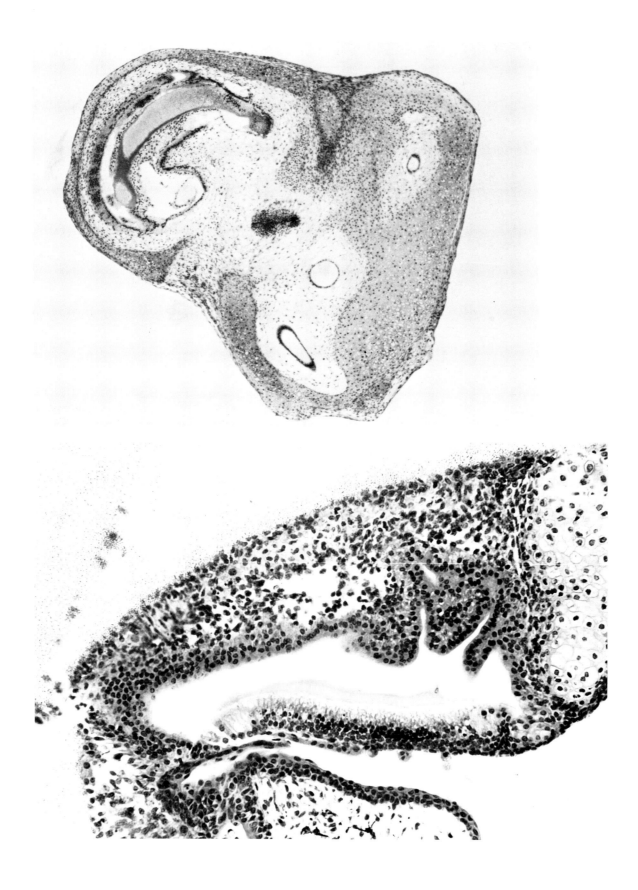

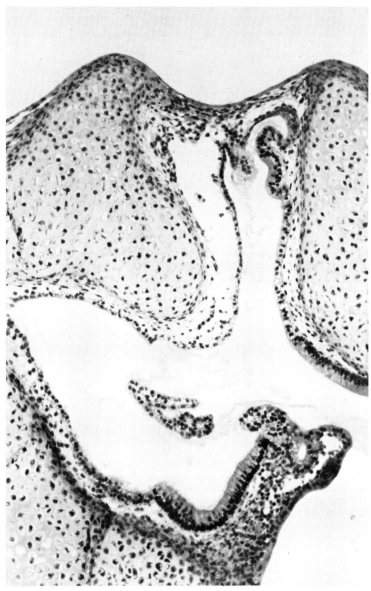

Figure 3.12 *Note macula (at bottom) formed by columnar (ciliated) epithelium resting on a layer of cuboidal basal cells. There are neural elements (confirmed electron microscopically) in the fibrous tissue.* ×180

◄ **Figure 3.10** *(top) Survey picture of 12-day gestation period mouse embryo otocyst grown in organ culture for additional 10 days; note cochlea (left) and cross-sections of the semicircular canals (right); also cartilaginous otic capsule and developing stapes (between cochlea and vestibule).* ×60

◄ **Figure 3.11** *(bottom) Mouse embryo otocyst of 12 days. Maintained for additional 8 days in vitro. The organ of Corti resting on the basilar membrane is covered by the tectorial membrane. There is a developing vascular stria (above) attached to the cartilaginous otic capsule.* ×225

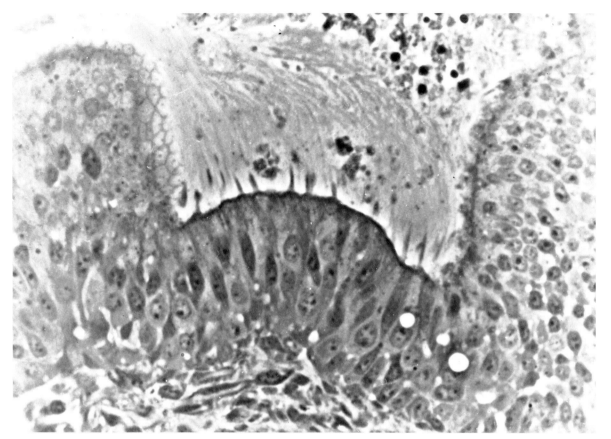

Figure 3.13 *Fowl embryo otocyst in tissue culture (12 days in vitro). Note crista-like structure lined by ciliated hair cells and covered by some filamentous material forming a cupula-like structure (phase contrast microscopy). (From (1974)* Pathology of the Ear, *Oxford: Blackwell, reproduced by kind permission of publishers) ×576*

(a)

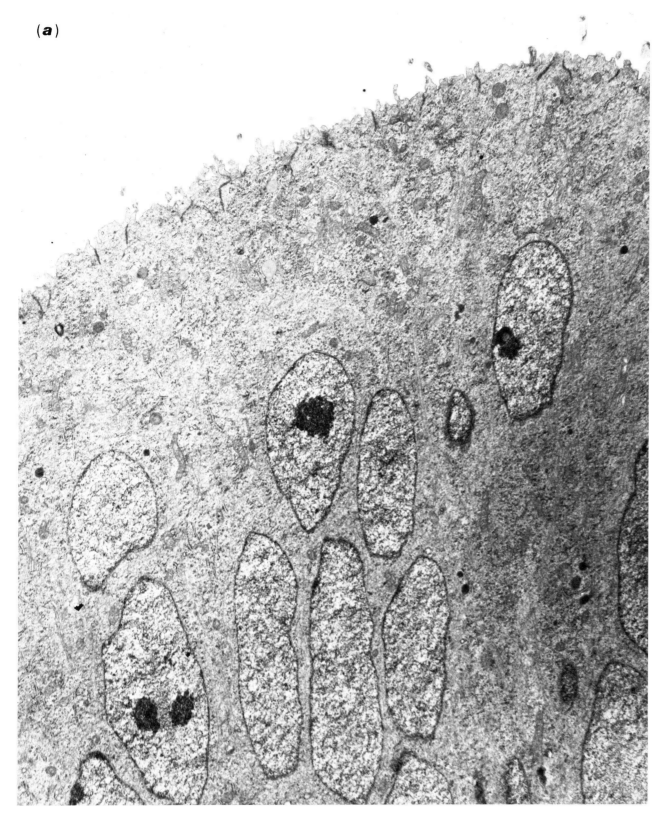

Figure 3.14 *(a) 8-day-old fowl embryo otocyst showing the developing surface epithelium with indistinct cell membranes. There is no obvious nerve tissue within the surface epithelium at this early stage of development. Original magnification ×10 500 reduced to 90%.*

(**b**)

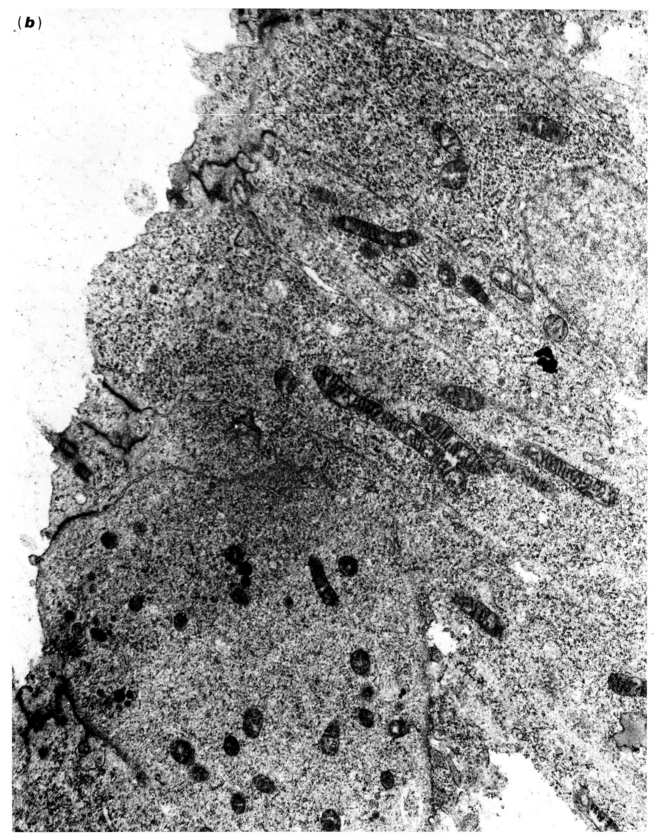

Figure 3.14 *(b) Similar area showing well-defined cells in the developing fowl embryo otocyst. Note rudimentary kinocilium. Original magnification ×17 500 reduced to 90%*

32

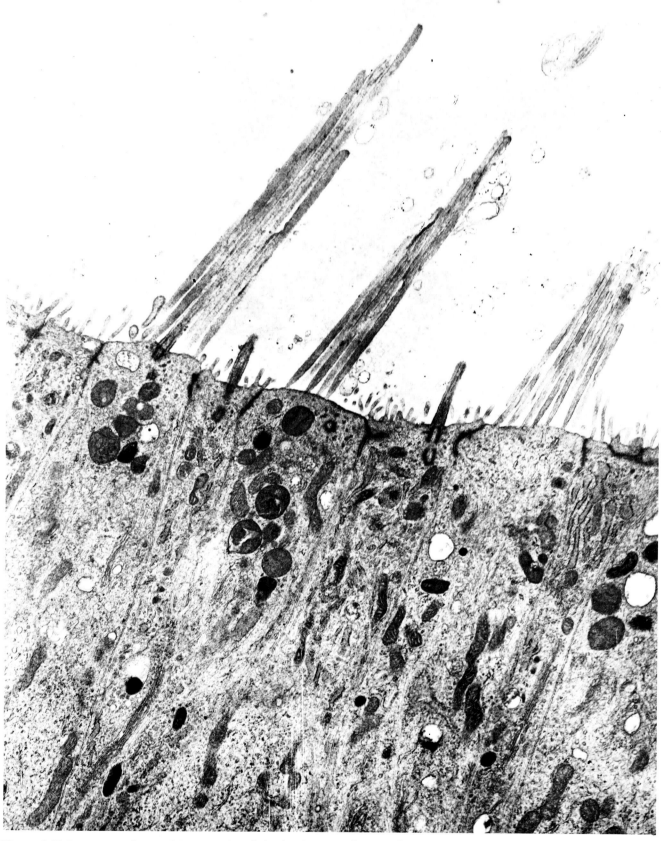

Figure 3.15 *Sensory area of organ of Corti type (papilla basilaris) showing hair cells endowed with long stereocilia. Note several rudimentary kinocilia. Original magnification ×14 000 reduced to 92%*

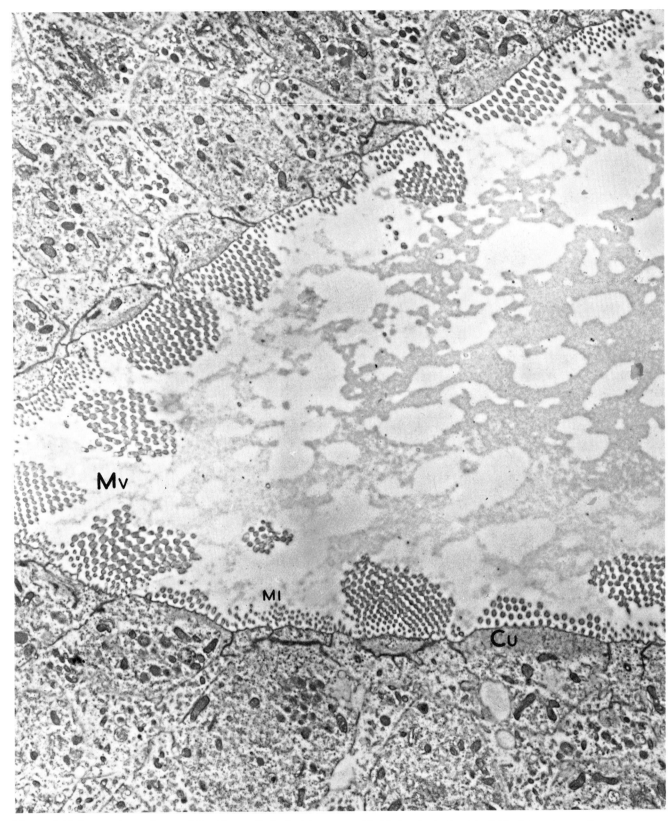

Figure 3.16 *Survey picture of organ of Corti (or papilla basilaris) of chick embryo otocyst in tissue culture (16 days in vitro). Note bunches of cross-cut macrovilli (Mv) or stereocilia arising from the cuticle (Cu) and mitochondria (Mi). Original magnification ×9325 reduced to 92%*

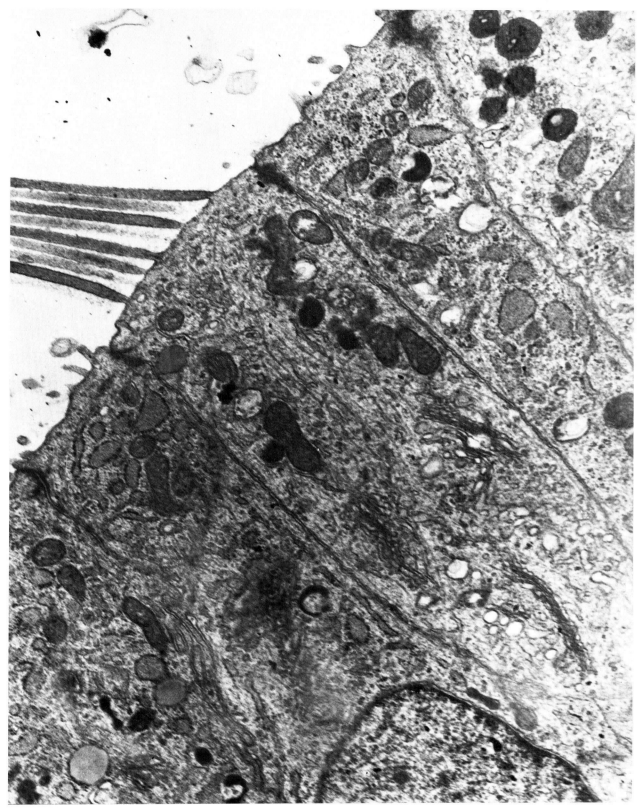

Figure 3.17 *Hair cell of organ of Corti (or papilla basilaris) with long stereocilia. Note Golgi complex in the cytoplasm. Original magnification ×26 250 reduced to 88% (From (1974)* Pathology of the Ear, *London: Blackwell, reproduced by kind permission of publishers.)*

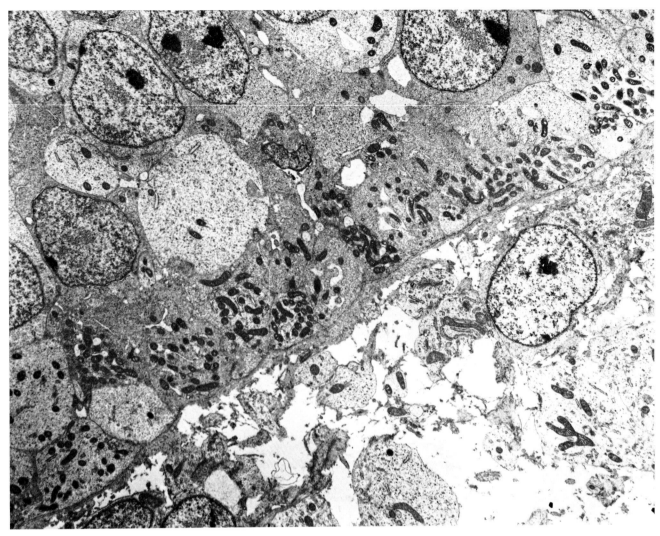

Figure 3.18 *Basal cells of sensory area containing large numbers of mitochondria. Note large number of nerves. Original magnification ×70 000 reduced to 72%*

Figure 3.19 *Fowl embryo otocyst in a 12-day-old tissue culture. Detail of a fully-developed sensory area resembling the macula. Note type II hair cells – lined by dark cells – with stereocilia and kinocilia; also rudimentary kinocilium on a dark cell. Original magnification ×15 000 reduced to 91%* ▶

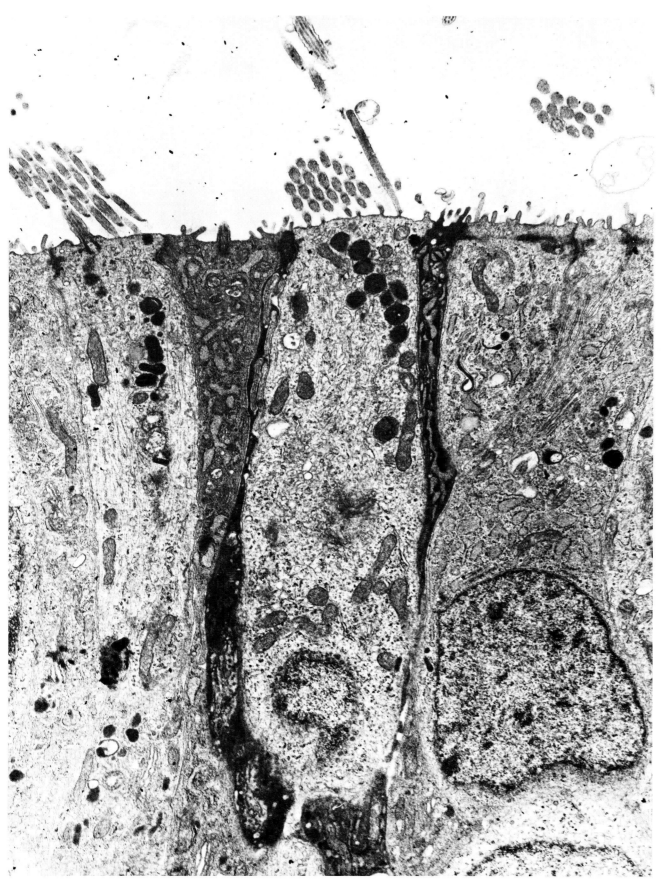

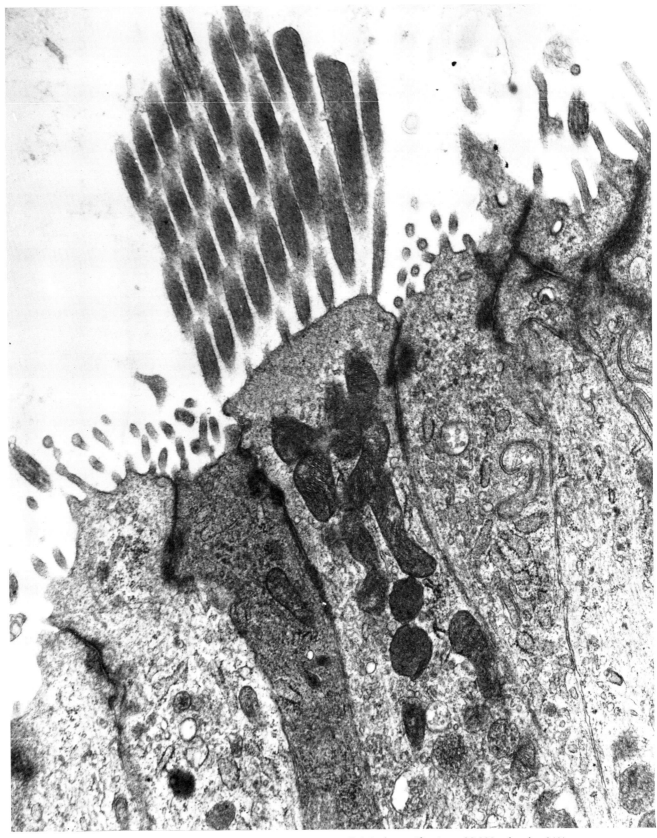

Figure 3.20 *Tissue culture of fowl embryo otocyst showing the macula. Hair cell with obliquely-cut stereoocilia and kinocilia.*

Original magnification ×28 000 reduced to 91%

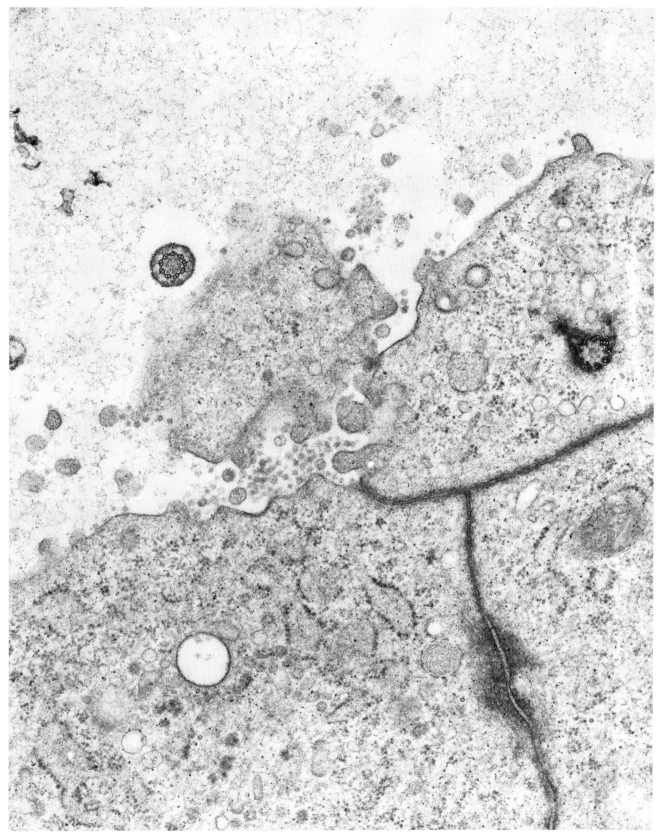

Figure 3.21 *Surface of hair cells with cross-cut stereocilia and kinocilium. Note base of kinocilium in the cuticle with triple tubules and anchoring process. Original magnification ×35 000 reduced to 91%*

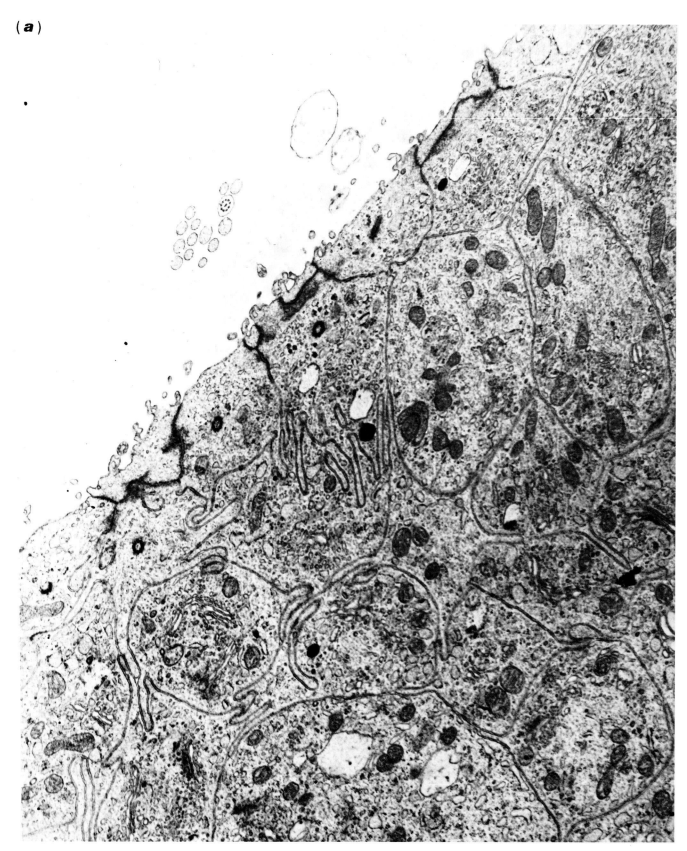

Figure 3.22 *(a) Obliquely cut cells of the developing macula. Note centrioles in several surface cells and cross-cut kinocilia. Original magnification ×17 500 reduced to 92%.*

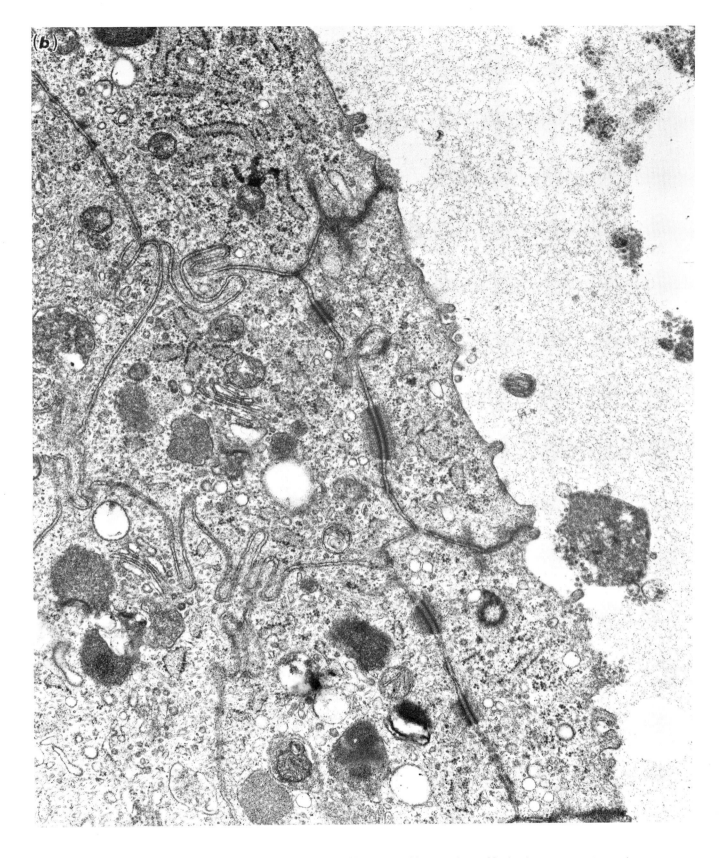

Figure 3.22 *(b) Note well-formed desmosomes and centriole in the cells (12-day-old tissue culture of fowl embryo otocyst). Original magnification ×25 000 reduced to 92%*

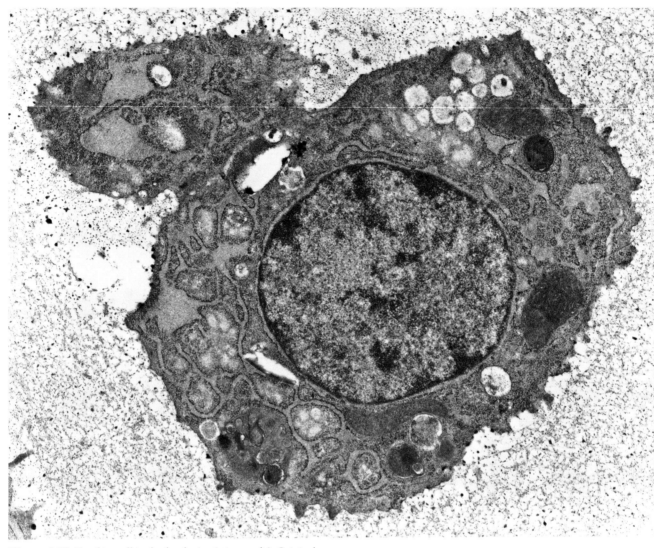

Figure 3.23 *Cartilage cell in the developing 'otic capsule'. Original magnification ×28 000 reduced to 80%*

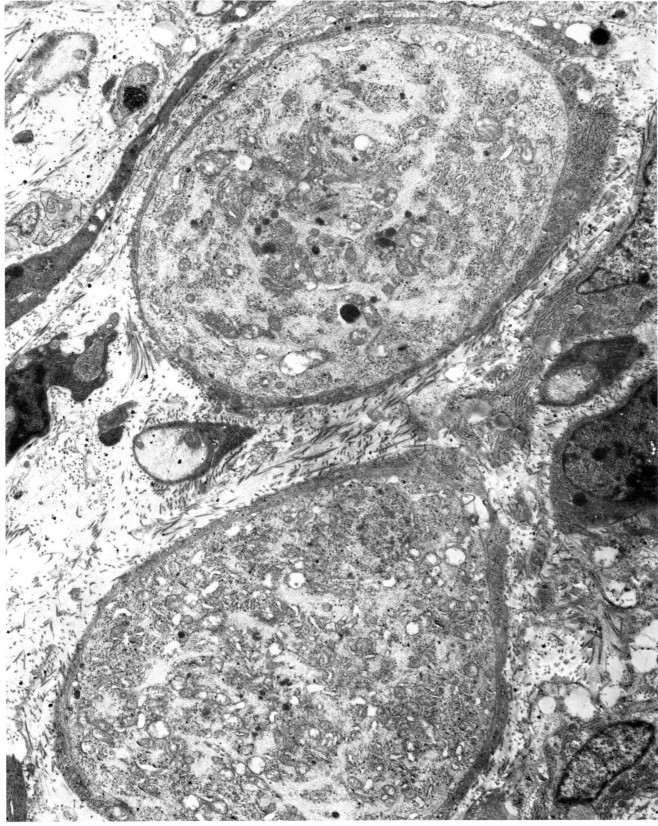

Figure 3.24 *Two filamentous neurons of the spiral ganglion fowl embryo otocyst (12 days in vitro). The neurons have a developing myelin sheath, multiple Nissl bodies, mitochondria, Golgi complexes and many scattered dense bodies. Original magnification ×7000 reduced to 92%*

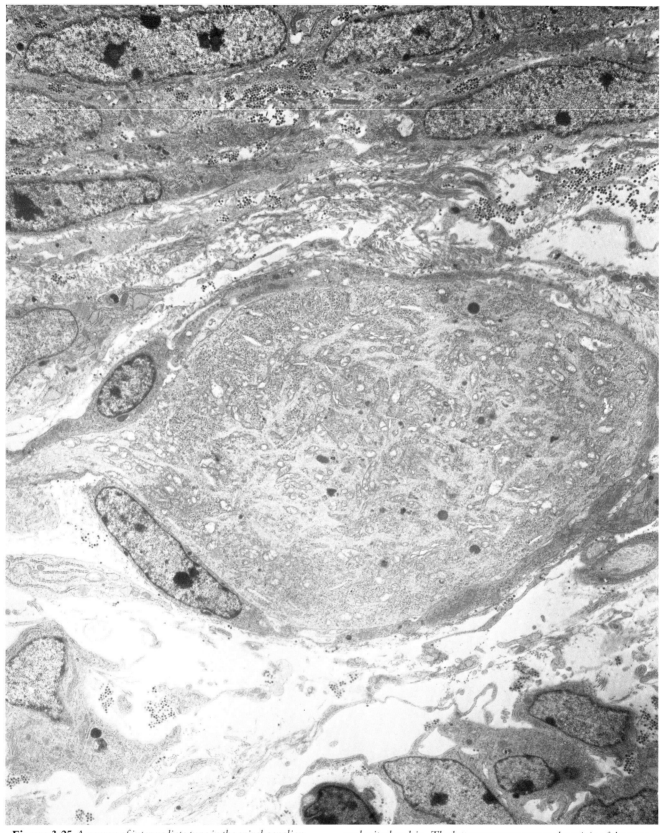

Figure 3.25 *A neuron of intermediate type in the spiral ganglion fowl embryo otocyst (12 days in vitro). Note the initial segment and the axon. The neuron contains numerous Nissl bodies, dense bodies and mitochondria. The latter appear to stop at the origin of the axon. Note nuclei of the satellite Schwann cells 'embracing' the axon. Original magnification ×7000 reduced to 92%*

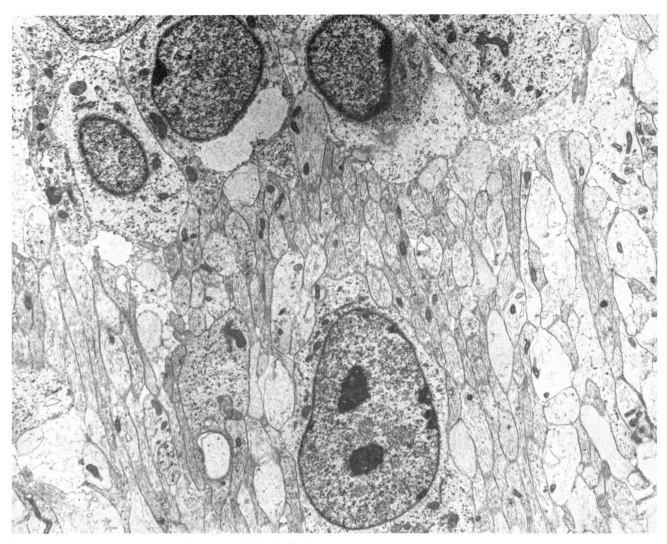

Figure 3.26 *Multiple nerve axons advancing on the basal cells of a sensory area (duck embryo otocyst in tissue culture). Original magnification × 10 500 reduced to 72%*

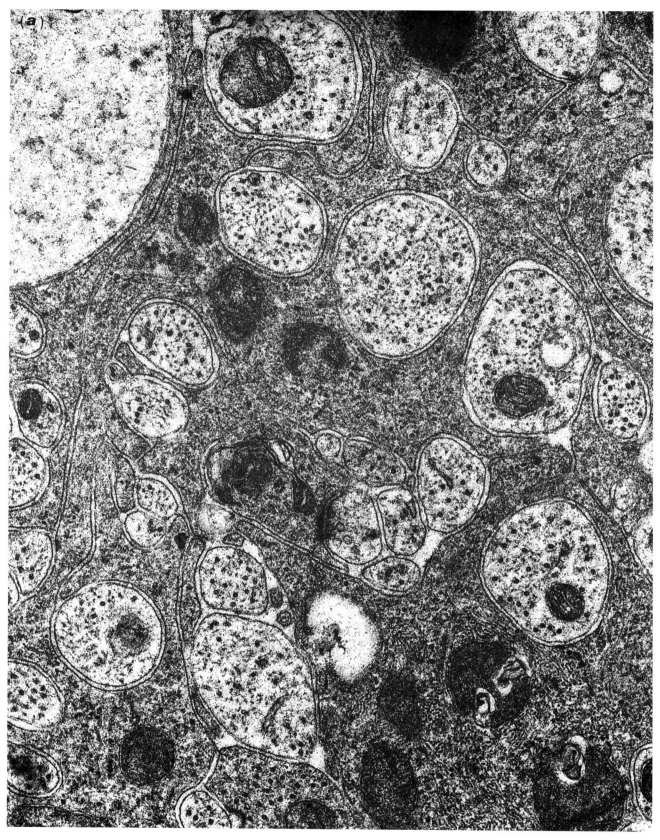

Figure 3.27 *(a) Nerve bundle composed of multi-axonal fibres (cross-cut). Note neurotubules and neurofilaments in the axon. Fowl embryo otocyst 12 days in vitro. Original magnification ×35 000 reduced to 90%.*

(**b**)

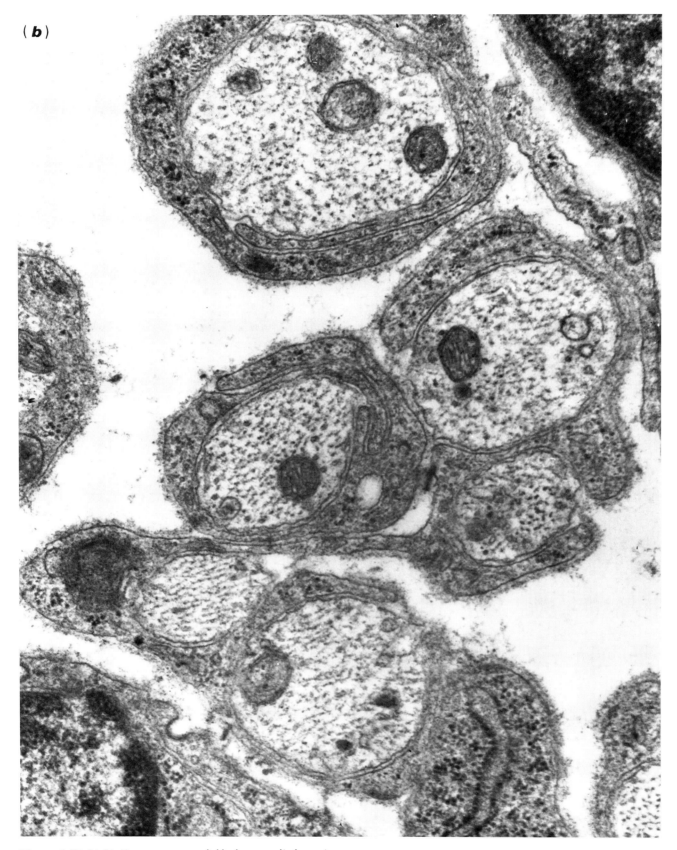

Figure 3.27 *(b) Similar axons surrounded by loose myelin layers in the mouse otocyst. Original magnification ×70 000 reduced to 90%*

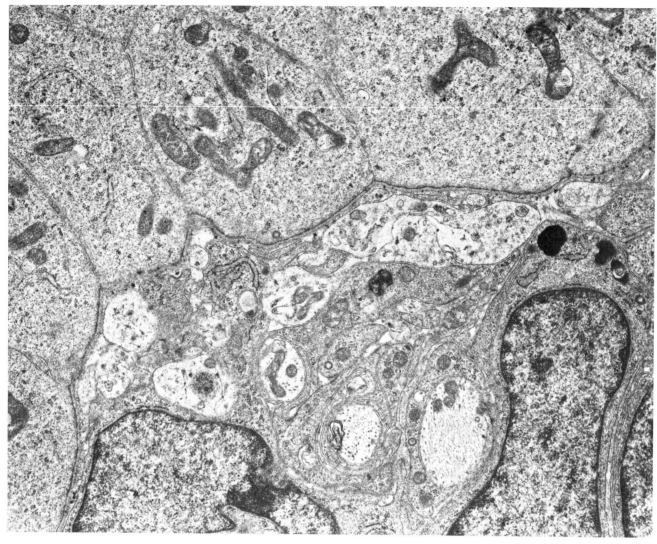

Figure 3.28 *Non-myelinated nerves near basal lamina (fowl embryo otocyst – 12 days in vitro). Many axons contain neurofilaments and mitochondria. Some are surrounded by loose myelin lamellae. Note basal lamina and basal cells containing many mitochondria. Original magnification ×32 300 reduced to 73%*

Figure 3.29 *Non-myelinated axons penetrating the basal lamina into the neuroepithelium. Note neurofilaments and mitochondria. Duck embryo otocyst 14 days in vitro. Original magnification ×10 500 reduced to 92%* ▶

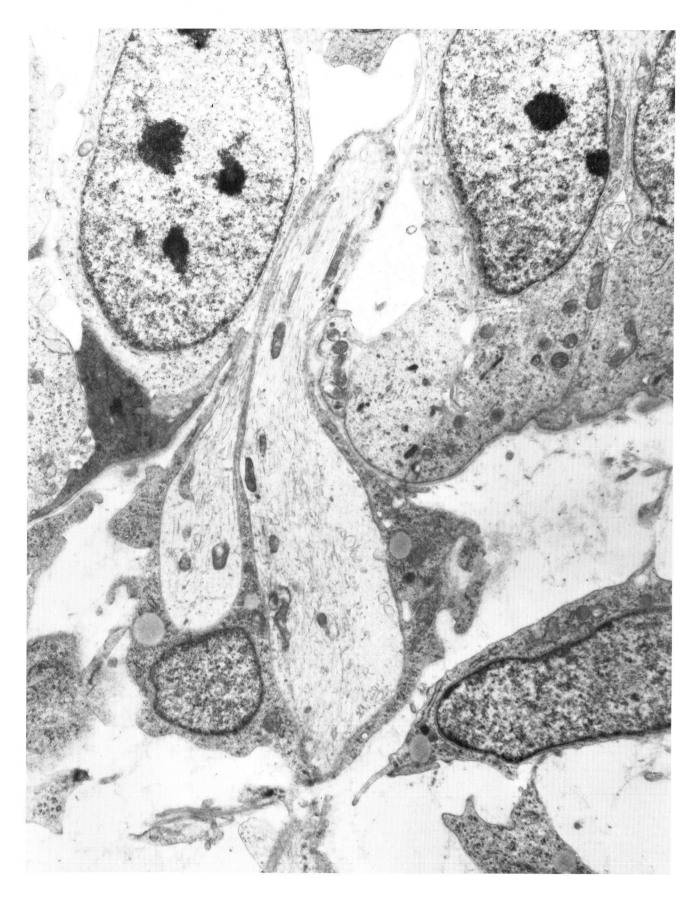

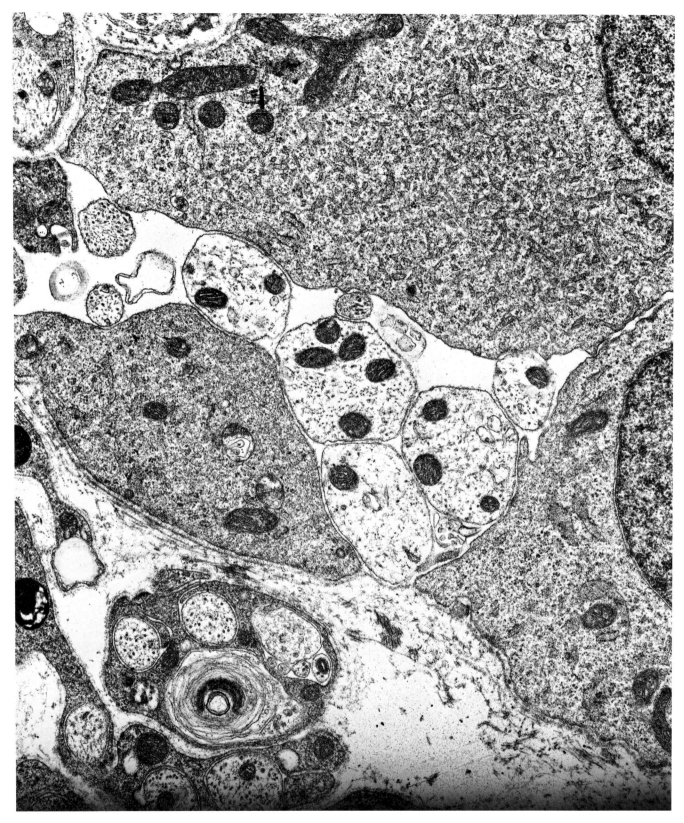

Figure 3.30 *Non-myelinated nerve fibres within and underneath the basal lamina. The axons contain mitochondria, neurotubules and neurofilaments. A Schwann cell has enveloped four axons, one of which is surrounded by several myelin layers. Original magnification ×32 300 reduced to 92%*

50

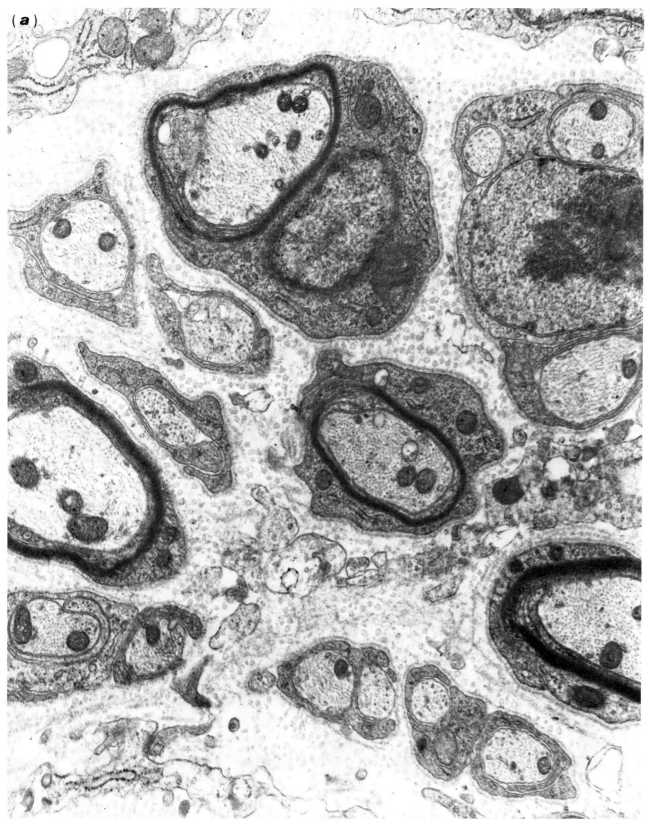

Figure 3.31 *(a) Myelinated and non-myelinated nerves in the developing fowl embryo otocyst in tissue culture (12 days in vitro). The nerve fibres are surrounded by Schwann cells. Note basal lamina around the Schwann cells. Original magnification ×28 000 reduced to 89%.*

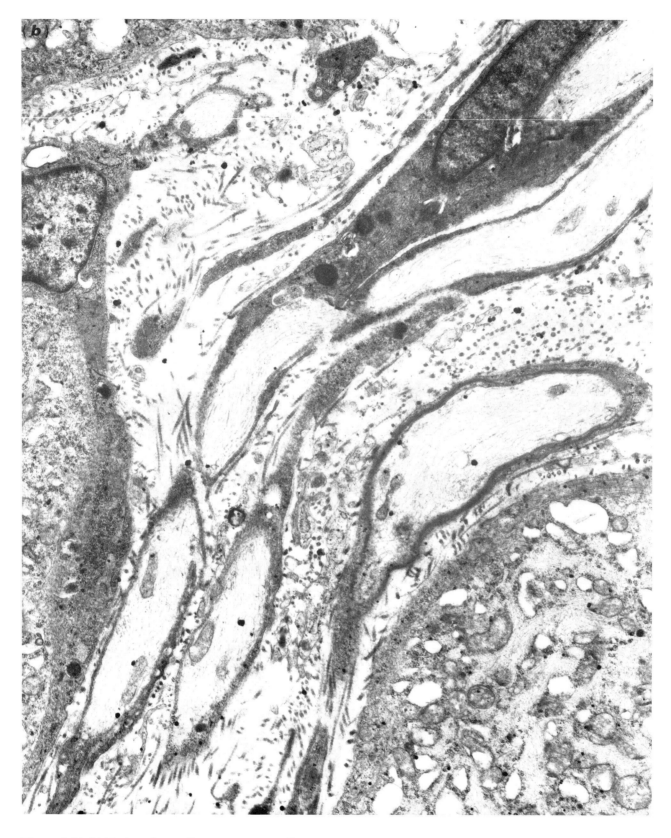

Figure 3.31 *(b) Myelinated nerve fibres containing neurofilaments running between two neurons. Original magnification ×17 500 reduced to 89%*

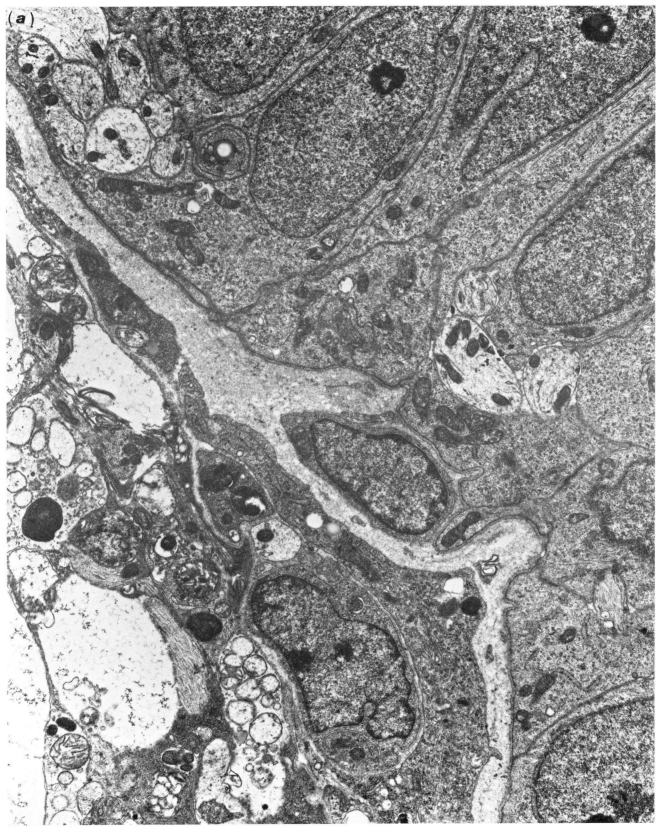

Figure 3.32 *(a) Non-myelinated nerves spreading among the supporting cells of the papilla basilaris. Original magnification ×28 000 reduced to 90%.*

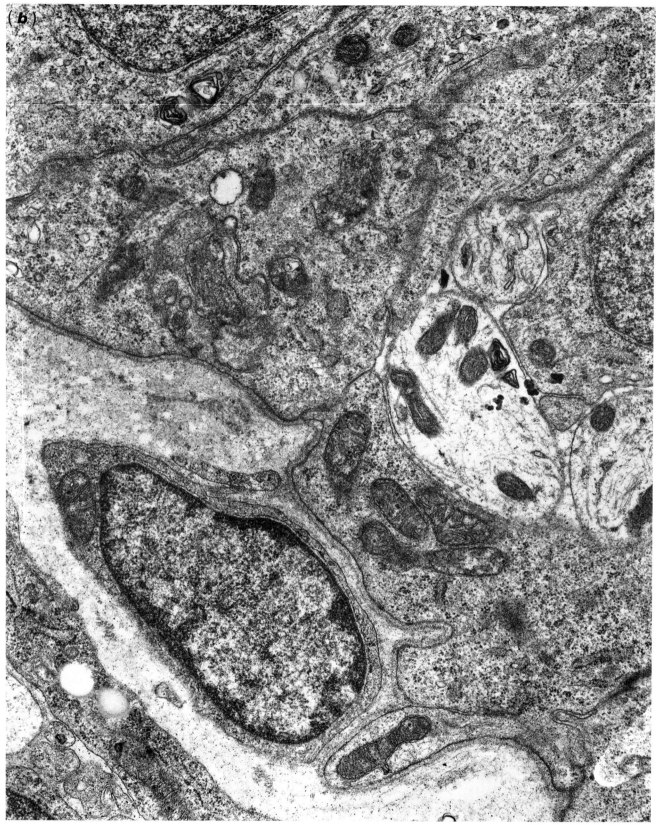

Figure 3.32 *(b) Detail of (a) showing cross-cut nerve axons between supporting basal cells. Original magnification ×35 000 reduced to 90%*

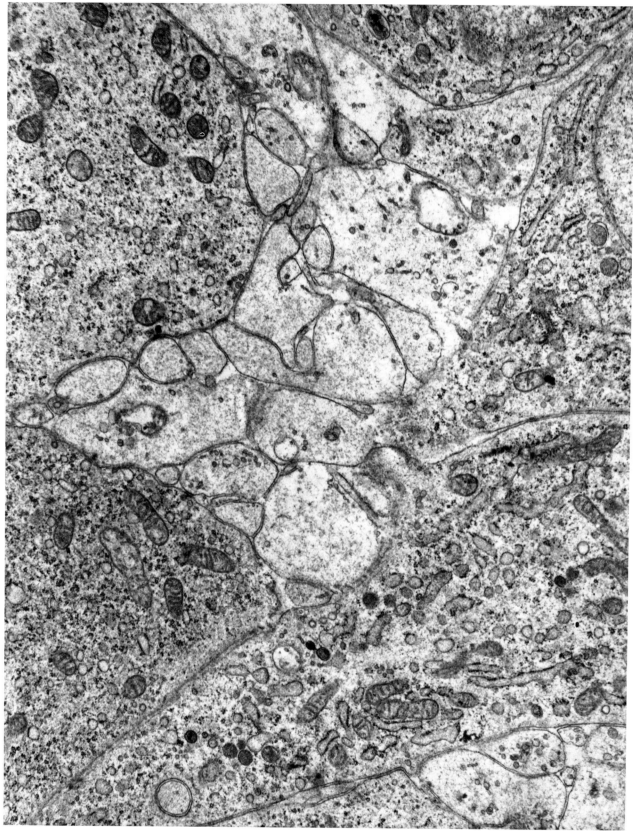

Figure 3.33 *10-day-old embryo tissue culture of chick embryo showing base of hair cell with numerous nerve endings. Note moderately granular endings of efferent type. Original magnification ×24 000 reduced to 88%*

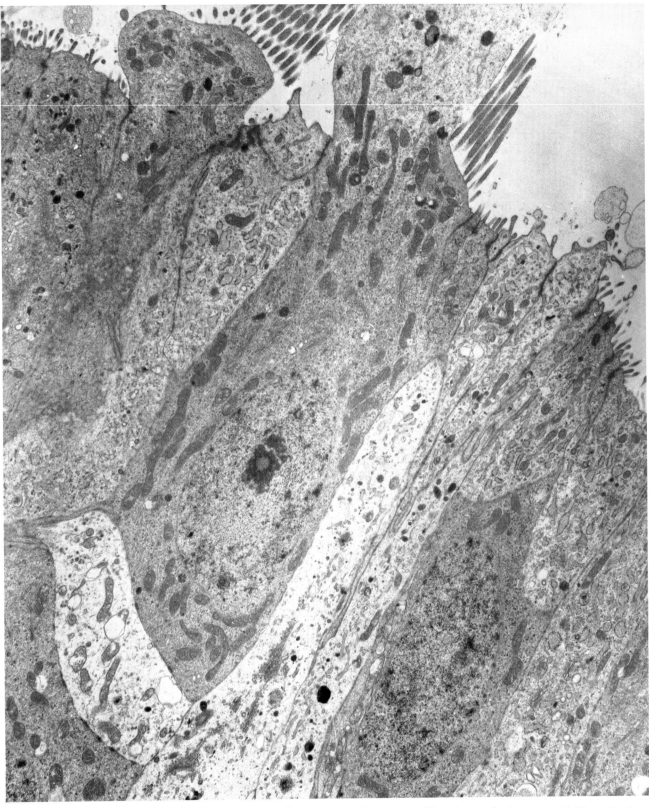

Figure 3.34 *12-day-old culture of chick embryo otocyst – note hair cells with nerves branching at their base. Well-developed cell membrane, stereocilia and organelles (compare with Figure 3.14). Original magnification ×8750 reduced to 90%*

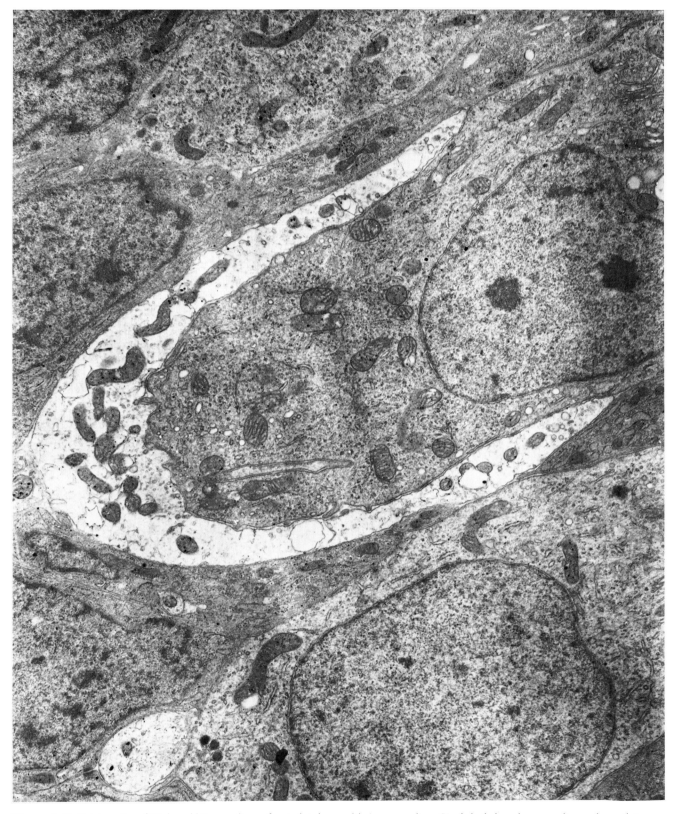

Figure 3.35 *Crista region of 12-day-old tissue culture of an isolated fowl embryo otocyst showing well-differentiated cup-shaped (calyx-like) nerve ending. Single bud-shaped nerve ending underneath type II cell. Original magnification × 14 000 reduced to 90%*

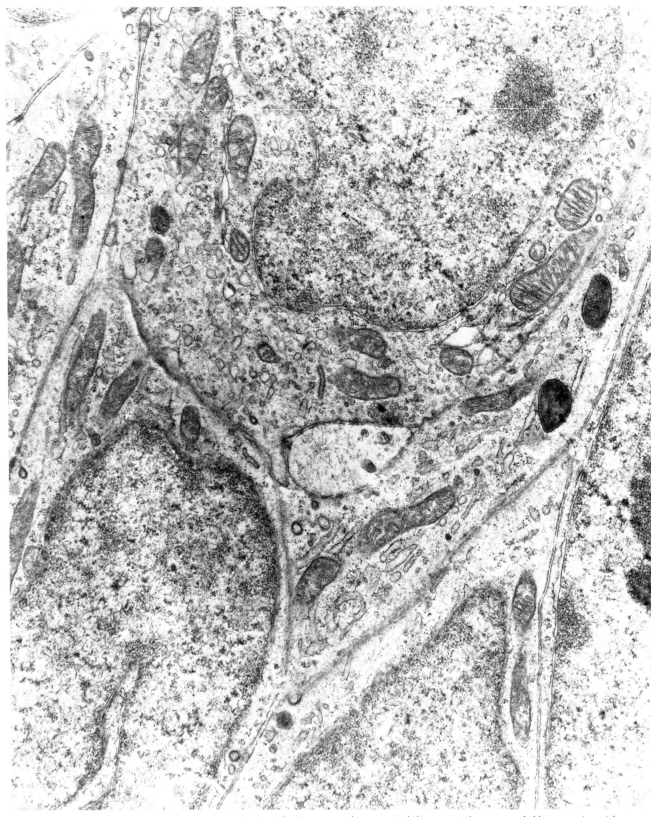

Figure 3.36 *Crista region in 12-day-old culture of isolated fowl embryo otocyst. Base of hair cell with bud-shaped nerve ending and dense osmiophilic synaptic bar surrounded by synaptic vesicles. Original magnification ×28 000 reduced to 92%*

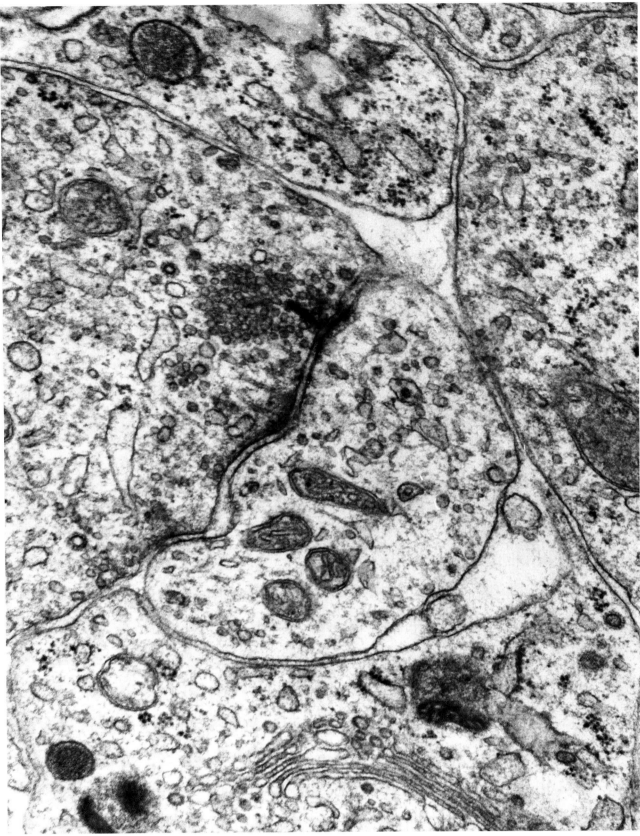

Figure 3.37 *Crista region in 12-day-old culture of isolated fowl embryo otocyst. Base of hair cell with dense osmiophilic synaptic bar* *surrounded by synaptic vesicles. Original magnification ×70 000 reduced to 90%*

References

Cajal, Ramon y. (1960) *Studies on Vertebrate Neurogenesis* (translated by L.Guth). Springfield, Illinois: Thomas.

Cohen, G. M. and Fermin, C. D. (1978) The development of hair cells in the embryonic chick's basilar papilla. *Acta Otolaryngologica*, **86**, 342–358.

Fell, H. B. (1929–1930) The development *in vitro* of the isolated otocyst of the embryonic fowl. *Arch. Exp. Zellforsch.*, **71**, 69.

Fell, H. B. and Robison, R. (1929) The growth and development and phosphatase activity of embryonic avian femora and limb buds cultivated *in vitro*. *Biochemical Journal*, **23**, 767.

Friedmann, I. (1959) Electron microscopic observations on *in vitro* cultures of the isolated fowl embryo otocyst. *Journal of Biophysics, Biochemistry and Cytology*, **5**, 263.

Friedmann, I. (1968) The chick embryo otocyst in tissue culture: a model ear. *Journal of Laryngology and Otology*, **82**, 185.

Friedmann, I. (1969) Innervation of the developing fowl embryo otocyst. *Acta Otolaryngologica (Stockholm)*, **67**, 224.

Friedmann, I. and Bird, E. S. (1961) The effect of ototoxic antibiotics on the isolated chick embryo otocyst. *Journal of Pathology*, **81**, 81.

Friedmann, I., Hodges, M. G. and Riddle, P. N. (1977) Organ culture of the mammalian and avian embryo otocyst. *Annals of Otology, Rhinology and Laryngology*, **86**, 371.

Harrison, G. R. (1907) Observations on the living developing nerve fibre. *Anatomical Record*, **1**, 116.

Hensen, V. (1864) The development of the nerves (in German). *Virchow's Archiv*, **31**, 51.

Hilding, D. A. (1969) Electron microscopy of the developing hearing organ. *Laryngoscope*, **79**, 1991.

Holley, R. W. (1975) Control of growth of mammalian cells in cell culture. *Nature (London)*, **258**, 487.

Lawrence, M. and Merchant, D. I. (1953) Tissue culture techniques for the study of the isolated otic vesicle. *Annals of Otology, Rhinology and Laryngology*, **62**, 770.

Meier, S. (1978) Development of the embryonic chick otic placode. *Anatomical Record*, **191**, 459–478.

Orr, M. F. (1975) A light and electron microsopic study of the embryonic chick otocyst: the effects of Trypsin and Ca- and Mg-free salt solution. *Developmental Biology*, **47**, 325–340.

Proctor, C. A. and Proctor, B. (1967) Understanding hereditary nerve deafness. *Archives of Otolaryngology*, **85**, 45.

Sher, A. (1971) The embryonic and post-natal development of the inner ear of the mouse. *Acta Otolaryngologica (Stockholm)*, **220**, 5.

Sobkowicz, H. M., Rose, J. E., Scott, G. L., Kuwada, S., Hind, J. E., Oertel, D. and Slapnick, S. M. (1980) Neuronal growth in the organ of Corti in culture. In *Tissue Culture in Neurobiology*, Eds. E. Giacobini *et al.*, pp. 253–273. New York: Raven Press.

Van De Water, T. (1976) Effects of removal of the stato-acoustic ganglion complex upon the growing otocyst. *Annals of Otology, Rhinology and Laryngology*, **85**, Supplement 33, 1.

Van De Water, T. R. and Ruben, R. J. (1973) Quantification of the 'in vitro' development of the mouse embryo inner ear. *Annals of Otology, Rhinology and Laryngology*, **82**, 19–21.

Van De Water, T. R., Heywood, P. and Ruben, R. J. (1973) Development of sensory structures in organ cultures of the 12th and 13th gestation day mouse embryo inner ears. *Annals of Otology, Rhinology and Laryngology*, **4**, 3–17.

Yamashita, T. and Vosteen, K. M. (1975) Tissue culture of the organ of Corti and the isolated hair cells from the newborn guinea pig. *Acta Otolaryngologica (Stockholm)*, **73**, 330.

4
The Ultrastructure of the Developing Organ of Corti of the Mouse in Culture

Hanna M. Sobkowicz   Jerzy E. Rose   Grayson L. Scott   Jon M. Holy

Note: In all Figures, the abbreviation Ih indicates inner hair cells; the abbreviations Oh-1, 2, or 3 indicate outer hair cells of the corresponding row.

Introductory remarks

Sobkowicz, Bereman and Rose (1975) have shown that the organ of Corti of the mouse, excised at or before birth, may survive explantation and continue to develop organotypically *in vitro* for some time. It is well known that all afferent fibres originate in the spiral ganglion; it is also generally assumed that all efferent fibres, which supply the sensory cells, originate in the region of the superior olives. There is evidence that, in the mouse, the efferent innervation develops postnatally; the efferents reach the inner hair cells around the second day after birth, and the outer hair cells about 1 week later. Thus the innervation of the organ of Corti in culture is especially intriguing, for an organ explanted at or before birth is deprived of any efferent inflow of central origin throughout its entire development.

Although the organ of Corti may undoubtedly differentiate and develop in isolation, a question to be answered is, to what extent do the various elements grow and mature in culture?

In this chapter we shall show that the ultrastructure of numerous components of the explant may remain for some days, or even weeks, remarkably similar to that of the intact animal. Most illustrations depict cultures of different ages. Unless otherwise stated, the culture has been explanted at birth. It was thought useful to include some electron micrographs obtained from intact animals, to stress the similarities and differences between normal and cultured tissues.

The organ of Corti and spiral ganglion can be excised together from the fetal, newborn, or early postnatal mouse. The cochlea of the fetal mouse is small enough to be explanted as a whole, but the larger newborn or

postnatal cochleas must be divided into several sectors. If the cochlear duct remains uninjured, the organ retains its structural organization in culture for several weeks. *Figure 4.1* presents an overall view of an 8-day culture excised from a newborn mouse and stained with silver. Apart from the injury of transection, the cochlear duct has remained intact. Spiral ganglion cells are assembled under the limbus, and the peripheral fibres of the closest cells collect into radial bundles. Most of the radial fibres stop at the level of the inner hair cells but some continue as tunnel fibres and form the outer spiral bundles, which innervate the outer hair cells (for details concerning the development and maintenance of the innervation of the organ of Corti in culture, see Sobkowicz, Bereman and Rose, 1975; Sobkowicz *et al.*, 1980, 1982; Rose, Sobkowicz and Bereman, 1977; Sobkowicz and Rose, 1983).

Hair Cells and Supporting Cells in Culture
Some Observations on Live Tissue

The use of the Maximow slide assembly and of an optically transparent substrate permits one to view live cells at high magnifications.

Figure 4.2 shows a live tectorial membrane. As reported (Sobkowicz, Bereman and Rose, 1975), the tectorial membrane develops and it can be maintained in an organotypic culture for over 10 days. In cultures derived from newborn animals and viewed the day after explantation, the tectorial membrane can be detected in polarized light as a layer of amorphous cloudy material overlying the cells of the inner spiral sulcus. During the first few days in culture, thin, needle-like radially oriented fibres gradually appear. The membrane extends towards the outer hair cell region and its development proceeds apicalwards. The same developmental sequence has been observed by Lim (1977) in the cat.

It is our impression that the tectorial membrane does not become attached to the outer edge of the hair-cell region. Around 12–14 days in culture, regressive changes occur, resulting in the retraction of the membrane along the edge of the limbus.

The fibrillar structure of the tectorial membrane was inferred by Iurato (1967) from his study of fresh and fixed specimens. According to Kronester-Frei (1978), the tectorial membrane of the mouse consists of two types of fibrils: straight unbranched (about 110 Å in diameter; 11 nm) and coiled branched (about 150–200 Å in diameter; 15–20 nm). The straight fibrils form bundles that are seen in the light microscope, whereas coiled branched fibrils appear as amorphous substance.

In culture, the arrangement of live hair cells and supporting cells can be viewed in different optical planes. *Figure 4.3* shows an unusual side-view of the outer hair cell region; *Figure 4.4* presents the traditional view of the reticular lamina as seen from above.

Ultrastructural Survey of the Developing Organ

Despite the trauma of explantation, there may be surprisingly few, if any, signs of degeneration within the hair-cell region (*see Figure 4.6*). Counts of inner hair cells, in silver-stained cultures, yield a mean value of 130 cells per millimetre length of basilar membrane (Sobkowicz, Bereman and Rose, 1975). This value compares rather closely with measurements of Retzius (1884), whose corresponding figures are: 104 inner hair cells per one millimetre for man, 111 for the cat and 107 for the rabbit. It appears that, in mammals, the number of hair cells is a function of the length of the organ. The larger numbers in our material are probably the result of denser cell packing in the smaller cochlea of the newborn, as compared with that of the adult.

The appearance of the fetal or newborn organ of Corti in a young culture may be quite similar to that of the animal of a corresponding age (*Figures 4.5* and *4.6*). Most cells in the organ are tightly packed, and the hair cells and supporting cells do not interlace fully. Consequently, the sensory cells may partly appose each other. Moreover, the inner hair cells, supported only in part by border and inner phalangeal cells, may sometimes be seen in apposition either to the inner spiral sulcus cells or to the inner pillars. The inner spiral sulcus is as yet multicellular (*Figure 4.5*); cell reduction to a single layer will occur, both in the intact mouse and in culture, in the course of further development. Small extracellular spaces may be present already in the newborn cochlea. Such pockets develop first among the supporting cells, at a level close to the basilar membrane. In the adult animal, the inner and outer hair cells are inclined toward the tunnel of Corti. In the newborn cochlea, the pillars are apposed to each other and there is no tunnel of Corti; as a result the hair-cell arrangement is fairly columnar (*Figure 4.6*).

A significant advance in the organotypic development of the cochlea occurs in culture. A comparison of *Figures 4.6* and *4.9* permits some estimate of the progress *Figure 4.7* shows an apical turn in a 14-day culture. The outer pillar apposes the head of the inner pillar at an angle, and the tunnel of Corti has begun to form. Pockets of fluid are present around the outer hair cells. Processes of Deiters' cells supply specialized support for the outer hair cells and ascending outer spiral fibres; the latter cross the lower part of the tunnel.

The intermediate zone of the hair cells, immediately below the cuticular plate, contains mitochondria, free ribosomes, smooth and rough endoplasmic reticulum, and microtubules. Multivesicular bodies and lysosomes are present in the Golgi region. The round or oval nuclei often are indented (*see Figure 4.6*, 0h-3).

The infranuclear zone of the hair cells may be narrow and contains mitochondria that tend to group at the sites of the nuclear indentations. Microtubules, microfilaments, and smooth and rough endoplasmic reticulum also are present within the infranuclear region of the hair cells. In some cells, however, in which the organelles are thinly distributed, free ribosomes and smooth endoplasmic reticulum dominate the organelle population.

Microtubules and microfilaments are already present in the hair cells at birth. Their number increases postnatally, especially in the infranuclear zone and just below the cuticular plate.

Differentiation of the outer hair cells in culture is similar to that in the intact animal. In contrast to the outer hair cell, the inner hair cell retains some features of a relatively undifferentiated cell. The most striking feature is the smooth outline of the receptor pole; in the intact mouse, the afferent endings deeply indent the cell surface. The smooth membrane outlines observed in culture (*see Figure 4.15*) may be the outcome of a partial deafferentation of the inner hair cells, due to neuronal loss that follows the explantation (Sobkowicz, Bereman and Rose, 1975).

Kinocilia

In the hair cell of the mouse, both the cuticular plate and the stereocilia develop largely postnatally. *Figure 4.10* illustrates the apical zone of an outer hair cell in a 14-day culture. The cuticular plate is wedge shaped; the hair cell is always oriented so that the widest portion of the plate faces the modiolus. The cuticular plate contains stereocilia, which are arranged nearly parallel to each other. Each hair consists of a dark root, a slender neck, and a free segment. The root may be split in the cuticular plate. Both stereocilia and the cuticular plate contain actin filaments (Flock *et al.*, 1981).

A single cilium is inserted into the cuticular-free region of the apical zone. This is the kinocilium, whose free segment in cross-section shows a typical 9+2

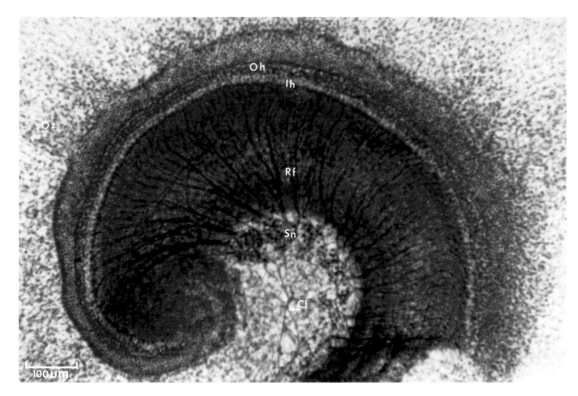

Figure 4.1 *An 8-day organotypic culture of the organ of Corti. The apical tip is at the lower left; the line of transection at the lower right. The cells of the outgrowth zone (Oz) surround the explant. Spiral neurones (Sn) give rise both to the radial (Rf) and to the central (Cf)* *fibres. Holmes silver stain. Partial polarization (Modified from Sobkowicz, Bereman and Rose, 1975, by kind permission of publishers.)*

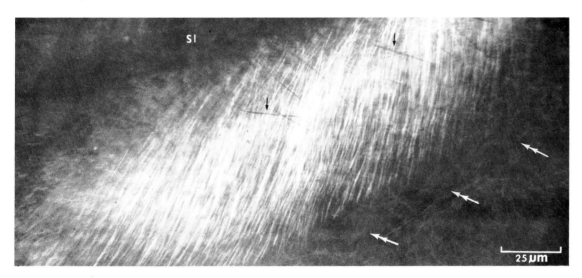

Figure 4.2 *A live tectorial membrane in polarized light in an 11-day culture of a fetal (15–16 days of gestation) organ of Corti. Needle-like fibrils run radially from the edge of the limbus (SI) toward the outer edge of the outer hair cell region, whose approximate* *location is indicated by the double-headed arrows. Note the occasional fibrillar bundles crossing the radial fibrils (arrows). The reticular plate of the same culture is shown in Figure 4.4*

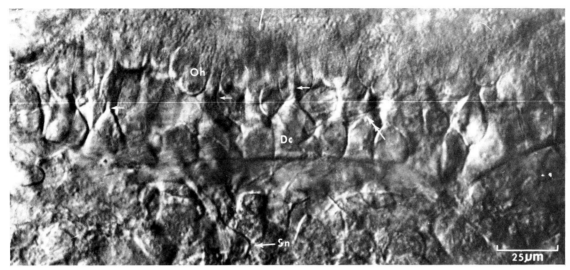

Figure 4.3 *A side view of the outer hair cell region in a live 10-day culture of a fetal (15–16 days of gestation) cochlea. The wide bases of Deiters' cells (Dc) delineate the basilar membrane; their narrow processes (arrows) extend to the reticular plate. Outer hair cells are visible in the upper part of the Figure. In the lower portion of the photograph, a neurone (Sn) emits a fibre which appears to penetrate the basilar membrane. A double-headed arrow points to spirally running fibres at the base of the outer hair cells. Nomarski differential interference*

Figure 4.4 *The reticular plate of a live 11-day culture of a fetal (15–16 days of gestation) cochlea. Stereocilia identify the sensory cells. The orderly alignment of the inner and outer hair cells forms a geometric pattern. The tectorial membrane of the same culture is shown in Figure 4.2. Nomarski differential interference*

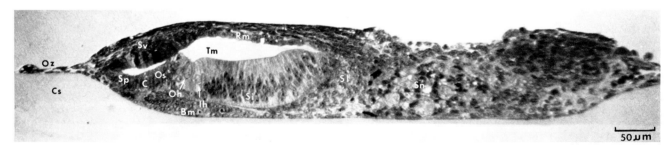

Figure 4.5 *A cross-section of an explant of the apical turn, 48 hours after explantation. Bm = Basilar membrane; C = Claudius' cells; Cs = collagen substrate; Os = outer spiral sulcus; Oz = outgrowth zone; Rm = Reissner's membrane; Si = inner spiral sulcus cells; Sl = spiral limbus; Sn = spiral neurons; Sp = spiral prominence; Sv = stria vascularis; Tm = tectorial membrane. Semi-thin section, Richardson's stain. Nomarski differential interference*

64

three largely denervated cultures, where no nerve endings were present within the region studied. This finding raises the question whether the denervation may influence the production of annulate lamellae in the receptor cells.

Filamentous Bodies

An unusual organelle was found in four inner hair cells in a 24-day culture. The organelle is a filamentous body not bound by a membrane (*Figure 4.15*). Reconstruction based on serial sections (over 64) of one such body revealed it to be a rod over 5 μm in length. In some sections, the organelle lies very close to the nucleus in a trough-like invagination of the nuclear membrane.

Intracytoplasmic filamentous bodies not bound by a membrane are only rarely reported. Bands of dense filaments are characteristic of Purkinje cells of the Syrian hamster (Morales and Duncan, 1967). Mountford

(1964) described filamentous spindles in the receptor-bipolar synapses in the retina of the albino guinea-pig. Some of these spindles resemble the organelle shown in *Figure 4.15*. Takahashi and Hama (1965) found round filamentous bodies in the ciliary ganglion of the chick. Somewhat irregular filamentous bodies have been observed by Elfvin (1963) in the cervical sympathetic ganglion of the cat.

In contrast to the rare occurrence of *intracytoplasmic* fibrillar inclusions, the *intranuclear* filamentous structures, sometimes called the rodlets of Roncoroni (Ramón y Cajal, 1952) are well known. The fibrillar rod found in our material most closely resembles the intranuclear rodlets described by Siegesmund, Dutta and Fox (1964) in some populations of nerve cells in the squirrel monkey and rabbit.

The rodlets are also known to occur in the bipolar and ganglion cells in the retina of the rabbit and rat (Magalhaes, 1967); in the neurones of the hippocampus

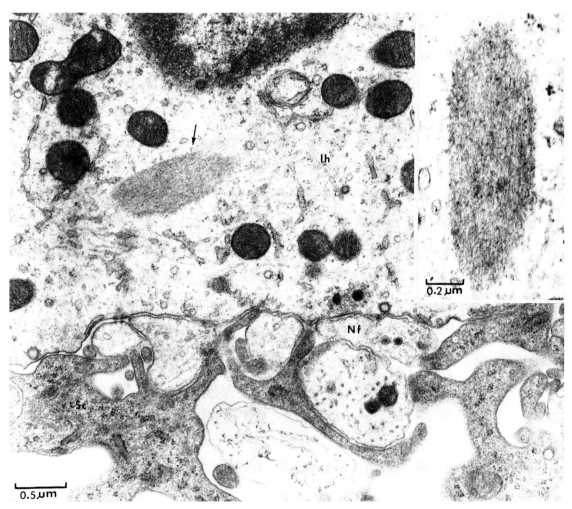

Figure 4.15 *A section through a long fibrillar rodlet (arrow) in an inner hair cell of a 24-day culture. The insert shows the filamentous content of the rodlet under high magnification. Reconstruction from serial sections indicates that this rodlet was at least 5 μm long. Two synaptic ribbons and a coated vesicle are present at the apposition of a*

nerve ending (Nf). Note the dense-core vesicles within the ending. Sc = Supporting cell. Apex.

Note: Figures 4.16–4.23 show different views of an organelle encountered in the cells of the outer spiral sulcus and Hensen's cells of older cultures.

and thalamus of the mouse (Chandler, 1966); in the sympathetic ganglion cells of the chick in culture (Masurovsky *et al.*, 1970); and in the ependymal cells of the lateral ventricle of the rat (Hirano and Zimmerman, 1967). In addition, the intranuclear rodlets may occur in about 10 per cent of some malignant glioma cells (Robertson and MacLean, 1965). In the sympathetic ganglion cells of the chick, the size and frequency of occurrence of the rodlets may be enhanced by deuterium (Murray and Benitez, 1967, 1968).

On rare occasions, intranuclear rodlets have been observed to pass from the nucleus into the cytoplasm (Holmgren, 1899; Hirano and Zimmerman, 1967; Popoff and Stewart, 1968; Masurovsky *et al.*, 1970). The rod observed by us was located next to the nucleus in many sections, but there was no evidence of it penetrating the nuclear envelope.

Cisternal Intercalated Bodies

We have encountered in the organ of Corti another organelle (*Figures 4.16–4.23*) that appears to have been noticed only once before, in a guinea-pig oocyte (Adams and Hertig, 1964). The organelle was found in the cells of the inner and outer spiral sulcus. It has been encountered in older cultures (14–34 days *in vitro*) and in the intact animal older than the fourteenth postnatal day. More than one organelle may occur in the same cell.

The organelle consists of interconnected cisternae of endoplasmic reticulum and of elongated rod-like elements that are intercalated with the cisternae. The cisternal membranes are predominantly smooth, but some may display numerous ribosomes.

The rod-like elements are approximately 0.1 μm wide and vary in length. They are membrane bound and contain irregularly distributed electron-dense material. The rods of adjacent rows are staggered and form a hexagonal pattern (*Figures 4.16, 4.17* and *4.23*). The rods appear to be connected with the cisternae (*Figure 4.19*). Parasagittal sections through the organelle show that the cisternae are interconnected with each other (*Figures 4.20* and *4.21*). In an oblique parasagittal section, the connecting channels form a diagonal pattern, lending the organelle a checkered appearance.

Mitochondria are commonly seen in close proximity to the organelle. It is our impression that the rod-like structures may be derived from mitochondria. Close alignment of endoplasmic reticulum to structures that possibly are modified mitochondria often are seen

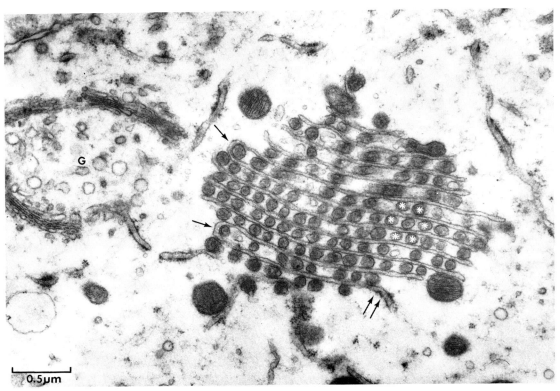

0.5μm

Figure 4.16 *A transverse section of the organelle reveals cisternae of smooth endoplasmic reticulum and electron-dense rod-like structures (compare with Figure 4.22) that are sandwiched between the cisternae and lie parallel to their plane (compare with Figure 4.22). The cisternae of smooth endoplasmic reticulum are of a constant width, lie parallel to each other, and appear to form 'u'-like loops (arrows). The rod-like structures, shown in cross-section, are membrane bound and contain irregularly distributed electron-dense material. Note the staggered arrangement of the rodlets giving rise to a hexagonal array (asterisks). A double arrow points to the transition between rough and smooth endoplasmic reticulum. Presumably an outer spiral sulcus cell. G = Golgi complex. Apex. 24-day culture*

Figure 4.18). In about 100 organelles observed so far, however, no clear evidence pointing to the origin of the rod-like structures was obtained.

The location of the organelle within the cell varies. It often was found near the Golgi complex or cell membrane. It was seen in the outgrowth zone in cells forming fluid pockets. Possibly, the organelle is involved in the production or transport of endolymph.

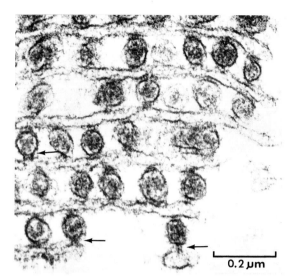

Figure 4.19 *An underexposed print, at higher magnification, of the lower right part of the organelle shown in Figure 4.17. The rod-like structures are membrane bound and display an electron-dense granular content. Note the apparent connections between the rod-like structures and the cisternae (arrows)*

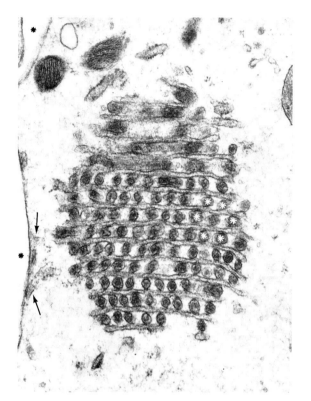

Figure 4.17 *A section through the intercalated body in a cell in the outgrowth zone of a 22-day culture. The organelle lies near a wide extracellular space, which is probably a fluid pocket (black asterisks). Some cisternae appear to be in contact with the cell membrane (arrows). Note the shift in rodlet orientation in the upper part of the organelle. White asterisks indicate the hexagonal pattern of the rodlets. Base. Same magnification as in Figure 4.16*

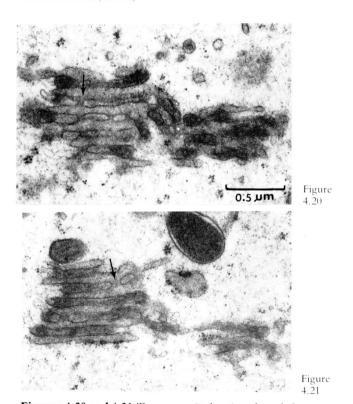

Figure 4.20

Figure 4.21

Figures 4.20 and 4.21 *Two parasagittal sections through the same organelle. Note the luminal continuity between the individual cisternae (arrows). A cell in the outer spiral sulcus. Base. 14-day culture. The scale in Figure 4.20 also applies to Figures 4.21–4.23*

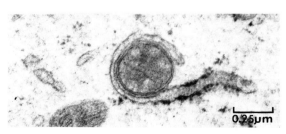

Figure 4.18 *A higher magnification of a round structure, presumably of mitochondrial origin, entrapped by endoplasmic reticulum. The round body shows a double membrane but no cristae. Note the transition of rough endoplasmic reticulum into smooth reticulum on contact with the mitochondrion-like structure. The structure was found in close proximity to the organelle shown in Figure 4.22. Outer spiral sulcus cell in a 14-day culture. Base*

71

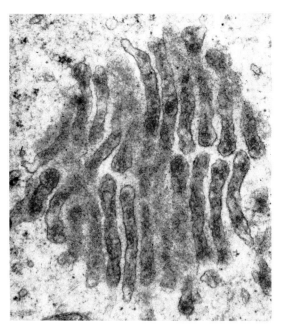

Figure 4.22 *A longitudinal section through a layer of the rod-like structures. The rods are about 1 μm long and 0.1 μm wide. The rods form two rows; the rods of the upper row appear to alternate with those of the lower one. They are membrane bound and contain irregularly distributed electron-dense material. The rods may show either club-like or tapering ends. Presumably an outer spiral sulcus cell. Base. 14-day culture. Same magnification as in Figure 4.20*

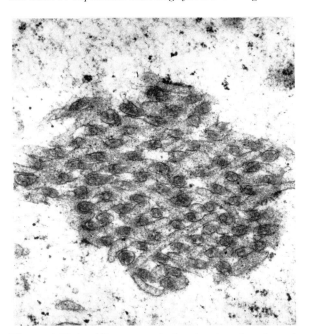

Figure 4.23 *An oblique parasagittal section through an intercalated body showing a diagonal pattern of interconnected channels. Notice the elegant geometrical pattern of smooth endoplasmic cisternae alternating with rod-like elements. Outer spiral sulcus cell. Base, 14-day culture. Same magnification as in Figure 4.20.*

Afferent and Efferent Patterns

The relations of sensory cells to the supporting cells and nerve endings in the developing cochlea differ substantially from those in the mature animal. In the adult, the hair cells are separated from each other by supporting cells, and the basal poles of the receptors are surrounded either entirely (inner hair cells) or partly (outer hair cells) by nerve endings. Both the inner and outer hair cells receive afferent and efferent innervation; the efferent endings are especially prominent in junctions with the outer hair cells. In the mouse, much of the innervation develops postnatally during the first 2–3 weeks. Only afferent endings are present in the organ before birth. The first vesiculated efferent fibres enter the inner spiral bundle during the second postnatal day; the first efferent endings approach the outer hair cells around the ninth postnatal day. During the first 2–3 weeks – before the final pattern is established – the sensory and supporting cells develop transient appositions with the incoming nerve endings. It is noteworthy that, according to behavioural tests, the mouse begins to hear around the twelfth postnatal day. The development of hearing continues at a fast pace until the fourteenth to sixteenth postnatal day (Alford and Ruben, 1963; Ehret, 1976), at which time the structure of the organ appears largely mature.

Early Neurosensory Contacts

As mentioned before, in the cochlea of the newborn mouse, the supporting cells may still envelop the infranuclear surface of the hair cell. *Figure 4.24* shows an outer hair cell surrounded entirely by Deiters' cells. Such relations are transient and are subsequently interrupted by the ingrowing nerve fibres (*Figures 4.25* and *4.26*). The mechanism by which the tip of a growing nerve ending separates the membranes of adjacent cells is unknown; with the exception of specialized junctions, the contacts between cells may not be binding, or a growing nerve ending may elaborate a proteolytic enzyme. In any case, the afferent endings in the organ of Corti seem to travel freely among the supporting cells.

The initial apposition of the nerve fibre to the hair cell is exploratory: a growing tip surrounds most of the receptor surface and emits filopodia. *Figure 4.26* illustrates a growth cone at the receptor pole of an outer hair cell in a 1-day-old mouse. Thin filopodia insinuate themselves between the hair cell and supporting cells, apparently interrupting the contacts between them.

After reaching the target cell, the fibre may retract its filopodia and cease to grow in the area of contact (*Figure 4.27*). Some fibres, however, may resume growth in a new direction: the nerve ending in *Figure 4.26* contains a collection of growth vesicles located at some distance from the hair-cell membrane – indicating an incipient growth away from the cell. The change in the direction of growth, following contact with the hair cell, appears

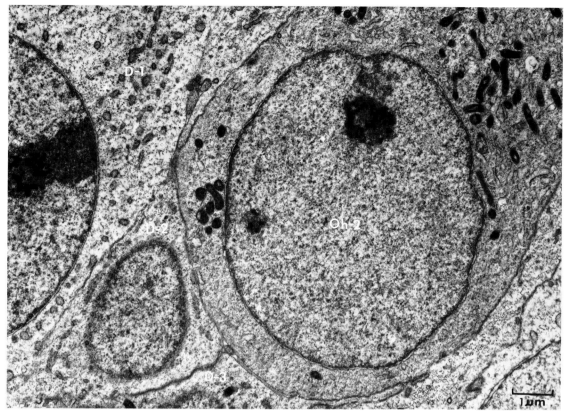

Figure 4.24 *An outer hair cell surrounded by three Deiters' cells (D-1,2,3) in a 2-day culture. Note the dense cell packing and lack of* *extracellular fluid spaces. Mid-turn*

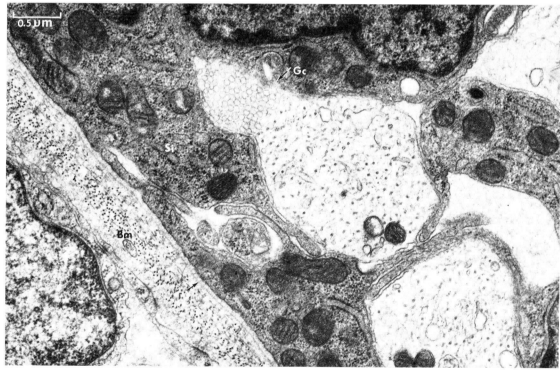

Figure 4.25 *A growth cone (Gc) of an afferent fibre among the inner spiral sulcus cells (Si). The arrow at the lower left points to the* *basement membrane. Bm = Basilar membrane. Apex. 6-day culture*

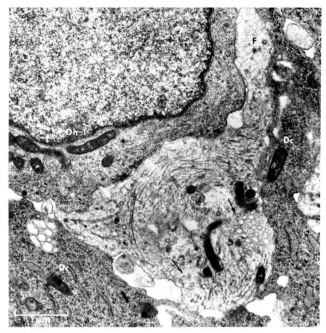

Figure 4.26 *Early apposition of a growth cone of an afferent fibre to an outer hair cell. Note the filopodium (F) burrowing between the hair cell and supporting cell (Dc). The location of growth vesicles (arrows) suggests growth away from the cell. Apex. 1-day-old intact mouse. Magnification reduced to 81% on reproduction*

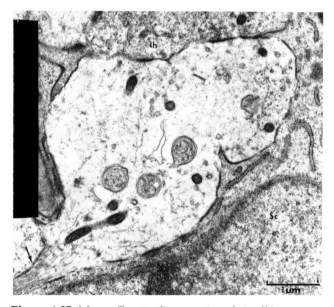

Figure 4.27 *A large afferent ending on an inner hair cell in a young (2 days in vitro) culture. The slender shaft of the nerve fibre can be seen to the lower left (arrow). Sc = Supporting cell. Apex. Magnification reduced to 81% on reproduction*

to be characteristic of the outer and inner spiral fibres. This implies that the spirally running fibre approaches the hair cells sequentially, and the radially oriented end-collaterals are formed later.

Deposition of clear and dense-core vesicles along the hair-cell membrane appears to follow the apposition of the nerve fibre. *Figure 4.28* shows the lower pole of an outer hair cell in a 1-day-old mouse. A row of vesicles overlies the hair-cell membrane along the area of nerve-fibre apposition. Such alignment of vesicles is a transient event that occurs prior to the formation of synaptic specializations. Numerous dense-core vesicles in the early afferent synapses in the guinea-pig have been noted by Thorn (1975). These vesicles may be akin to those which, as Pfenninger *et al.* (1969) suggest, carry material involved in the formation of presynaptic densities.

Newly formed afferent endings are large, and each occupies an extensive receptor surface (*see Figure 4.27*). They eventually are either replaced by or transformed into a series of small synaptic endings. *Figures 4.29* and *4.30* illustrate small multiple nerve endings surrounding the receptor pole of an outer hair cell in culture and in the intact animal, respectively. The developmental sequence in the differentiation of the neurosensory contacts (*Figures 4.24–4.30*) is similar in both the intact and cultured cochlea.

Afferent Synapses

In the cochlea of the mouse, afferent synapses are formed mainly postnatally, and their development proceeds for at least 14 days after birth. A conspicuous organelle of the afferent synapse is the synaptic ribbon (*Figures 4.31–4.33*). Synaptic ribbons (electron-dense bodies surrounded by synaptic vesicles) were first described by Smith and Sjöstrand (1961). They occur in sensory cells of many species (for a review, see Sobkowicz *et al.*, 1982).

Synaptic ribbons are already present in the newborn mouse, and they continue to develop in culture (Scott *et al.*, 1976; Sobkowicz *et al.*, 1982). In the very young intact animal and in culture, most ribbons are spheroid. This form is thought to be phylogenetically older (Wersäll, Gleisner and Lundquist, 1967; Favre and Sans, 1979) – a notion consonant with our observations that during development the roundish forms are largely replaced by plate ribbons (Sobkowicz *et al.*, 1982).

Spheroid ribbons are usually attached to two triangular presynaptic densities. In *Figure 4.33 (a)* and *(b)*, the dense body extends two processes toward the presynaptic membrane. Thin rodlets connecting both processes with presynaptic densities can be seen at left in *Figure 4.33 (a)* and at right in *Figure 4.33 (b)*. A single layer of synaptic resides surrounds the dense body of the ribbon (*Figure 4.32 (a)–(c)*). The presynaptic and postsynaptic membranes are electron dense, the cleft is

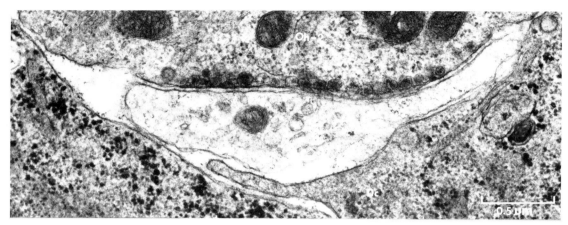

Figure 4.28 *Alignment of dense-core vesicles along the membrane of an outer hair cell, following nerve fibre apposition. Dc = Deiters' cell. Mid-turn. 1-day-old* intact *mouse*

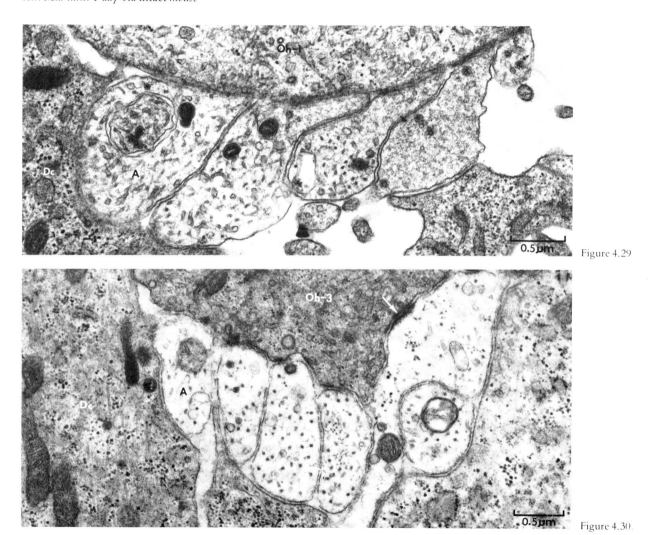

Figure 4.29

Figure 4.30

Figures 4.29 and 4.30 *Multiple afferent endings (A) at the receptor poles of outer hair cells. Figure 4.29: Apex. 14-day culture. Figure 4.30: Mid-turn. 9-day* intact *mouse. The arrow points to slightly asymmetrical densities at the apposition of the nerve fibre to the hair-cell membrane. Dc = Deiters' cell*

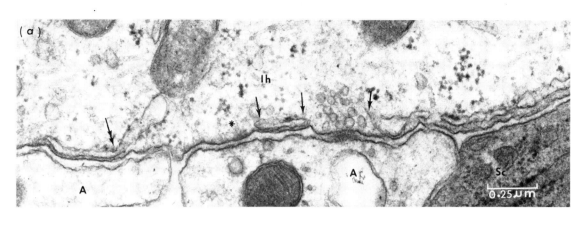

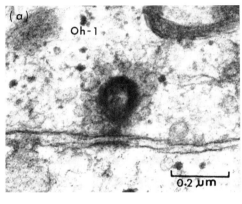

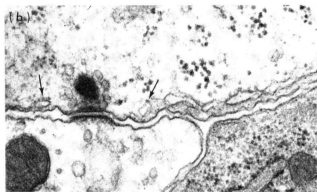

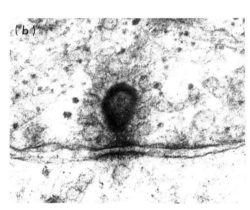

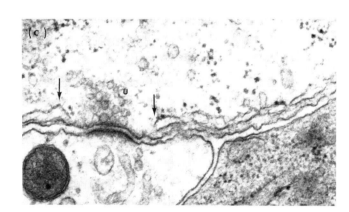

Figure 4.31 (a) and (b) Two consecutive sections through a ring-like ribbon in an outer hair cell in a 6-day culture. The ribbon is attached to a single presynaptic density. Apex. The scale in (a) also applies to (b)

Figure 4.32 (a–c) Three consecutive sections through a ribbon synapse in a 14-day culture. The scale in (a) also applies to (b) and (c). Synaptic vesicles surround the electron-dense body of the ribbon. The asterisk in (a) indicates the site of another incoming ribbon. Subsurface cisternae extend over the areas of hair cell afferent fibre (A) apposition but are interrupted (arrows) by the presynaptic complexes. Note the transition of rough endoplasmic reticulum into a subsurface cisterna (double-headed arrow). Sc = Supporting cell. Apex

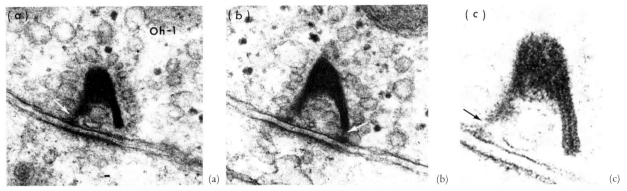

Figure 4.33 (a–c) Two consecutive sections through a laminated ribbon in an outer hair cell in a 6-day culture. (c) shows, under higher magnification, an underexposed print of the same section shown in(a); note the laminated substructure of the ribbon. The ribbon is attached through thin rodlets to two discrete presynaptic densities (arrows). Apex. The scale in 4.31(a) applies to 4.33(a) and (b); 4.33(c) is an enlargement, by about ×2, of 4.31(a)

of uniform width, and the postsynaptic density is coextensive with the presynaptic ribbon complex. The slight elevation of the postsynaptic surface of the nerve fibre (*Figure 4.32(c)*) and the attachment to a single presynaptic density (*Figure 4.31*) are both signs of synapse maturation that occur in culture.

The dense body of the ribbon may appear granular or show a laminated substructure. The lamellar substructure is easier to detect in underexposed prints (*Figure 4.33 (c)*). In plate ribbons of adolescent mice, the lamellae run parallel to the long axis of the ribbon; there are at least three electron-dense lamellae per ribbon. The lamellar substructure of the synaptic ribbons has been noted in the cochlear hair cells of the guinea-pig (Saito, 1980) and in the vestibular hair cells of the cat (Favre and Sans, 1979).

The function of synaptic ribbons is unknown. There are reasons, however, to suppose that in some sensory cells the ribbons may store neurotransmitter or its precursors (Thornhill, 1972; Osborne and Thornhill, 1972; Wagner, 1973; Monaghan, 1975; Osborne, 1977). According to Spadaro, deSimone and Puzzolo (1978), the number of ribbons in the retina varies with the circadian cycle, the average life span of a ribbon being about 8 hours. A laminated structure indicates a newly formed organelle, whereas granularity of the dense body implies an aged ribbon. This suggests that synaptic ribbons may be temporary structures.

The centre of some ribbons is less electron dense than the periphery is, often lending the ribbon a ring-like appearance (*Figure 4.31*). This may be another feature related to the life cycle of the ribbon. The size of the lighter region varies, and in some ribbons the dense ring is incomplete. Ribbons of this kind have been described in the cochlear and vestibular hair cells of various mammals (Smith and Sjöstrand, 1961; Engström, Bergström and Ades, 1972; Kimura, 1975; Bodian, 1978, 1980; Favre and Sans, 1979; Liberman, 1980; Saito, 1980); but usually little attention has been paid to such observations. There is also some evidence that the dense body of the ribbon may change its position within the

cell (Flock, Jorgensen and Russell, 1973). Thus, it seems feasible that ribbons that differ in their appearance and location represent organelles in different functional states.

Efferent Fibres in Culture

As far as is known, the organ of Corti in culture is innervated exclusively by cells of the spiral ganglion. *Surprisingly, however, nerve endings containing synaptic-sized vesicles may sometimes be seen in culture in conjunction with the inner hair cells.* These endings (*Figure 4.34*) closely resemble some efferents of the inner spiral bundle in the intact mouse (*Figure 4.35*). The fibres, both in culture and in the intact animal, contain islands of synaptic vesicles and occasional dense-core vesicles but are otherwise indistinguishable from afferent fibres. The synaptic vesicles may aggregate at the synaptic site or lie free in the axoplasm. *Figure 4.36* illustrates an efferent ending forming an annular synapse with an afferent spine in the region of the inner hair cell in the intact mouse. The ending is filled with synaptic vesicles that are densely packed around the presynaptic membrane. A few dense-core vesicles are present. In the

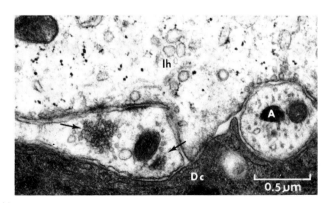

Figure 4.34 A vesiculated ending (arrows), seemingly forming an efferent synapse with an inner hair cell in a 14-day culture. A = Afferent ending; Dc = Deiters' cell. Apex

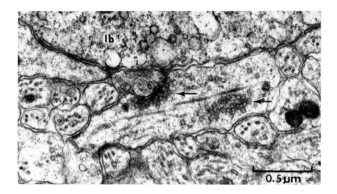

Figure 4.35 *An efferent fibre containing two islands of vesicles (arrows). The left arrow points to a synapse formed on a spine of an afferent fibre. Base. 6-day intact mouse*

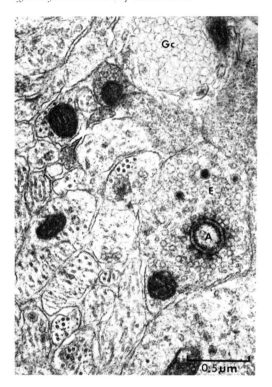

intact mouse, such endings were first seen in the inner spiral bundle in the basal turn on the second postnatal day. They may occur in conjunction with afferent fibres as well as with the inner hair cells; in the latter, a postsynaptic cisterna overlies the hair-cell membrane.

The occasional presence of vesiculated fibres in culture implies that, in the intact animal, *some efferent fibres actually originate in the spiral ganglion.*

Outer Hair Cells and Their Efferents

Early neurosensory relationships are similar in both the inner and outer hair cells. The large influx of efferent fibres that occurs during the second postnatal week, however, causes major changes in the innervation of the outer hair cells in the intact animal. *Figures 4.37* and *4.38* illustrate young efferent endings at the receptor poles of the outer hair cells. The endings are small but may be identified by islands of synaptic vesicles.

The incoming efferent fibres apparently replace a good many of the afferents (Shnerson, Devigne and Pujol, 1982). The elimination of numerous afferent endings and synapses in the outer hair cells is a puzzling developmental event. A substantial decrease in the numbers of ribbon-containing afferent synapses is already apparent in the intact animal at the end of the first postnatal week – well before the arrival of the efferent fibres to the outer hair cell region (Sobkowicz *et al.*, 1982). Moreover, a similar decrease has been

◀ **Figure 4.36** *An efferent ending (E) forming an annular synapse with an afferent spine (A) in the inner spiral bundle. Note the dense-core vesicles within the efferent ending. Gc = Growth cone. Mid-turn. 6-day-old intact mouse*

Figure 4.37 *A young efferent ending (E) at the receptor pole of an outer hair cell. The ending appears to be replacing the afferent endings (A). Note the small size of the ending and the absence of a postsynaptic cisterna (compare with Figures 4.42–4.46). Dc = Deiters' cell. Mid-turn. 9-day-old intact mouse*

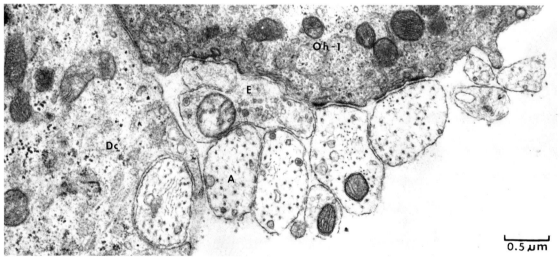

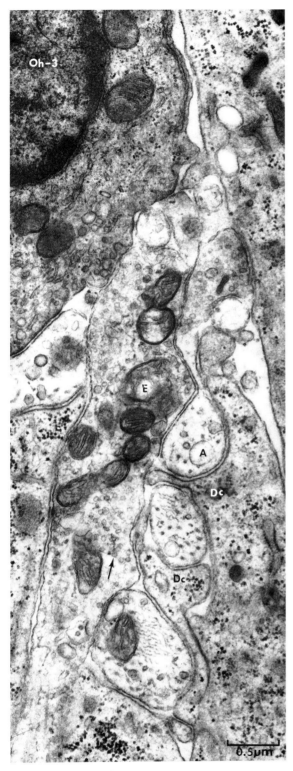

Figure 4.38 *A long slender efferent fibre (E) just reaching an outer hair cell. Note the synaptic vesicles within the efferent fibre (arrow). Note also the processes of a Deiters' cell (Dc) supporting the nerve fibres. A = Afferent fibre. Base. 9-day* intact *mouse*

observed in culture, providing further evidence that the elimination of the afferent endings and synapses in the outer hair cells is not (at least not directly) due to the incoming efferents. *Figure 4.37* depicts an efferent fibre seemingly replacing afferent endings at the base of an outer hair cell in the intact mouse. The micrograph could imply a competitive process between the efferent and afferent innervation. *Figure 4.39*, however, illustrates that in the absence of efferents the receptor area of the outer hair cell may remain almost entirely free of endings (compare with *Figure 4.41*).

As mentioned before, the first efferent endings in the mouse enter the region of the outer hair cells about the ninth postnatal day. At that time, the endings and the areas of their apposition to the hair cells are small (*Figures 4.37* and *4.38*). In the nerve ending, synaptic vesicles are dispersed and few mitochondria are present. Postsynaptic cisternae may be absent in the hair cell. Large differentiated endings are first seen about the fourteenth postnatal day, and their development continues, we believe, until at least the eighteenth day.

Efferent endings innervating the outer hair cells vary greatly in the number and packing density of vesicles: some are almost devoid of vesicles, whereas others are densely packed (*Figures 4.42–4.46*). The vesicles tend to be focally distributed and form a single or a series of clusters along the presynaptic membrane. Focal accumulations of vesicles, similar to those described here, have been noted by Gulley and Reese (1977) in the chinchilla. One is tempted to assume that variations in the groupings of vesicles reflect variations in synaptic activity.

Subsurface and Postsynaptic Cisternae

The development of subsurface cisternae in the hair cells occurs postnatally. Subsurface cisternae develop in both the inner and outer hair cells (*Figures 4.46* and *4.49*) but are more prominent in the latter. According to Le Beux (1972b), neuronal subsurface cisternae are continuous with the endoplasmic reticulum. Our data likewise suggest a connection between the hair cell subsurface cisternae and endoplasmic reticulum (*see Figure 4.31(a)*).

The alignment of mitochondria along the subsurface cisternae is a hallmark of the outer hair cell. *Figures 4.39* and *4.41* show a maturing hair cell in culture and in the intact mouse, respectively. *Figure 4.41* suggests that mitochondria not only cover the walls of the cell but may form a roof over the upper portion of the nucleus. An accumulation of mitochondria often is found below the Golgi region (*Figures 4.39* and *4.41*). The characteristic alignment of mitochondria along the cell membrane is not apparent in the young cultured cell shown in *Figure 4.39*.

(continued on page 82)

Figures 4.39 and 4.40 *An outer hair cell of a 12-day culture. Most of the cell membrane is lined by subsurface cisternae. The receptor pole is largely devoid of nerve endings; the three remaining small afferent endings (A) appear to be limited to the only surface of the cell that is free from a subsurface cisterna. Figure 4.40 shows the endings under higher magnification. An accumulation of mitochondria is present below the Golgi complex (G) in Figure 4.39. The arrows point to the mitochondria that lie in an immediate supranuclear position. Apex*

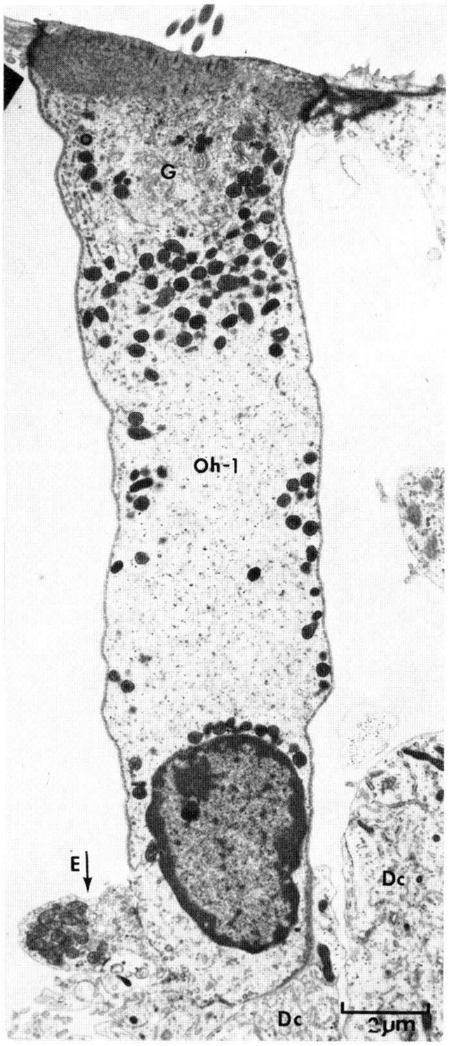

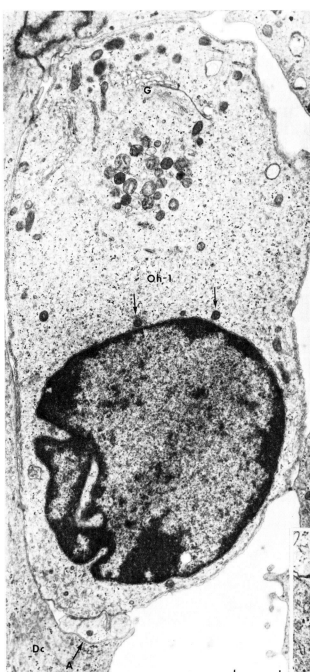

Figure 4.39

Figure 4.40

Figure 4.41 *An outer hair cell of a 14-day intact mouse. The cell is elongated and is largely surrounded by fluid spaces. The cell membrane is lined by subsurface cisternae. Mitochondria tend to align along the subsurface cisternae; they also roof the upper surface of the nucleus. Note an accumulation of mitochondria below the Golgi complex (G; compare with Figure 4.39). The arrow points to a large efferent ending (E). Dc = Deiters' cell. Apex*

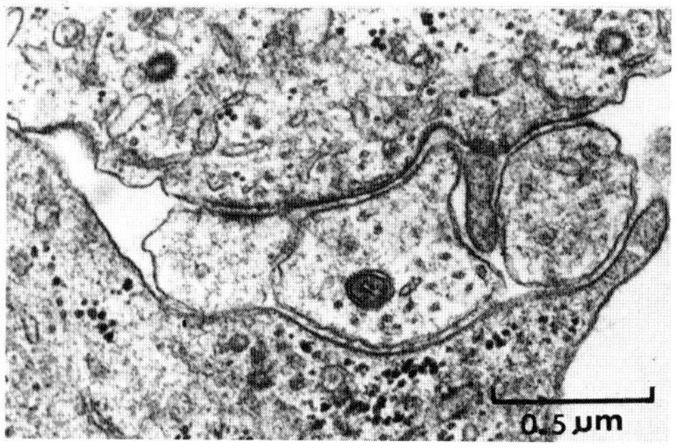

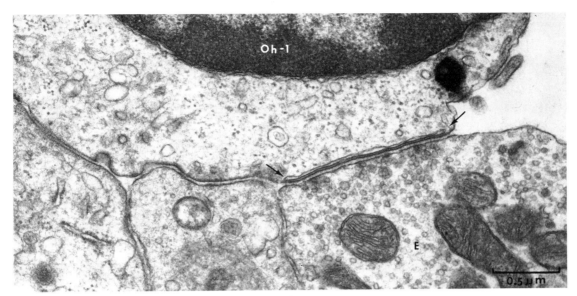

Figure 4.42 *A large efferent ending (E) forms a synapse with an outer hair cell in a 23-day-old* intact *mouse. Synaptic vesicles fill the ending and aggregate into discrete groups along the presynaptic membrane. Several mitochondria are present in the ending. A postsynaptic cisterna (arrows) is coextensive with the apposition of the presynaptic membrane. The electron-dense body at right resembles a synaptic ribbon, but the lack of vesicles makes the identification uncertain. Apex*

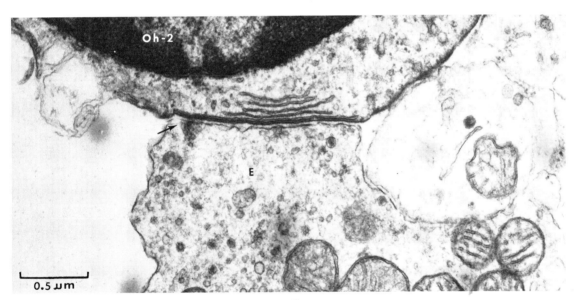

Figure 4.43 *A stack of postsynaptic cisternae accompany an efferent synapse. Note the paucity of synaptic vesicles in the ending. The arrow points to a discrete group of vesicles at the presynaptic membrane. Outer hair cell. Apex. 21-day* intact *mouse. E = Efferent ending*

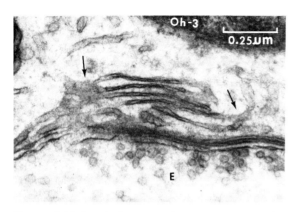

Figure 4.44 *A stack of postsynaptic cisternae in an outer hair cell. The electron micrograph suggests continuity between the postsynaptic cisternae and smooth endoplasmic reticulum (arrows). Well-defined groups of synaptic vesicles form 'hot spots' along the presynaptic membrane. Mid-turn. 18-day* intact *mouse. E = Efferent ending*

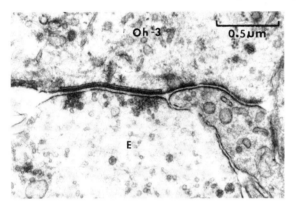

Figure 4.45 *Two focal aggregates of synaptic vesicles along the presynaptic membrane in an efferent synapse. 18-day-old* intact *mouse. Mid-turn. E = Efferent ending*

In the intact animal, the subsurface cisternae line the hair-cell membrane with the exception of the areas of apposition of afferent fibres. Upon contact with efferent fibres, the cisterna flattens and straightens abruptly, forming a flat postsynaptic cisterna (*Figure 4.46*). The insert of *Figure 4.46* shows the transition of the subsurface cisterna to a flattened postsynaptic cisterna. Continuity between subsurface and postsynaptic cisternae was first described by Kimura (1975).

Occasionally, stacks of postsynaptic cisternae may be observed in the adolescent mouse (*Figures 4.43* and *4.44*). Stacks of nearly parallel cisternae sometimes are present throughout the entire infranuclear zone of the outer hair cell; the innermost cisterna may nearly border the nuclear membrane. Some of these multiple postsynaptic cisternae appear to be continuous with the endoplasmic reticulum (*Figure 4.44*). We are not aware of such formations being reported in the hair cells, and they may be peculiar to the developmental period. Stacks of flattened subsurface cisternae have been described in the vestibular ganglion cells and in some cortical neurons (Rosenbluth, 1962b).

One could speculate that the postsynaptic cisterna is a response of the sensory cell to the apposition of the efferent fibre. However, cisternae quite similar to the postsynaptic cisternae of the intact animal were observed in culture in conjunction with afferent fibres (*Figure 4.47*). This observation implies that at least *some postsynaptic specializations of the sensory cells may develop regardless of the type of apposed fibre.*

In the intact animal, the areas of hair-cell afferent nerve fibre apposition usually lack subsurface cisternae. In culture, however, a subsurface cisterna may continue over the area of afferent fibre apposition (*see Figure 4.32(a)*). This implies that the afferent fibre does not have a decisive influence on the extent of subsurface cisternae. The cisternae are invariably interrupted by presynaptic complexes. It is our impression that, at least in culture, subsurface cisternae may play a role in the elimination of afferent endings. *Figure 4.39* depicts an outer hair cell that retained only three small afferents. They are restricted to the only receptor surface free of a subsurface cisterna (*see Figure 4.40*). The extensive infranuclear receptor area – lined by a subsurface cisterna – presumably remains denuded because no efferents are present.

In culture, flattened cisternae, similar to postsynaptic cisternae, occasionally are seen in hair cells that appose each other (*Figure 4.48*) or in hair cells apposed to supporting cells (*Figure 4.49*). No such formations were seen in the developing intact mouse.

Symmetrical appositions of flattened subsurface cisternae have been described by Kumegawa, Cattoni and Rose (1968) in the cells of human epidermoid carcinoma and by Tennyson and Pappas (1968) in the epithelial cells of the choroid plexus of the rabbit. Transient symmetrical appositions of flattened subsurface cisternae have been observed in spinal ganglia of the developing chick (Weis, 1968) and in cortical pyramidal cells of the golden hamster (Buschmann, 1975). In both reports, symmetrical subsurface cisternae were seen only for a short time during development when the nerve cells were in direct contact with each other. The cisternae disappeared following wrapping of the nerve cells by the supporting cells.

(*continued on page 85*)

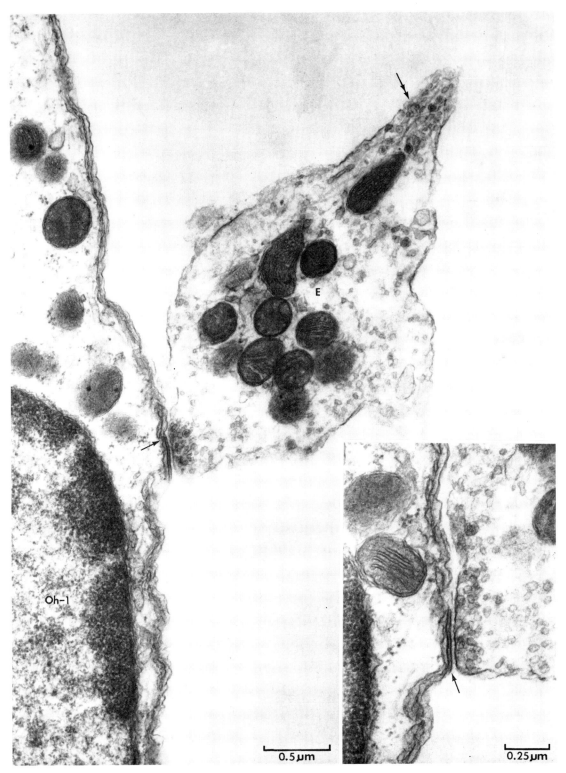

Figure 4.46 *An efferent synapse in a 22-day-old intact mouse. The insert shows the same synapse, under higher magnification, a few sections away. The single-heaaded arrows point to an abrupt transition of a subsurface cisterna into a flat postsynaptic cisterna that follows the apposition of the efferent ending. Note the accumulation of synaptic vesicles and a solitary dense-core vesicle at the presynaptic membrane of the fibre. A similar population of vesicles is present within the shaft of the fibre (double-headed arrow). E = Efferent ending. Apex*

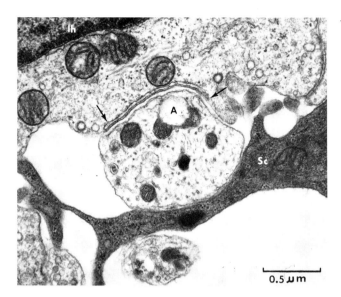

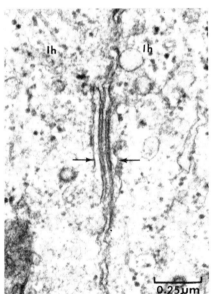

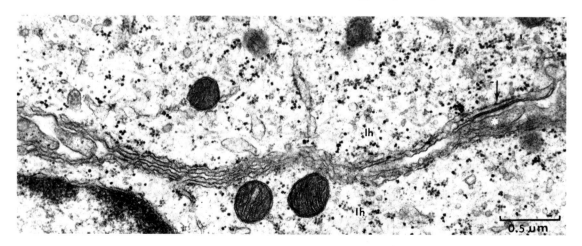

Figure 4.47 *A postsynaptic-like cisterna (arrows) in conjunction with an afferent fibre (A) in a 12-day culture. Note that the distance between the outer cisternal membrane and the cell membrane is narrow and of a uniform width. Note also the characteristic apposition of the mitochondrion to the inner surface of the cisterna. Sc = Supporting cell. Apex*

Figure 4.48 *Symmetrical flat subsurface cisternae (arrows) closely aligned along the membranes of 2 apposing inner hair cells. The cisternae are facing each other in a mirror-like fashion. The cell membranes and the outer cisternal membranes are straight, parallel, and electron dense. A few ribosomes are attached to the inner cisternal membranes. Apex. 6-day culture*

Figure 4.49 *(below) Subsurface cisternae in two apposing inner hair cells in an 11-day denervated culture obtained from a 3-day-old mouse. A tongue of supporting cell cytoplasm (asterisks) partly separates the hair cells. The membranes of the subsurface cisternae are wavy; their inner membranes show attached ribosomes. Note two mitochondria apposed to the inner surface of the cisterna of the lower cell. The arrow points to a flat cisterna, similar to the cisterna in Figure 4.48, formed by the upper hair cell along the apposition to the supporting cell. Apex*

Spiral Ganglion Cells

Bipolar spiral ganglion cells may withstand injury to either or both processes and still survive in culture (*Figures 4.50* and *4.51*). The number of surviving cells varies greatly in different specimens. One reason for the variations in cultures derived from postnatal animals is the extent of the direct injury during explantation. In contrast to the behaviour in culture, an injury *in situ* is almost always followed by neuronal death (Wittmaack, 1936; Weaver and Neff, 1947; Schuknecht and Woellner, 1955; Ruben, Hudson and Chiong, 1962). Although there are other examples of survival in culture of neurons known to die after axotomy *in situ* (Guillery, Sobkowicz and Scott, 1968; Sobkowicz, Guillery and Bornstein, 1968; Sobkowicz, Bleier and Monzain, 1974; Sobkowicz *et al.*, 1973, 1974), the plasticity of the spiral ganglion neuron is remarkable. A spiral ganglion cell surviving in culture either remains bipolar or becomes a monopolar cell by shedding one of its processes. Usually it is the central – injured – process that is discarded. The central fibre, if present, is always free growing; the peripheral fibre may be maintained by the cell as either a synaptically engaged or a free-growing fibre.

The availability of the hair-cell region is critical for the normal growth of peripheral fibres. In organotypic culture, the growth of peripheral fibres is limited and may result in the acquisition of synaptic contacts with appropriate hair cells. This restricted growth apparently follows the normal developmental pattern. The outcome is quite different when the target cell is missing: the free growth is fast, continuous, uncontrolled and unlimited in amount (Rose *et al.*, 1977).

Figure 4.52 shows two fetal spiral neurons in a 2-day culture. Both cells are wrapped by a single layer of satellite-cell cytoplasm. The upper cell has a large eccentric nucleus with two nucleoli located close to the nuclear membrane, as was noted in the developing intact and cultured spinal ganglion cells of the rat (Sobkowicz *et al.*, 1973). The nuclear membrane facing the bulk of the cytoplasm is slightly indented. Immediately beneath it, rough endoplasmic reticulum and free ribosomes are present. The central portion of the cell contains the Golgi apparatus, mitochondria, centrioles, microtubules and microfilaments (*Figures 4.52* and *4.55*). This region corresponds to the fibrillogenous zone of Held (Held, 1905; Tennyson, 1965; Sobkowicz *et al.*, 1973) and is involved in the formation of neurotubules and neurofilaments. In developing cells, the axon originates in this zone. Rough endoplasmic reticulum and free ribosomes occupy the periphery of the cell.

In older cells, the fibrillogenous zone loses its affinity for silver, the nucleus acquires a more central position (*see Figure 4.50*), and rough endoplasmic reticulum is more evenly distributed (*see Figure 4.56*). Centrioles may move toward the periphery of the cell (*see Figure 4.56*).

On one occasion, a ciliated neuron was seen in culture (*Figures 4.53* and *4.54*). The cilium measured about

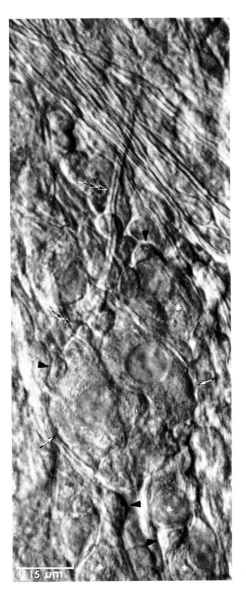

Figure 4.50 *Spiral ganglion cells and their myelinated processes in a live 21-day culture. In the lower part of the photograph, two fairly large neurones are in focus (arrows); some smaller, slightly out-of-focus nerve cells are marked by asterisks. The nerve cell at left appears to emit an axon (double-headed arrows), which joins the bundle of other myelinated fibres. Satellite cells closely appose the nerve cell bodies (arrowheads). Apex. Nomarski differential interference*

7.5 µm before it disappeared from the plane of the section. Ciliated neurons are reported only rarely (for review, see Lafarga *et al.*, 1980).

Many spiral ganglion cells in culture are densely packed with filaments (*Figures 4.53* and *4.54*). Cells with a considerable content of rough endoplasmic reticulum, however, also are seen (*Figures 4.56* and *4.58*). Filamentous and granular cells (the latter with a preponderance of rough endoplasmic reticulum) have been described in the spiral ganglion of the rat by Rosenbluth (1962a). He stressed, however, the existence of a considerable overlap between the two types. The filamentous cells constitute only a small population of the spiral ganglion cells in the cat (Spoendlin, 1973), guinea-pig (Kellerhals, Engström and Ades, 1967), and man (Ota and Kimura, 1980). At present, we do not have enough material to compare with confidence the ultrastructure of the developing spiral ganglion cells in culture with that of the intact animal.

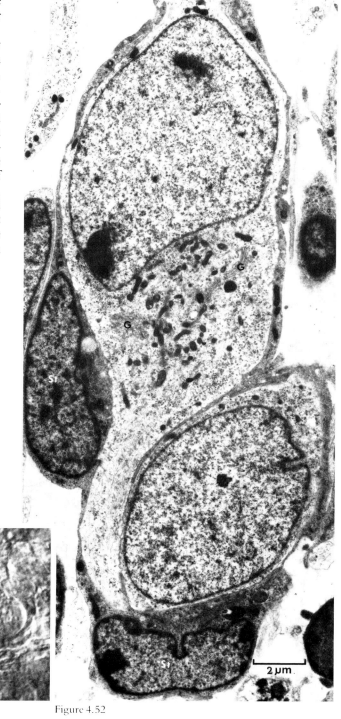

Figure 4.51

Figure 4.52

Figure 4.51 *A live spiral ganglion cell in a 22-day culture. Note the centrally located nucleolus within the nucleus. The neuronal cytoplasm displays a distinct granulation. The cell appears to emit an axon at the left (arrow). Base. Nomarski differential interference. Same magnification as Figure 4.50*

Figure 4.52 *An electron micrograph of two spiral ganglion cells in a 2-day culture of a fetal (15 days of gestation) cochlea. Each cell displays a prominent nucleus; the nucleus of the upper cell is eccentric. Golgi complexes (G), mitochondria, and neurofibrils occupy the central, fibrillogenous, zone of the cytoplasm. Rough endoplasmic reticulum and free ribosomes are located at the periphery of the fibrillogenous zone. Satellite cells (St) surround the young neurons*

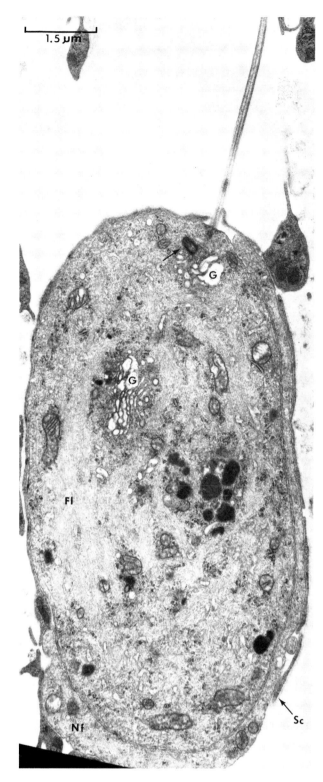

Figure 4.54 *A higher magnification of the region of anchorage of the cilium shown in Figure 4.53 (two sections away). The arrow points to a rootlet, anchoring the basal body (B). Note the characteristic association of the basal body with the Golgi complex*

Figure 4.53 *A ciliated neuron in a 29-day culture. Note the filamentous content of the neuron (Fl). A long, narrow process of a growing nerve fibre (Nf) is apposed to the cell. Both are partly wrapped by satellite-cell cytoplasm (Sc). The arrow points to a centriole. G = Golgi complex. Apex*

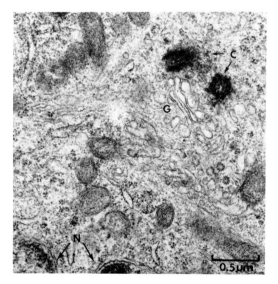

Figure 4.55 *Centrioles (C) in the paranuclear region of a 6-day-old intact mouse. N = nucleus. Mid-turn*

Desmosomal-like Connections

In culture, desmosomal-like connections are common in the spiral ganglion. We observed such connections between Schwann cells, between Schwann cells and nerve fibres, and between satellite cells and neuronal somas.

Desmosomal-like connections involving the spiral ganglion cells are of special interest. *Figure 4.59* illustrates a junction between processes of two spiral ganglion cells. One of the processes emerges from an axon hillock, whereas the other is a gradually tapering process: the former is probably a central axon, the latter may be a peripheral fibre. *Figure 4.61* depicts two desmosomal junctions between a nerve ending and a spiral neuron; *Figure 4.62* illustrates a connection between a small nerve fibre and a neuronal soma.

Synapses adjacent to desmosomal connections were described in the primate spiral ganglion by Kimura and Ota (1981) on small filamentous neurons. The presynaptic nerve fibres were unmyelinated and could be traced to the intraganglionic spiral bundle, which is known to contain efferent nerve fibres of the olivocochlear bundle.

We did not observe typical axo-somatic synapses in our material. On one occasion, however, a synaptic connection was noted between two nerve fibres in the spiral ganglion (*Figures 4.58* and *4.60*). The presynaptic fibre in *Figure 4.60(a)* contains synaptic vesicles. Their appearance precedes synaptic contact (*Figure 4.62(b)*) which in turn ends in symmetrical desmosomal-like densities (*Figure 4.60(c)*). This axo-axonal synapse was observed in the same culture in which vesiculated endings were found in contact with the inner hair cells (*see Figure 4.34*). These observations could imply that the spiral ganglion contains nerve cells that are pre-synaptic to either the hair cells or spiral ganglion cells, or both.

Myelination

Spiral ganglion cells are among the nerve cells that may acquire myelin ensheathment (for a review, *see* Beal and Cooper, 1976; Cooper and Beal, 1977; Braak, Braak and Strenge, 1977). The spiral neuron obtains its myelin sheath from a satellite cell; the outermost layer of the myelin is continuous with the satellite-cell cytoplasm. A basement membrane covers the outer surface of the satellite cell. The number of lamellae varies, and they may form loose or compact myelin (*see Figures 4.56 –4.58*). The loose lamellae may be interconnected by desmosomes (*see Figure 4.57*), which often are arranged in stacks (Rosenbluth, 1962a; Ota and Kimura, 1980).

Both peripheral and central fibres myelinate in culture. Central fibres, if present, grow continuously, and the process of myelination continues as long as the fibre grows. Initially, the Schwann cell appears to recruit spiral fibres, collecting them into compact bundles. *Figure 4.63* shows the outgrowth zone of a 12-day culture; a long process of a Schwann cell appears to grasp three newly grown central fibres. *Figure 4.64* depicts two Schwann cells, each surrounding a compact bundle of nerve fibres. This 1-day culture probably reflects the relations that existed between the spiral fibres and Schwann cells at the time of explantation. *By grouping neighbouring fibres, the Schwann cells apparently initiate the formation of the tonotopic organization of the peripheral and central bundles.*

(continued on page 92)

Figure 4.56–4.58 *A partly myelinated nerve cell (Sn) and its satellite cell (St) in a 14-day culture. Apex*

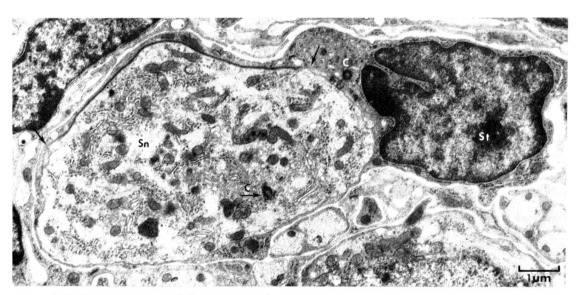

Figure 4.56 *The two upper arrows indicate the extent of compact myelin. Note the wealth of rough endoplasmic reticulum in the neuron. Both the nerve cell and satellite cell contain a centriole (C)*

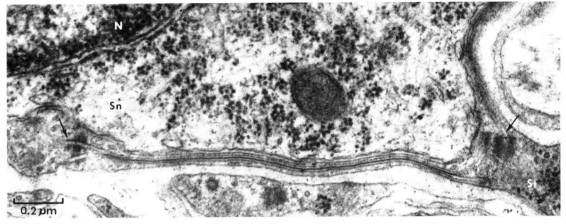

Figure 4.57 *A higher magnification of several layers of compact myelin. On both ends, the lamellae of compact myelin change to loose cytoplasmic lamellae that are connected to each other by desmosomal junctions (arrows). N = Nucleus*

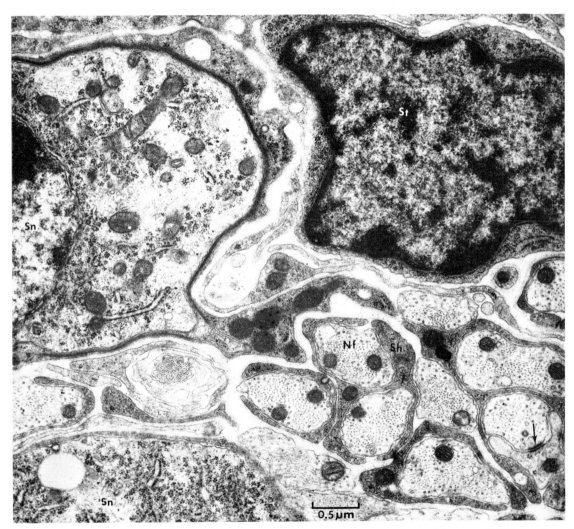

Figure 4.58 *The compact myelin in this region covers most of the cell surface (compare with Figure 4.56). In the lower right part of the micrograph, a number of nerve fibres (Nf) is present: some naked and some wrapped by Schwann-cell (Sh) cytoplasm. The arrow points to a desmosomal-like connection that in serial sections proved to be a synapse (see Figure 4.60)*

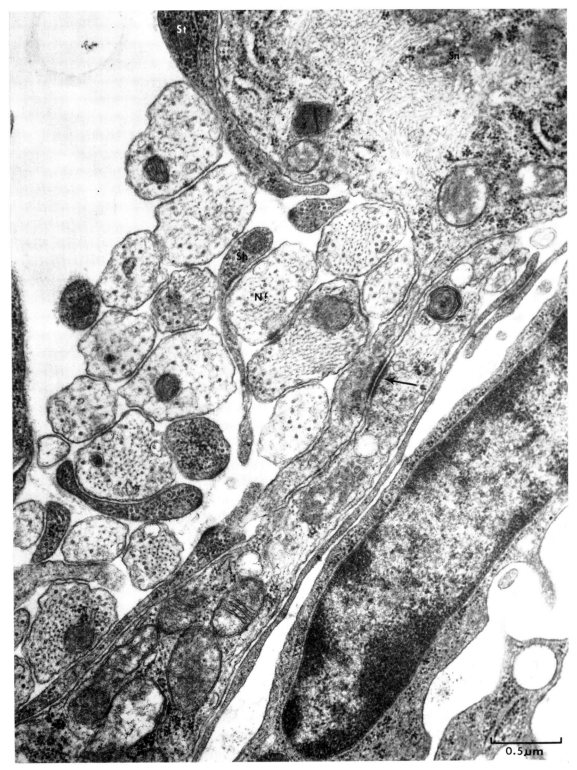

Figure 4.59 *A desmosomal-like connection (arrow) between two spiral fibres. Note that the slender central fibre narrows at the axon hillock and shows a paucity of organelles; the peripheral fibre widens towards its soma (towards the left) at which point mitochondria and free ribosomes become abundant. Small symmetrical densities can be seen between the labelled fibre (Nf) and the fibre below it. Nf = Nerve fibre; Sh = Schwann cell; Sn = spiral neuron; St = satellite cell. Apex. 29 days in vitro*

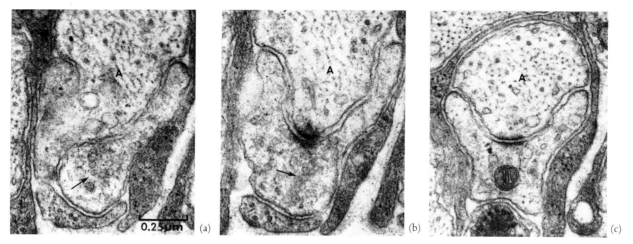

Figure 4.60 *Three sections selected from a serial through the synapse shown in Figure 4.58. The scale in (a) applies to (b) and (c), and also to 4.61 and 4.62. Apex. 14 days in vitro. (a) A vesiculated lower ending apposes an afferent nerve profile (A). (b) A few sections away, synaptic densities are apparent between the fibres.*

(c) In another section, the symmetrical densities resemble a desmosomal-like connection (compare with Figures 4.59, 4.61, and 4.62). The lower fibre does not show vesicles and appears similar to the upper one. Apex. 14 days in vitro

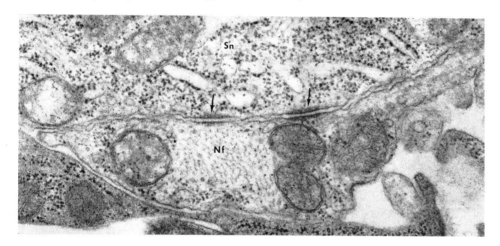

Figure 4.61 *Two desmosomal-like connections (arrows) between a nerve ending (Nf) and the soma of a spiral neuron (Sn). Note the* *rough endoplasmic reticulum in the neuron. Apex. 29 days in vitro. Same magnification as in Figure 4.60*

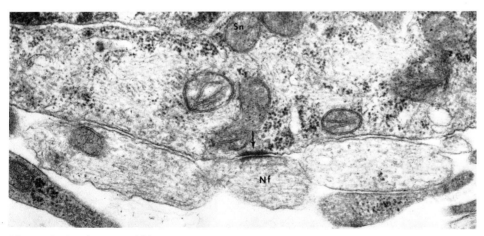

Figure 4.62 *A desmosomal-like connection between a nerve fibre and a soma of the spiral neuron (arrow). Note that this connection differs from the ones shown in Figure 4.61: the 'cleft' is narrower and* *contains electron-dense material. Apex. 29 days in vitro. Same magnification as in Figure 4.60*

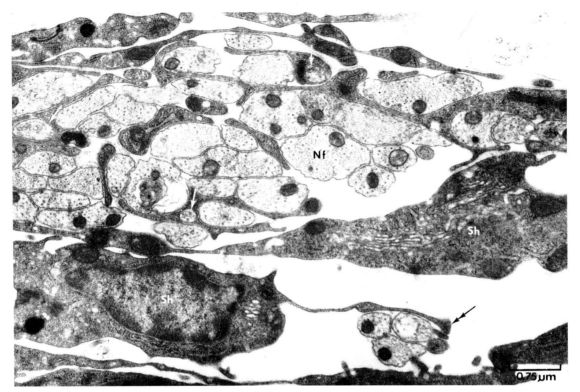

Figure 4.63 *A cross-section through the outgrowth zone of a 12-day culture. In the lower part of the photograph, a Schwann cell recruits growing nerve fibres. Note the long process of the cell cytoplasm that has reached three small nerve fibres, and is about to encircle them (double-headed arrow). In the upper part of the Figure, probably a* *different Schwann cell wraps a bundle of growing nerve fibres (Nf). Its long cytoplasmic processes invade the bundle and split it into smaller fasciculi. Some fibres are already entirely surrounded by the Schwann cell cytoplasm (arrows). Sh = Schwann cell. Apex*

The next step toward myelination is the invasion of the neuronal bundle by processes of Schwann cell cytoplasm (*see Figures 4.59* and *4.63*). Finally, each nerve profile is fully encircled by Schwann cell cytoplasm (*see Figure 4.58*). The acquisition of Schwann cell wrapping by the processes of spiral neurons is similar to that described for peripheral nerves (Peters and Muir, 1959; Peters, 1961; Peters and Vaughn, 1970; Ochoa, 1971; Webster, 1971, 1975).

During the growth of peripheral nerve fibres in the intact animal, the successive stages of the Schwann-cell wrapping are related to the diameter of the growing nerve fibre (Peters and Muir, 1959; Peters, 1961; Peters and Vaughn, 1970). According to these reports, the diameter of naked fibres is very small. When the fibres reach a diameter of 0.2–0.5 μm, Schwann-cell cytoplasm invades the bundle and begins to wrap individual fibres. In the 1-day culture shown in *Figure 4.64*, however, the nerve bundle contains naked fibres of different diameter: the smallest is about 0.3 μm, the largest about 1.5 μm.

Before the onset of myelination, each fibre at any point is apposed by only one Schwann cell. *Figure 4.65* shows a Schwann cell apposed to a myelinated fibre in a live culture. *Figure 4.66* shows the same relationship in an electron micrograph. A diameter of about 2 μm is the

critical size a peripheral axon must attain before its myelination (Peters and Vaughn, 1970). In contrast to the growth of peripheral nerves in the intact animal, our material in culture indicates that fibres of similar diameter may remain naked, they may show individual wrapping by Schwann cell cytoplasm, or they may myelinate (*see Figures 4.58, 4.59, 4.63, 4.64, 4.66* and *4.69*). *Figures 4.67–4.69* show details of the myelin structure that developed in culture.

Acknowledgements

This work was supported by the National Institutes of Health Grants NS 15061, NS 12732 and NS 5P3OHDO3352. We thank Ms J. Subervi, for culturing the tissues; Ms S. Slapnick, for providing many of the electron micrographs; and Ms J. Lichtenstein and Mr C. Levenick for invaluable help with the preparation of the figures. Special thanks are due to Professors D. Green, H. Ris and G. M. ZuRhein for their interest and helpful advice in interpreting our observations on some of the organelles studied.

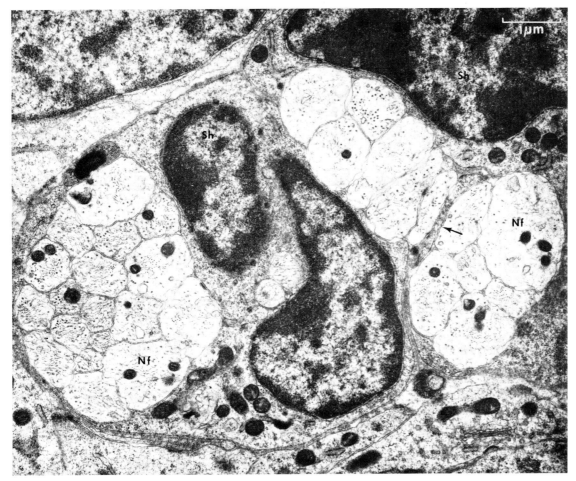

Figure 4.64 *Two Schwann cells (Sh), each wrapping a bundle of naked spiral fibres (Nf), in a 1-day culture. The diameter of the naked fibres ranges from about 0.3 to 1.5 μm. Note the tongue of Schwann cell cytoplasm (arrow) invading the upper bundle and separating it into two fasciculi. Apex*

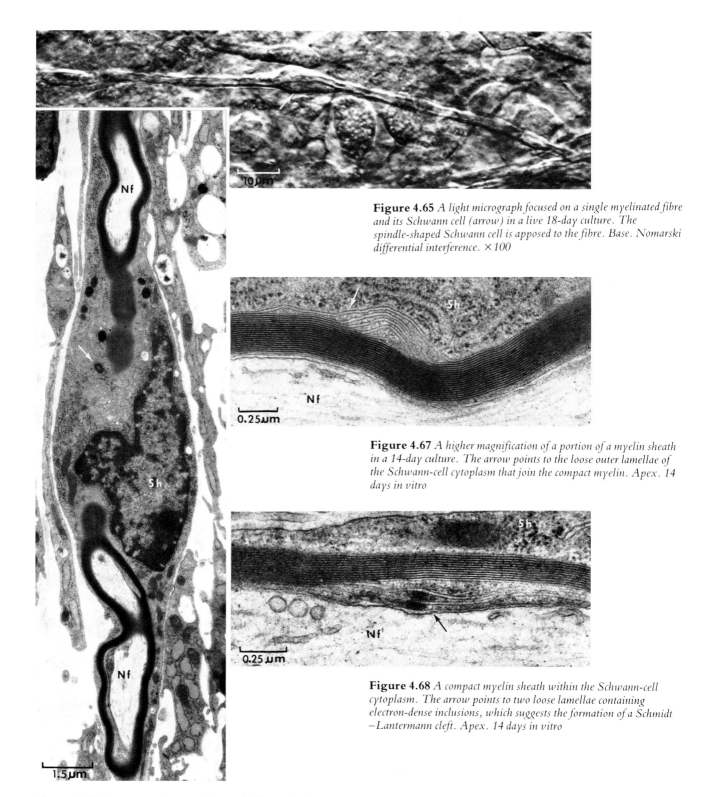

Figure 4.65 *A light micrograph focused on a single myelinated fibre and its Schwann cell (arrow) in a live 18-day culture. The spindle-shaped Schwann cell is apposed to the fibre. Base. Nomarski differential interference. ×100*

Figure 4.67 *A higher magnification of a portion of a myelin sheath in a 14-day culture. The arrow points to the loose outer lamellae of the Schwann-cell cytoplasm that join the compact myelin. Apex. 14 days in vitro*

Figure 4.68 *A compact myelin sheath within the Schwann-cell cytoplasm. The arrow points to two loose lamellae containing electron-dense inclusions, which suggests the formation of a Schmidt–Lantermann cleft. Apex. 14 days in vitro*

Figure 4.66 *A low-power electron micrograph, illustrating the relationship between a spiral fibre (Nf) and myelinating Schwann cell (Sh) (compare with Figure 4.65). The lamellae of compact myelin are surrounded by the cytoplasm of the Schwann cell. The arrow points to a centriole. Apex. 14 days in vitro*

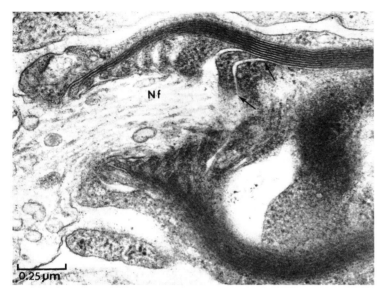

Figure 4.69 *A longitudinal section through the node of Ranvier. Note the transition from the laminae of compact myelin to loose cytoplasmic lamellae, anchored to the axonal surface (arrows). As in the intact animal, the innermost lamella ends first and the outermost one extends farthest. Apex. 14 days in vitro*

References

Adams, E. C. and Hertig, A. T. (1964) Studies on guinea-pig oocytes. I. Electron microscopical observations on the development of cytoplasmic organelles in oocytes of primordial and primary follicles. *Journal of Cell Biology*, **21**, 397–427.

Alford, B. R. and Ruben, R. J. (1963) Physiological, behavioral and anatomical correlates of the development of hearing in the mouse. *Annals of Otology, Rhinology and Laryngology*, **72**, 237–247.

Anzil, A. P., Herrlinger, H. and Blinzinger, K. (1973) Nucleolus-like inclusions in neuronal perikarya and processes: phase and electron microscope observations on the hypothalamus of the mouse. *Zeitschrift für Zellforschung und mikroskopische Anatomie*, **146**, 329–337.

Beal, J. A. and Cooper, M. H. (1976) Myelinated nerve cell bodies in the dorsal horn of the monkey (*Saimiri sciureus*). *American Journal of Anatomy*, **147**, 33–48.

Bodian, D. (1978) Synapses involving auditory nerve fibers in primate cochleas. *Proceedings of the National Academy of Sciences of the United States of America*, **75**, 4582–4586.

Bodian, D. (1980) Presynaptic bodies of auditory hair cells in old world monkeys. *Anatomical Record*, **197**, 379–386.

Braak, E., Braak, H. and Strenge, H. (1977) The fine structure of myelinated nerve cell bodies in the bulbus olfactorius of man. *Cell and Tissue Research*, **182**, 221–233.

Buschmann, T. (1975) Membrane organelles in developing hamster frontal cortex. In *33rd Annual Proceedings of the Electron Microscopy Society of America*, ed. G. W. Bailey, pp. 316–317.

Chandler, R. L. (1966) Intranuclear structures in neurones. *Nature (London)*, **209**, 1260–1261.

Comings, D. E. and Okada, T. A. (1972) The chromatoid body in mouse spermatogenesis: evidence that it may be formed by the extrusion of nucleolar components. *Journal of Ultrastructure Research*, **39**, 15–23.

Cooper, M. H. and Beal, J. A. (1977) Myelinated granule cell bodies in the cerebellum of the monkey (*Saimiri sciureus*). *Anatomical Record*, **187**, 249–256.

Desclin, J. C. and Colin, F. (1980) The olivocerebellar system. II. Some ultrastructural correlates of inferior olive destruction in the rat. *Brain Research*, **187**, 29–46.

Doolin, P. F., Barron, K. D. and Seber, A. (1967) Annulate lamellae in cat lateral geniculate neurons. *Anatomical Record*, **159**, 219–230.

Eddy, E. M. (1974) Fine structural observations on the form and distribution of nuage in germ cells of the rat. *Anatomical Record*, **178**, 731–758.

Eddy, E. M. (1975) Germ plasm and the differentiation of the germ cell line. *International Review of Cytology*, **43**, 229–280.

Ehret, G. (1976) Development of absolute auditory thresholds in the house mouse (*Mus musculus*). *Journal of the American Audiology Society*, **1**, 179–184.

Elfvin, L. -G. (1963) The ultrastructure of the superior cervical sympathetic ganglion of the cat. *Journal of Ultrastructure Research*, **8**, 403–440.

Engström, H., Ades, H. W. and Hawkins, J. E. (1962) Structure and functions of the sensory hairs of the inner ear. *Journal of the Acoustical Society of America*, **34**, 1356–1363.

Engström, H., Bergström, B. and Ades, H. W. (1972) Macula utriculi and macula sacculi in the squirrel monkey. *Acta Otolaryngologica (Stockholm)*, Supplement, **301**, 75–126.

Favre, D. and Sans, A. (1979) Morphological changes in afferent vestibular hair cell synapses during postnatal development of the cat. *Journal of Neurocytology*, **8**, 765–775.

Flock, Å., Kimura, R., Lundquist, P. -G. and Wersäll, J. (1962) Morphological basis of directional sensitivity of the outer hair cells in the organ of Corti. *Journal of the Acoustical Society of America*, **34**, 1351–1355.

Flock, Å., Jorgensen, M. and Russell, I. (1973) The physiology of individual hair cells and their synapses. In *Basic Mechanisms in Hearing*, ed. A. R. Møller, pp. 273–306. New York: Academic Press.

Flock, Å., Cheung, H. C., Flock, B. and Utter, G. (1981) Three sets of actin filaments in sensory cells of the inner ear. Identification and functional orientation determined by gel electrophoresis, immunofluorescence and electron microscopy. *Journal of Neurocytology*, **10**, 133–147.

Friedmann, I. and Bird, E. S. (1967) Electron microscopic studies of the isolated fowl embryo otocyst in tissue culture. *Journal of Ultrastructure Research*, **20**, 356–365.

Grillo, M. A. (1970) Cytoplasmic inclusions resembling nucleoli in sympathetic neurons of adults rats. *Journal of Cell Biology*, **45**, 100–117.

Guillery, R. W., Sobkowicz, H. M. and Scott, G. L. (1968) Light and electron microscopical observations of the ventral horn and ventral root in long term cultures of the spinal cord of the fetal mouse. *Journal of Comparative Neurology*, **134**, 433–476.

Gulley, R. L. and Reese, T. S. (1977) Freeze-fracture studies on the synapses in the organ of Corti. *Journal of Comparative Neurology*, **171**, 517–544.

Hamori, J. and Lakos, I. (1980) Ultrastructural alterations in the initial segments and in the recurrent collateral terminals of Purkinje cells following axotomy. *Cell and Tissue Research*, **212**, 415–427.

Held, H. (1905) Die Entstehung der Neurofibrillen. *Neurologisches Centralblatt*, **24**, 706–710.

Herman, M. M. and Ralston, H. J. (1970) Laminated cytoplasmic bodies and annulate lamellae in the cat ventrobasal and posterior thalamus. *Anatomical Record*, **167**, 183–196.

Hirano, A. and Zimmerman, H. M. (1967) Some new cytological observations of the normal rat ependymal cell. *Anatomical Record*, **158**, 293–302.

Holmgren, E. (1899) Weitere Mitteilungen über den Bau der Nervenzellen. *Anatomischer Anzeiger*, **16**, 388–397.

Iurato, S. (1967) *Submicroscopic Structure of the Inner Ear*. Oxford: Pergamon Press.

Kellerhals, B.,Engström, H. and Ades, H. W. (1967) Die Morphologie des Ganglion spirale Cochleae. *Acta Otolaryngologica (Stockholm)*, Supplement **226**, 5–78.

Kessel, R. G. (1968) Annulate lamellae. *Journal of Ultrastructure Research*, Supplement **10**, 1–82.

Kimura, R. S. (1975) The ultrastructure of the organ of Corti. *International Review of Cytology*, **42**, 173–222.

Kimura, R. S. and Ota, C. Y. (1981) Nerve fiber synapses on primate spiral ganglion. In *Abstracts of the Fourth Midwinter Meeting Association For Research In Otolaryngology*, St. Petersburg Beach, Florida, p. 82.

Kronester-Frei, A. (1978) Ultrastructure of the different zones of the tectorial membrane. *Cell and Tissue Research*, **193**, 11–23.

Kumegawa, M., Cattoni, M. and Rose, G. G. (1968) Electron microscopy of oral cells *in vitro*. II. Subsurface and intracytoplasmic confronting cisternae in strain KB cells. *Journal of Cell Biology*, **36**, 443–452.

Lafarga, M., Hervas, J. -P., Crespo, D. and Villegas, J. (1980) Ciliated neurons in supraoptic nucleus of rat hypothalamus during neonatal period. *Anatomy and Embryology*, **160**, 29–38.

Le Beux, Y. J. (1971) An ultrastructural study of the neurosecretory cells of the medial vascular prechiasmatic gland, the preoptic recess and the anterior part of the suprachiasmatic area. I. Cytoplasmic inclusions resembling nucleoli. *Zeitschrift für Zellforschung und mikroskopische Anatomie*, **114**, 404–440.

Le Beux, Y. J. (1972a) An ultrastructural study of a cytoplasmic filamentous body, termed nematosome, in the neurons of the rat and cat substantia nigra. *Zeitschrift für Zellforschung und mikroskopische Anatomie*, **133**, 289–325.

Le Beux, Y. J. (1972b) Subsurface cisterns and lamellar bodies: particular forms of the endoplasmic reticulum in the neurons. *Zeitschrift für Zellforschung und mikroskopische Anatomie*, **133**, 327–352.

Le Beux, Y. J., Langelier, P. and Poirier, L. J. (1971) Further ultrastructural data on the cytoplasmic nucleolus resembling bodies or nematosomes. Their relationships with the subsynaptic web and a cytoplasmic filamentous network. *Zeitschrift für Zellforschung und mikroskopische Anatomie*, **118**, 147–155.

Liberman, M. C. (1980) Morphological differences among radial afferent fibers in the cat cochlea: an electron-microscopic study of serial sections. *Hearing Research*, **3**, 45–63.

Lim, D. J. (1977) Fine morphology of the tectorial membrane. Fresh and developmental. In *INSERM*, eds. M. Portmann and J. -M. Aran, **68**, 47–60.

Magalhaes, M. M. (1967) Intranuclear bodies in cells of rabbit and rat retina. *Experimental Cell Research*, **47**, 628–632.

Masurovsky, E. B., Benitez, H. H., Kim, S. U. and Murray, M. R. (1970) Origin, development, and nature of intranuclear rodlets and associated bodies in chicken sympathetic neurons. *Journal of Cell Biology*, **44**, 172–191.

Maul, G. G. (1977) The nuclear and the cytoplasmic pore complex: structure, dynamics, distribution, and evolution. *International Review of Cytology*, Supplement **6**, 76–186.

Monaghan, P. (1975) Ultrastructural and pharmacological studies on the afferent synapse of lateral-line sensory cells of the African clawed toad (*Xenopus laevis*). *Cell and Tissue Research*, **163**, 239–247.

Morales, R. and Duncan, D. (1967) A special type of filament in the Purkinje cells of the Syrian hamster. *Zeitschrift für Zellforschung und mikroskopische Anatomie*, **83**, 49–52.

Mountford, S. (1964) Filamentous organelles in receptor-bipolar synapses of the retina. *Journal of Ultrastructure Research*, **10**, 207–216.

Murray, M. R. and Benitez, H. H. (1967) Deuterium oxide: direct action on sympathetic ganglia isolated in culture. *Science*, **155**, 1021–1024.

Murray, M. R. and Benitez, H. H. (1968) Action of heavy water (D_2O) on growth and development of isolated nervous tissues. In *Growth of the Nervous System*, pp. 148–178. Boston: Little, Brown and Co.

Ochoa, J. (1971) The sural nerve of the human foetus: electron microscope observations and counts of axons. *Journal of Anatomy*, **108**, 231–245.

Osborne, M. P. (1977) Role of vesicles with some observations on vertebrate sensory cells. In *Synapses*, pp. 40–63. New York: Academic Press.

Osborne, M. P. and Thornhill, R. A. (1972) The effect of monoamine depleting drugs upon the synaptic bars in the inner ear of the bullfrog (*Rana catesbeiana*). *Zeitschrift für Zellforschung und mikroskopische Anatomie*, **127**, 347–355.

Ota, C. Y. and Kimura, R. S. (1980) Ultrastructural study of the human spiral ganglion. *Acta Oto-laryngologica*, **89**, 53–62.

Peters, A. (1961) The development of peripheral nerves in *Xenopus laevis*. In *Electron Microscopy in Anatomy*, pp. 142–159. London: Edward Arnold.

Peters, A. and Muir, A. R. (1959) The relationship between axons and Schwann cells during development of peripheral nerves in the rat. *Quarterly Journal of Experimental Physiology*, **44**, 117–130.

Peters, A. and Vaughn, J. E. (1970) Morphology and development of the myelin sheath. In *Myelination*, pp. 3–79. Springfield Il: Charles C. Thomas.

Pfenninger, K., Sandri, C., Akert, K. and Eugster, C. H. (1969) Contribution to the problem of structural organization of the presynaptic area. *Brain Research*, **12**, 10–18.

Popoff, N. and Stewart, S. (1968) The fine structure of nuclear inclusions in the brain of experimental golden hamsters. *Journal of Ultrastructure Research*, **23**, 347–361.

Ramón y Cajal, S. (1952) *Histologie du Système Nerveux de L'Homme et des Vertébrés*. Vol. **1**, p. 200. Consejo Superior de Investigaciones Científicas. Madrid: Instituto Ramón y Cajal.

Retzius, G. (1884) *Das Gehörorgan der Wirbelthiere. Das Gehörorgan der Reptilien, der Vögel und der Säugethiere*. Stockholm: Die Zentral-Druckerei.

Robertson, D. M. and MacLean, J. D. (1965) Nuclear inclusions in malignant gliomas. *Archives of Neurology*, **13**, 287–296.

Rose, J. E., Sobkowicz, H. M. and Bereman, B. (1977) Growth in culture of the peripheral axons of the spiral neurons in response to displacement of the receptors. *Journal of Neurocytology*, **6**, 49–70.

Rosenbluth, J. (1962a) The fine structure of acoustic ganglia in the rat. *Journal of Cell Biology*, **12**, 329–359.

Rosenbluth, J. (1962b) Subsurface cisterns and their relationship to the neuronal plasma membrane. *Journal of Cell Biology*, **13**, 405–421.

Ruben, R. J., Hudson, W. and Chiong, A. (1962) Anatomical and physiological effects of chronic section of the eighth nerve in cat. *Acta Oto-laryngologica*, **55**, 473–484.

Saito, K. (1980) Fine structure of the sensory epithelium of the guinea pig organ of Corti: Afferent and efferent synapses of hair cells. *Journal of Ultrastructure Research*, **71**, 222–232.

Santolaya, R. C. (1973) Nucleolus-like bodies in the neuronal cytoplasm of the mouse arcuate nucleus. *Zeitschrift für Zellforschung und mikroskopische Anatomie*, **146**, 319–328.

Schuknecht, H. F. and Woellner, R. C. (1955) An experimental and clinical study of deafness from lesions of the cochlear nerve. *Journal of Laryngology and Otology*, **69**, 75–97.

Scott, G. L., Sobkowicz, H. M., Bereman, B. and Rose, J. E. (1976) Ultrastructure of synaptic profiles in the developing organ of Corti in culture. *Neuroscience Abstracts*, **2**, 25.

Shnerson, A., Devigne, C. and Pujol, R. (1982) Age-related changes in the C57BL/6J mouse cochlea. II. Ultrastructural findings. *Developmental Brain Research*, **2**, 77–88.

Siegesmund, K. A., Dutta, C. R. and Fox, C. A. (1964) The ultrastructure of the intranuclear rodlet in certain nerve cells. *Journal of Anatomy*, **98**, 93–97.

Smith, C. A. and Sjöstrand, F. S. (1961) A synaptic structure in the hair cells of the guinea pig cochlea. *Journal of Ultrastructure Research*, **5**, 184–192.

Sobkowicz, H. M., Guillery, R. W. and Bornstein, M. B. (1968) Neuronal organization in long term cultures of the spinal cord of the fetal mouse. *Journal of Comparative Neurology*, **132**, 365–396.

Sobkowicz, H. M., Hartmann, H. A., Monzain, R. and Desnoyers, P. (1973) Growth, differentiation and ribonucleic acid content of the fetal rat spinal ganglion cells in culture. *Journal of Comparative Neurology*, **148**, 249–284.

Sobkowicz, H. M., Bleier, R., Bereman, B. and Monzain, R. (1974) Axonal growth and organization of the mamillary nuclei of the newborn mouse in culture. *Journal of Neurocytology*, **3**, 431–447.

Sobkowicz, H. M., Bleier, R. and Monzain, R. (1974) Cell survival and architectonic differentiation of the hypothalamic mamillary region of the newborn mouse in culture. *Journal of Comparative Neurology*, **155**, 355–376.

Sobkowicz, H. M., Bereman, B. and Rose, J. E. (1975) Organotypic development of the organ of Corti in culture. *Journal of Neurocytology*, **4**, 543–572.

Sobkowicz, H. M., Rose, J. E., Scott, G. L., Kuwada, S., Hind, J. E., Oertel, D. and Slapnick, S. (1980) Neuronal growth in the organ of Corti in culture. In *Tissue Culture in Neurobiology*, eds. E. Giacobini, A. Vernadakis and A. Shahar, pp. 253–275. New York: Raven Press.

Sobkowicz, H. M., Rose, J. E., Scott, G. L. and Slapnick, S. (1982) Ribbon synapses in the developing intact and cultured organ of Corti in the mouse. *Journal of Neuroscience*, **2**, 942–957.

Sobkowicz, H. M. and Rose, J. E. (1983) Innervation of the organ of Corti of the fetal mouse in culture. In *Development of Auditory and Vestibular Systems*, ed. R.Romand. New York: Academic Press (In press).

Söderström, K. -O. (1978) Formation of chromatoid body during rat spermatogenesis. *Zeitschrift für mikroskopisch-anatomische Forschung*, **92**, 417–430.

Söderström, K. -O. (1981) The relationship between the nuage and the chromatoid body during spermatogensis in the rat. *Cell and Tissue Research*, **215**, 425–430.

Spadaro, A., de Simone, I. and Puzzolo, D. (1978) Ultrastructural data and chronobiological patterns of the synaptic ribbons in the outer plexiform layer in the retina of albino rats. *Acta Anatomica (Basel)*, **102**, 365–373.

Spoendlin, H. (1973) The innervation of the cochlear receptor. In *Basic Mechanisms in Hearing*, ed. A. R. Møller, pp. 185–234. New York: Academic Press.

Takahashi, K. and Hama, K. (1965) Some observations on the fine structure of nerve cell bodies and their satellite cells in the ciliary ganglion of the chick. *Zeitschrift für Zellforschung und mikroskopische Anatomie*, **67**, 835–843.

Tennyson, V. M. (1965) Electron microscopic study of the developing neuroblast of the dorsal root ganglion of the rabbit embryo. *Journal of Comparative Neurology*, **124**, 267–318.

Tennyson, V. M. and Pappas, G. D. (1968) The fine structure of the choroid plexus: adult and developmental stages. In *Progress in Brain Research*, Vol. **29**, Brain Barrier Systems, eds. A. Lajtha and D. H. Ford, pp. 63–85. Amsterdam: Elsevier.

Thorn, L. (1975) Die Entwicklung des Cortischen Organs beim Meerschweinchen. *Advances in Anatomy, Embryology, and Cell Biology*, **51**, 7–97.

Thornhill, R. A. (1972) The effect of catecholamine precursors and related drugs on the morphology of the synaptic bars in the vestibular epithelia of the frog *(Rana temporaria)*. *Comparative and General Pharmacology*, **3**, 89–97.

Wagner, H. J. (1973) Darkness-induced reduction on the number of synaptic ribbons in fish retina. *Nature (New Biology)*, **246**, 53–55.

Weaver, E. G. and Neff, W. D. (1947) A further study of the effects of partial section of the auditory nerve. *Journal of Comparative Physiological Psychology*, **40**, 217–226.

Webster, H. de F. (1971) The geometry of peripheral myelin sheaths during their formation and growth in rat sciatic nerves. *Journal of Cell Biology*, **48**, 348–367.

Webster, H. de F. (1975) Development of peripheral myelinated and unmyelinated nerve fibers. In *Peripheral Neuropathy*, Vol. I, eds. P. J. Dyck, P. K. Thomas and E. H. Lampert, pp. 37–61. Philadelphia: W. B. Saunders.

Weis, P. (1968) Confronting subsurface cisternae in chick embryo spinal ganglia. *Journal of Cell Biology*, **39**, 485–488.

Wersäll, J., Gleisner, L. and Lundquist, P-G. (1967) Vestibular mechanisms: fine structure. Ultrastructure of the vestibular end organs. *Ciba Foundation Symposium*, 105–120.

Wischnitzer, S. (1970) The annulate lamellae. *International Review of Cytology*, **27**, 65–100.

Wittmaack, K. (1936) Über sekundäre Degeneration im Cochlearnerven und über die funktionelle und biologische Beziehung zwischen Cortischem Organ und Hörnerven. *Acta Oto-laryngologica*, **23**, 274–289.

The Cochlea

5
Sensory and Accessory Epithelia of the Cochlea

Robert S. Kimura

Introduction

The organ of Corti, originally described by Corti in 1851, is a papillary epithelial mound situated on the spirally arranged basilar membrane that runs along the coils of the cochlea. The structure is composed of sensory cells, supporting cells and nerve fibres. There are two types of sensory cells, inner and outer hair cells, with different cytological characteristics. The apical cytoplasm of the sensory cells is modified to cuticular plates to accommodate the stereocilia, the tops of which are attached to the tectorial membrane. The outer hair cells synapse both with afferent and with efferent nerve fibres, whereas the inner hair cells synapse with afferent nerve fibres and make contact with efferent varicosities. The cellular origins of the afferent fibres innervating the outer and inner hair cells are different, as are those of the efferent fibres. The afferent fibres innervate one type of sensory cell; they do not branch to innervate both types. The supporting framework for the sensory cells is provided primarily by the inner and outer pillar cells and Deiters' cells, all of which contain intracellular large tonofibrils extending from the cell bases to their apices where phalangeal processes establish the reticular lamina. The cell bodies of the inner hair cells are surrounded by inner phalangeal cells and border cells and those of the outer hair cells are surrounded by the fluid-filled Nuel's space.

With the advanced technology in optics and methods of specimen preparation, our knowledge of this receptor organ has increased remarkably in recent years. By the freeze-fracture technique, differences between the post-synaptic membranes of the afferent nerve endings in the inner and outer hair cells have been revealed (Gulley and Reese, 1977). Actin filaments within the sterocilia of the cochlear sensory cells have been identified by applying a special cytochemical technique (Flock et al., 1981). Reciprocal synapses have been observed on the outer hair cells of the human organ of Corti (Nadol, 1981). Using the horseradish peroxidase injection method, Kiang et al. (1982) have been able to trace the peripheral processes of small spiral ganglion cells to the bases of the outer hair cells. Scanning electron microscopy has been useful in determining the sensory cell population along the length of the cochlea and in demonstrating the changes in size and dimension of the surface epithelia as well as the internal structures of Corti's tunnel (Ades and Engström, 1972; Hunter-Duvar, 1978; Lim, 1980).

The aim of this chapter is to focus our attention on recent developments as well as to review the essential structures in the present perspective. For more detailed information the readers are referred to several other major publications on the fine morphology of the organ of Corti (Iurato, 1967; Smith, 1968; Spoendlin, 1970; Engström and Ades, 1973; Kimura, 1975). Most of the photographs shown in this chapter are taken from specimens from rhesus monkeys (*Macaca mulatta*), cynomolgus monkeys (*Macaca fascicularis*) and squirrel monkeys (*Saimiris sciureus*). The contributions of several scanning electron micrographs by Dr Ivan M. Hunter-Duvar and Dr David J. Lim are greatly appreciated by the author.

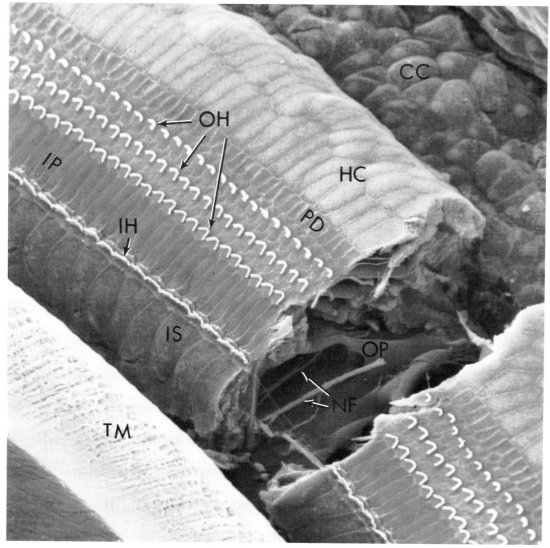

Figure 5.1 *A scanning electron micrograph of the organ of Corti of the chinchilla, Chinchilla laniger (courtesy of Dr I. Hunter-Duvar). The tectorial membrane (TM) has been retracted from the surface of the organ of Corti exposing the stereocilia of the inner hair cells (IH) and outer hair cells (OH). A fracture in the organ of Corti reveals the nerve fibres (NF) in Corti's tunnel, outer pillar cells (OP) and three rows of outer hair cells. Note that Hensen's cells (HC) are not attached to the outer hair cells but to the phalanges of Deiters' cells (PD). The specimen was prepared by a modified OTOTO method (Hunter-Duvar, 1978). IS = Inner sulcus cells; IP = head plates of inner pillar cells; HC = Hensen's cells; CC = Claudius cells*

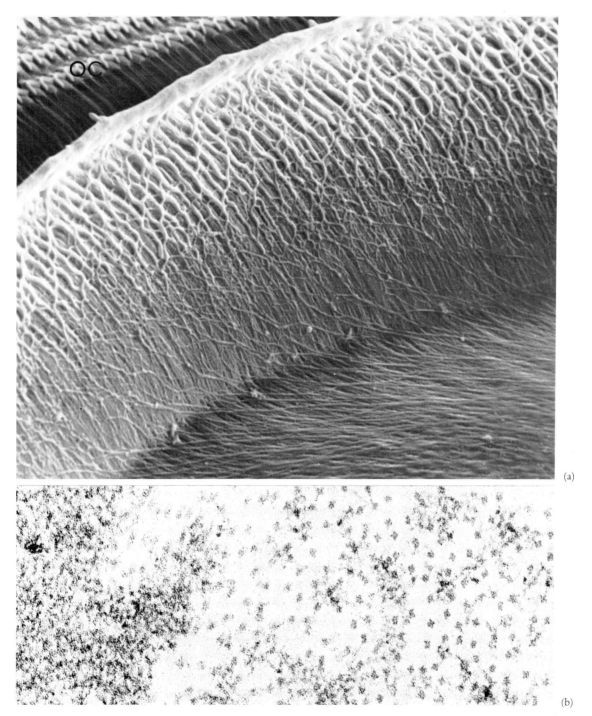

(a)

(b)

Figure 5.2(a) *Tectorial membrane of the chinchilla showing the fibrillar texture of the upper surface (courtesy of Dr I. Hunter-Duvar). The membrane has been lifted off the organ of Corti (OC) by the preparation method. The ridges in the membrane are more prominent towards the edge, marginal band. The fibrils of the membrane are oriented towards the apical and radial directions. The attachment of the tectorial membrane to the organ of Corti is a subject of controversy. By transmission electron microscopy, the tectorial membrane is found to be attached to the taller stereocilia of the outer hair cells (Kimura, 1966). By scanning electron microscopy other attachments are found at the phalanges of the outermost Deiters' cells (Lim, 1972), the Hensen's cells, and the supporting cells around the inner hair cells (Lawrence and Burgio, 1980). (b) A cross-section of the tectorial membrane close to the organ of Corti (M. fascicularis). The dense homogeneous ground substance and rectangular fibrils (130 Å (13 nm) across) are the main structural components. The tops of stereocilia (not shown here) are attached to the homogeneous ground substance by microfilaments. The coarse fibrils are similar in shape to those of the basilar membrane. ×146 200*

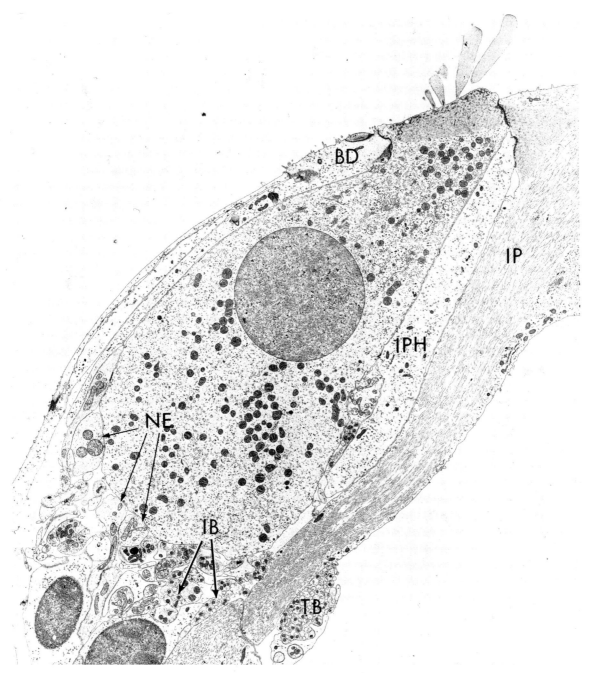

Figure 5.3 *Inner hair cell (M. fascicularis). The inner hair cell is flask shaped with a small apex and a large cell body. The long axis of the cell is inclined towards Corti's tunnel. Junctional complexes are formed at the cell apex whereas no membrane specialization is established along the sides with the inner phalangeal cell (IPH) or border cell (BD), both of which lack intracellular tonofibrils. The nucleus is located centrally or close to the cell apex, and its chromatin distribution is uniform. Mitochondria are more numerous in the subcuticular and infranuclear zones. Nerve fibres and nerve endings (NE) are located at the lower half of the cell body. IP = Inner pillar cell; TB = tunnel spiral nerve bundle; IB = inner spiral nerve bundle. ×11 500*

(a)

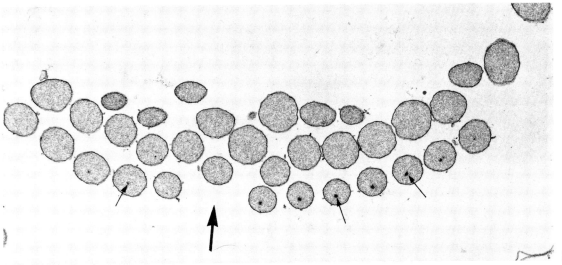

(b)

Figure 5.4(a) A scanning electron micrograph of the stereocilia of an inner hair cell of the chinchilla (courtesy of Dr I. Hunter-Duvar). The stereocilia are arranged in a few curved rows with tall ones facing the inner pillar cells. There is a considerable difference in height between the tall and short strereocilia. Note that the tips of the short stereocilia are flat, not round, and that they are slanted towards the tall ones. The large stereocilia of the inner hair cells are taller than those of the outer hair cells at the lower basal turn but are gradually exceeded by the stereocilia of the outer hair cells toward the apical turn (Lim, 1980). (b) A cross-section of the stereocilia of the inner hair cell (M. fascicularis). The stereocilia are arranged in a flat W-form with a notch (large arrow) where a kinocilium was located in a developmental stage of the organ of Corti. The number of stereocilia is generally smaller than those of the outer hair cells. The tops of tall stereocilia tend to be spaced widely apart and the interconnecting filamentous bridges between the cilia are rarely seen. The cores are shown in the tall cilia (small arrows). $\times 16\,400$

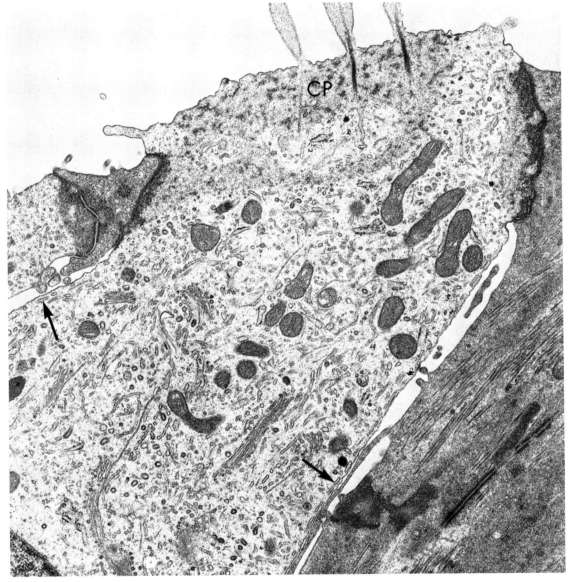

Figure 5.5 *Apical portion of an inner hair cell (M. fascicularis). The most prominent feature is an extensive network of tubules and some Golgi bodies. Coated vesicles, multivesiculated bodies, and lysosomes, even though not obvious or shown in this photograph, are also common in this area. Short subsurface cisternae (arrows) lie along the lateral surface of the plasma membrane. The stereocilia are attached vertically to the cuticular plate (CP) and their rootlets often extend into the cytoplasm. Note the cuticle-free zone to the right of the stereocilia where a basal body is often observed.* ×15 100

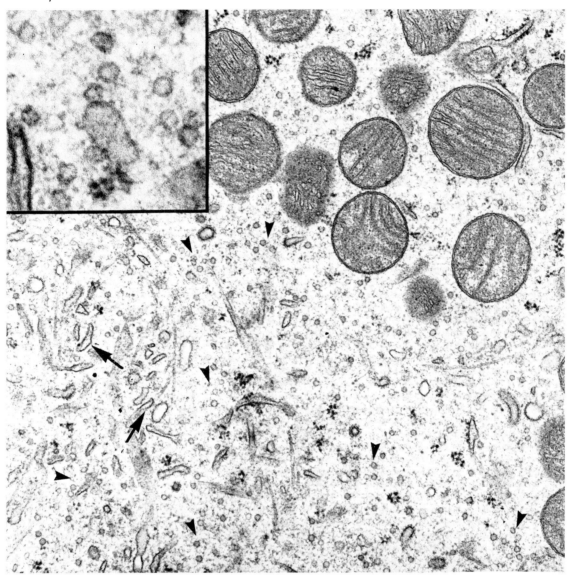

Figure 5.6 *Infranuclear zone of the inner hair cell (M. fascicularis). The cytoplasm is filled by a large number of small vesicles (arrowheads). The inset (×141 000) shows a high magnification of these vesicles. Their size (375 Å)(37.5 nm) diameter) is comparable with that of synaptic vesicles. There are some large tubules (arrows) and fine tubular fibrils. The rough endoplasmic reticulum is short and ribosomes are scanty. There are a few scattered coated vesicles. The mitochondria are round and are often grouped together around the nucleus. ×39 000*

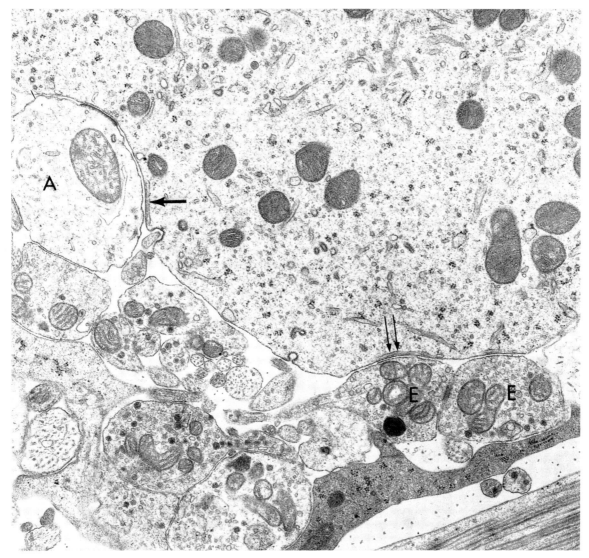

Figure 5.7 *Basal portion of an inner hair cell (M. fascicularis). The cytoplasm contains scattered mitochondria, a small number of large tubules, short rough endoplasmic reticulum, ribosomes in rosette form, and coated and non-coated vesicles. Two types of nerve terminals are shown in contact with the inner hair cell. The large terminal on the left (A), showing a thickening in the post-synaptic membrane and a few synaptic vesicles on the pre-synaptic side, is presumably an afferent nerve ending. Note the subsurface cisterna (single arrow) on the inner hair cell side opposite this nerve ending. On the right the two nerve varicosities (E) containing numerous vesicles, a few mitochondria, and dense-cored vesicles are presumably efferent. Note that there is a desmosome between these two varicosities. On the inner hair cell side, subsurface cisternae (double arrows) are aligned along one of these varicosities. There are no wedge-shaped electron-dense substances in the efferent nerve varicosities or vesicular aggregations in the nerve terminals of either type of nerve ending opposite the subsurface cisternae. Other nerve fibres are mostly efferent nerve fibres of the inner spiral bundle. Efferent nerve contacts on the inner hair cells are sparse, while afferent nerve terminals number as many as 20 per inner hair cell. ×18 000*

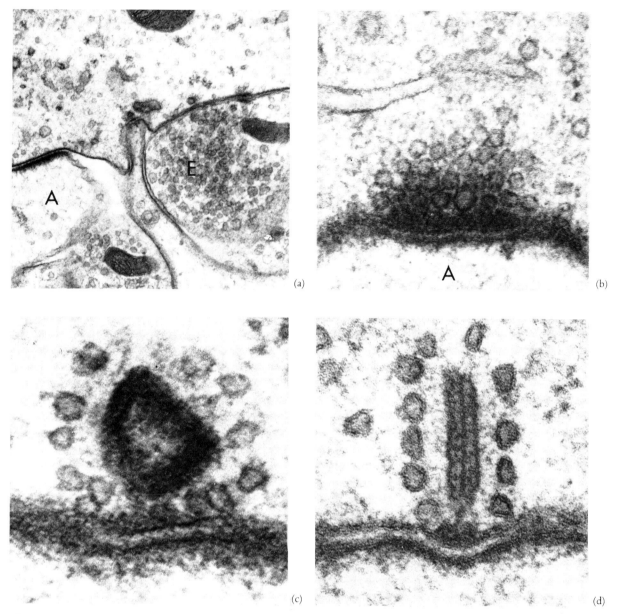

Figure 5.8 *Junctions between the nerve endings and inner hair cells. (a) On the right is a punctum adherens, a short desmosome, at the efferent nerve varicosity (E). On the left is a synapse of the afferent nerve ending (A), the most common type, with thickening of the post-synaptic membrane and an aggregation of synaptic vesicles at the pre-synaptic membrane (M. fascicularis). ×35 900. (b) A synapse of the afferent nerve ending (A) with a wider thickening of the pre-synaptic membrane and an aggregate of synaptic vesicles (M.* *fascicularis). ×117 900. (c) and (d) Synaptic bodies of the inner hair cells. The shapes and sizes of these bodies differ within the same and in different species. These bodies usually extend a slender dense bridge or bridges towards the pre-synaptic membrane and are surrounded by a series of synaptic vesicles. (c) (S. sciureus) shows a ring form (×199 100), and (d) (M. fascicularis) shows a rod form with a longitudinal cleavage line (×170 800)*

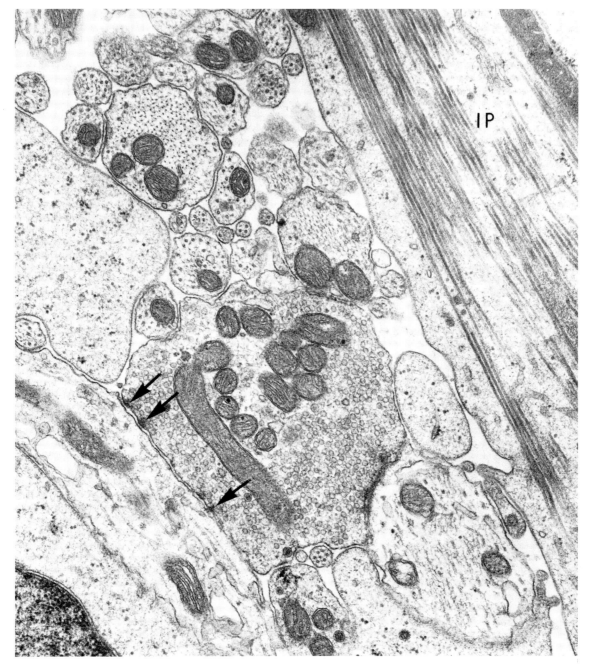

Figure 5.9 *Inner spiral bundle (M. fascicularis). This nerve bundle is located below the inner hair cells and close to the inner pillar cells (IP), and is composed mainly of efferent nerve fibres. They run in a spiral direction and make synaptic contact (arrows) with the afferent nerve fibres in the bundle or immediately below the nerve terminals of the inner hair cells. This is one of the main locations at which the efferent (inhibitory) fibres affect the neural activities of the afferent nerve fibres. The origin of the inner spiral bundle is predominantly from the small cells of the main ipsilateral superior olivary complex (Warr, 1978). ×33 300*

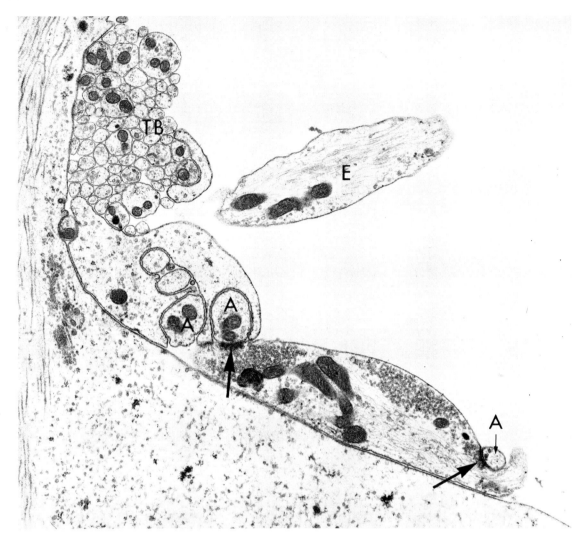

Figure 5.10 *Nerve fibres in Corti's tunnel (M. mulatta). The tunnel spiral bundle fibres (TB) run in a spiral direction and also head radially (E) toward the base of the outer hair cells in the middle of Corti's tunnel. A cross-section of these fibres shows more neural filaments than neural tubules. These nerve fibres are primarily efferent. Another group of nerve fibres travels radially towards the* *Deiters' cells at the base of the pillar cells, slightly above or even enclosed by these cells. These nerves are afferent (A) and comprise only 5 per cent of the total cochlear nerve fibres (Kimura, 1978; Spoendlin, 1979). Synapses (arrows) between the afferent and efferent nerve fibres are sometimes observed. ×18 000*

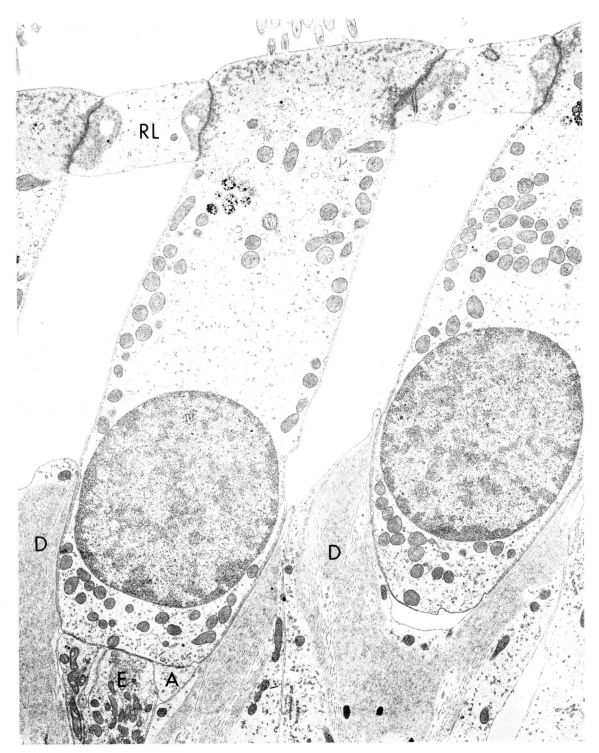

Figure 5.11 *Outer hair cell (S. sciureus). The cell is cylindrical in shape and is surrounded by Corti's fluid at the mid-portion. The long cell axis tilts toward Corti's tunnel and the caecum vestibulare. These hair cells are shorter at the basal turn and gradually increase in height towards the radial direction and apex of the cochlea. The cell apex is attached to the reticular lamina (RL) with a junctional complex formation, but at the base no such cytoplasmic modification is demonstrated with the Deiters' cells (D). The nucleus is located in the basal portion almost completely filling the cross-sectional area. Mitochondria are often localized in the subcuticular zone, periphery, and infranuclear zone. Short sub-surface cisternae run parallel to the lateral plasma membrane. At the inferior pole there are two types of nerve endings, afferent (A) and efferent (E). ×11 200*

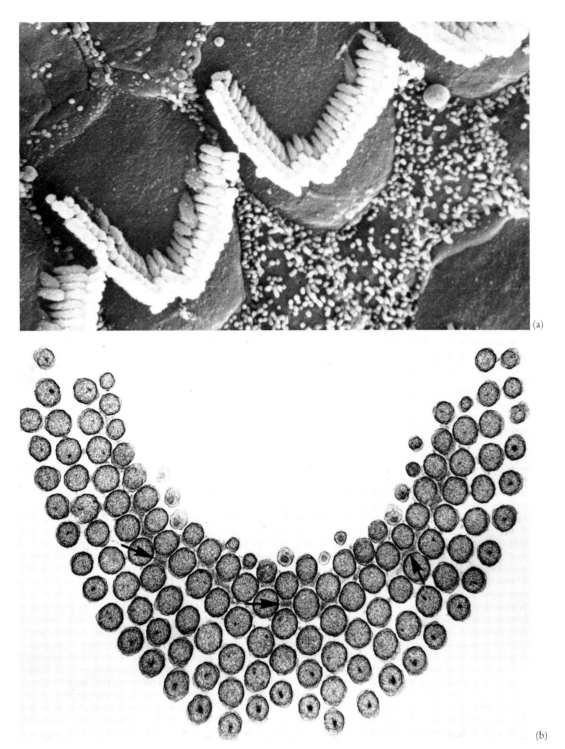

(a)

(b)

Figure 5.12(a) A scanning electron micrograph of the stereocilia of the outer hair cells of the chinchilla (courtesy of Dr D. Lim). The stereocilia are arranged in three rows in small animals and about six rows in primates, with the tall stereocilia located at the periphery. The height of the stereocilia increases from the inner to outer rows of the outer hair cells as well as from the base to the apex of the cochlea. Their height in the apical turn is about three times greater than that in the lower basal turn. The cell surface around the stereocilia of the outer hair cells is smooth, whereas in the supporting cells microvilli are numerous. ×400. (b) Cross-section of the stereocilia of the outer hair cell (human). The stereocilia are arranged in a W-form, the bases of which face the Hensen's cells. The number of stereocilia is large but their diameter is smaller than that of the inner hair cells. They are grouped closely together and interconnecting bridges (arrows) between the cilia are common. When displacement of the stereocilia occurs, from the shorter cilia towards the longer ones, excitation of the afferent nerve synapsing on the sensory cell occurs, and displacement in the opposite direction results in inhibition (Flock, 1965; Wersäll et al., 1965). Original magnification ×34 000 reduced to 92%

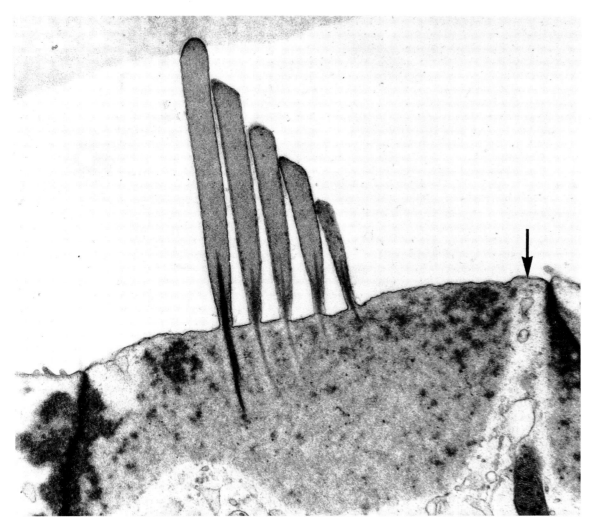

Figure 5.13 *Apical portion of the outer hair cell (S. sciureus). The tall stereocilia are attached to the tectorial membranes. The shorter cilia may have been embedded in the tectorial membrane in vivo. Although it is not shown clearly in this photograph, tall and short stereocilia are interconnected by microfilaments which are attached to thick aggregates of electron-dense substance midway between the cilia. By contrast, similar microfilaments with condensed midline substance have not been found between the stereocilia of the inner hair cells, though artefact cannot be ruled out. The stereocilia and rootlets are composed of longitudinally arranged actin filaments with functional polarization directed towards the cytoplasm (Flock et al., 1981). The cores of the stereocilia extend from the middle portion of the cilia into the cuticular plate as rootlets, some of which penetrate the cytoplasm. The cuticular plate is also composed of actin filaments which run parallel to the apical surface and are attached to the zonula adherens. The cuticle-free zone is not limited to the usual location near the tall stereocilia, but is found in other areas (arrow). ×18 400*

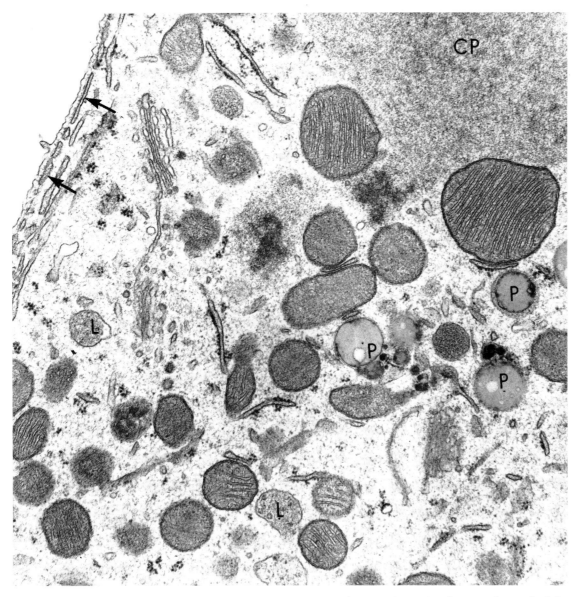

Figure 5.14 *Subcuticular zone of the outer hair cell (M. fascicularis). The mitochondria aggregated in this area are round and elongated, and some very large ones are located near the cuticular plate (CP). There are a few Golgi networks and elongated rough endoplasmic reticulum with a small number of ribosomes. Subsurface cisternae (arrows), either in single or multiple layers, are found underneath the plasmalemma. Another common feature is the presence of lysosomes (L) and pigment granules (P). This is a site of predilection for the early manifestation of cell injury. ×32 700*

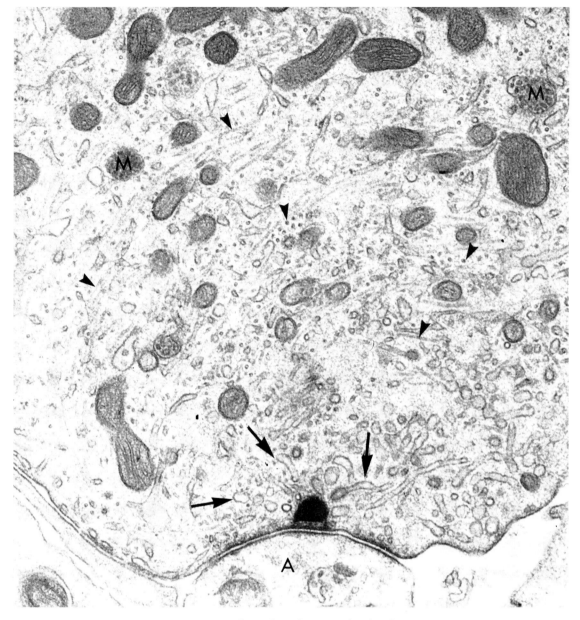

Figure 5.15 *Infranuclear zone of the outer hair cell (M. fascicularis). The prominent features are the presence of numerous crisscrossing long tubular fibrils (250 Å (25 nm) diameter)(arrowheads) and prominent large tubules (arrows). The large tubules located near the synaptic body may be involved in recycling the synaptic vesicles. A few multivesiculated bodies (M) and some coated vesicles are found. Ribosomes and rough endoplasmic reticulum are sparse. Note the thickened presynaptic membrane at the dense synaptic body opposite an afferent nerve ending (A). The origin of the afferent nerve endings below the outer hair cells is believed to be from the small pseudomonopolar unmyelinated spiral ganglion cells (Spoendlin, 1979; Kiang et al., 1982). ×33 300*

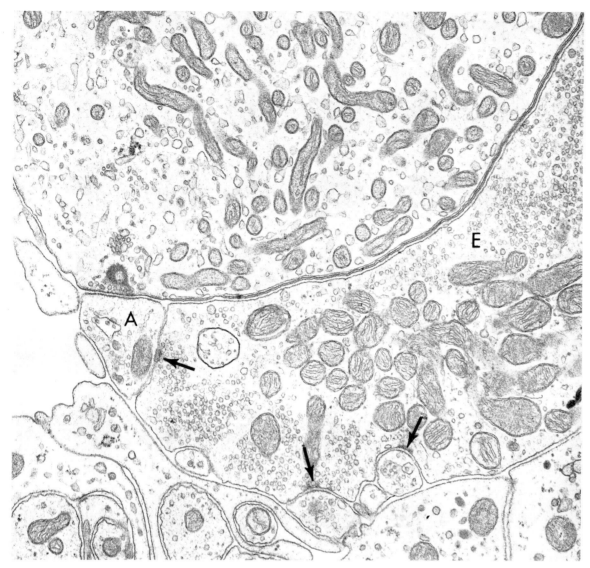

Figure 5.16 *Basal portion of the outer hair cell (M. mulatta). This portion is characterized by the presence of numerous tubules and mitochondria. A dense circular synaptic body surrounded by vesicles is shown adjacent to a small afferent nerve ending (A). In the monkey, the number of afferent nerve terminals is about 1–4 per outer hair cell in the basal turn. A long sub-synaptic cisterna is shown in the outer hair cell opposite a large efferent nerve ending (E). Efferent nerve synapses (arrows) are shown with an afferent nerve ending (A) and two nerve fibres. Few efferent nerve endings are found per outer hair cell in the basal turn but the number both of afferent and of efferent nerve endings varies considerably in different species. Serial sections show that efferent nerve endings may be missing from the outer hair cells. These efferent fibres of the outer hair cells originate predominantly in the large cells of the contralateral accessory superior olivary complex (Warr, 1978). The functional significance of the subsynaptic cisternae is not known. ×28 000*

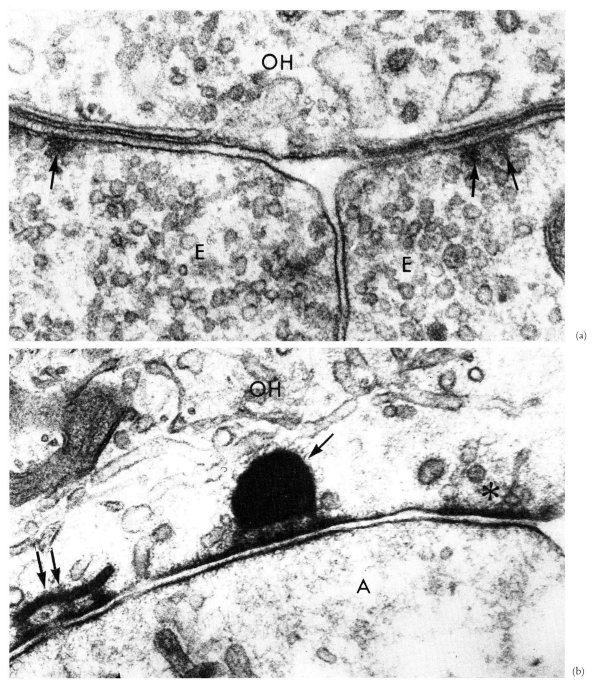

(a)

(b)

Figure 5.17 *(a) Efferent nerve synapses on the outer hair cell (S. sciureus). In the efferent nerve endings (E) opposite the subsynaptic cisternae of the outer hair cell (OH), some more or less wedge-shaped dense structures abut the pre-synaptic membranes (arrows). Synaptic vesicles of the efferent nerve endings are aggregated near the dense substance.* ×114 000. *(b) Afferent (A) nerve synapses on the outer hair cell (M. fascicularis). The synapses take a different form. The synaptic body may be spherical, half-dome (arrow), rod-shaped, plaque-shaped (double arrows) or irregular in shape. Another type of synapse (★) is a pre-synaptic membrane thickening with vesicular aggregations alongside. Puncta adherentes, though not shown in this figure, may also be present. Note numerous tubules in the hair cell (OH) side adjacent to the synaptic zone.* ×81 000

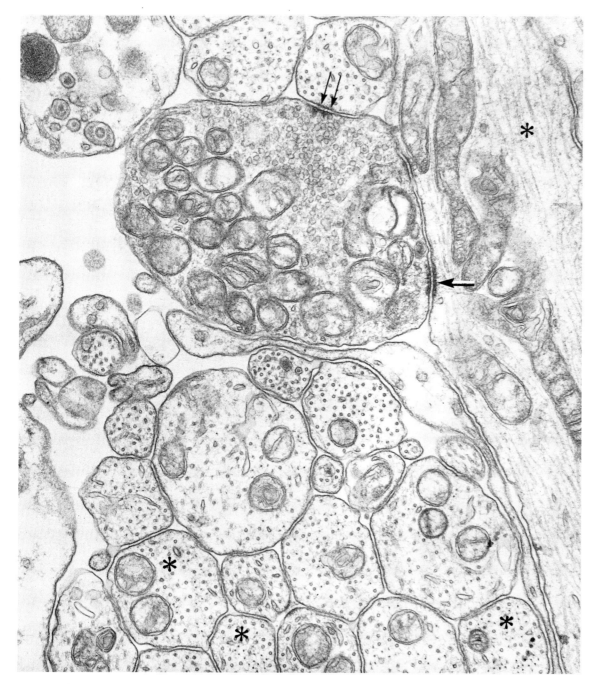

Figure 5.18 *Outer spiral bundle (human). In human specimens the outer spiral fibres often run in a large bundle, whereas in small animals they travel individually in a spiral direction. Within the bundle, synaptic contacts (double arrows) as well as desmosomes (single arrow) are often seen. The varicose nerve containing numerous vesicles and mitochondria is an efferent fibre coming from the olivocochlear bundle. Nerve fibres containing almost exclusively neurotubules are in the majority; however, a few fibres also contain a large number of neurofilaments (*). Differentiation of afferent and efferent nerve fibres based on a cross-sectional view is still obscure, though afferent fibres are more often seen in a tubular form, the efferent in a filamentous form.* ×34 800

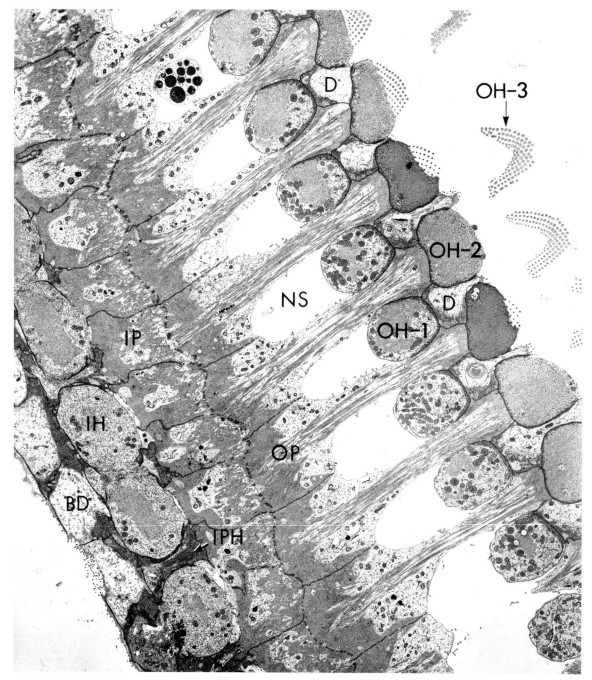

Figure 5.19 *Reticular lamina (M. fascicularis). The reticular lamina is a thick net-like plate formed by the supporting cells to hold the apical ends of the hair cells in place. The inner hair cells (IH) are supported by the inner pillar cells (IP), inner phalangeal cells (IPH) and border cells (BD). The outer hair cells (OH-1) of the inner row are held by the inner and outer (OP) pillar cells and Deiters' cells (D), the middle row (OH-2) by the outer pillar cells and Deiters' cells, and the outer row (OH-3) by the Deiters' cells. Owing to the curvature of the surface of the organ of Corti, not all cell junctions are shown in this photograph. NS = Nuel's space. ×2700*

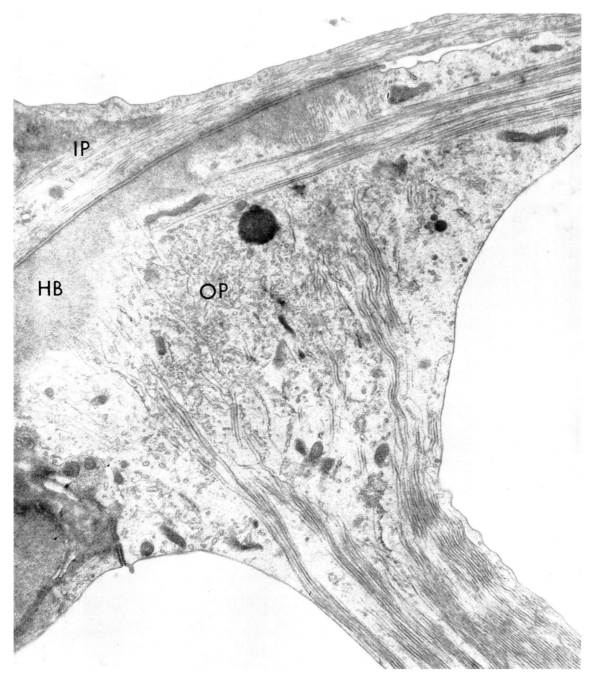

Figure 5.20 *The junction of the outer and inner pillar cells at their apices (human). The most remarkable feature is a series of desmosomes along the length of the cell junctions between the inner pillar (IP) and outer pillar (OP) cells. In the outer pillar cell one group of* tonofibrils runs parallel to the endolymphatic surface and a second group of tonofibrils runs towards the basilar membrane. The head body (HB) is a cementing body into which the tonofibrils are inserted. ×17 500

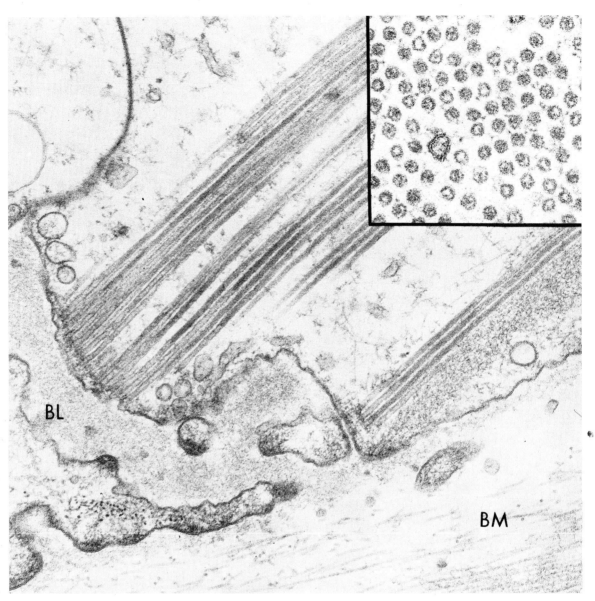

Figure 5.21 *Base of an outer pillar cell (human). The tonofibrils terminate a short distance from the plasmalemma. At the terminal portions of the fibrils an electron-dense substance fills the spaces between them. BM = Basilar membrane, a deep extension of the basal lamina (BL). ×58 900. Inset: A cross-section of tonofibrils (M. fascicularis). The fibrils are round (about 275 Å (27.5 nm) in diameter) and the insides are hollow and/or filled with an electron-dense substance. Smaller filaments are also shown between the tonofibrils. ×124 500*

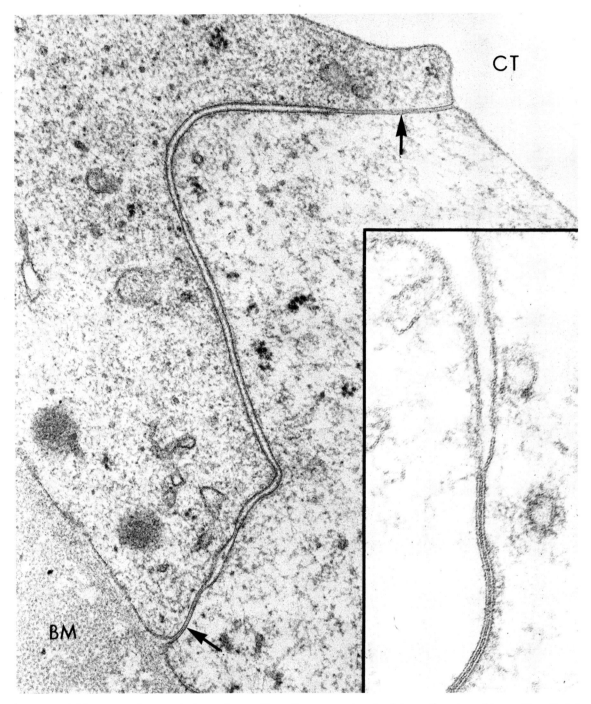

CT

BM

Figure 5.22 *The junction between the bases of outer pillar cells at Corti's tunnel (M. fascicularis). There is no typical tight junction from the basilar membrane (BM) side of the scala tympani to Corti's tunnel (CT). Only gap junctions are demonstrable (arrows)(Iurato et al., 1976). ×61 100. Inset: A high-magnification view of the gap junction between the outer and inner pillar cells facing Corti's tunnel (human). These junctions do not prevent the passage of ions or small molecules from one fluid space to another. The fluid in Corti's tunnel might be similar to perilymph. ×155 000*

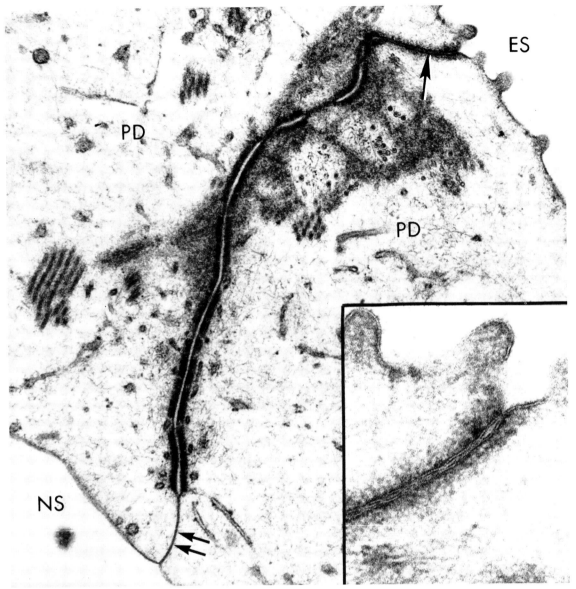

Figure 5.23 *The junction of the phalanges of Deiters' cells (PD) between the endolymphatic space (ES) and Nuel's space (NS) (M. fascicularis). A series of desmosomes is a prominent feature through almost the entire length of the junction. In the area adjacent to the desmosomes are tonofibrils in cross-section. At the endolymphatic surface is a tight junction (single arrow) and on the Nuel's space side is a gap junction (double arrows) (Iurato et al., 1976). ×42 000.*

The inset shows a high-magnification view of the tight junction. The outer leaflets of the unit membrane approximate each other at certain intervals and a cytoplasmic condensation is noted along its length. The tight junction at the endolymphatic surface side probably plays an essential role in maintaining the different chemical composition of the fluids in the organ of Corti and the endolymphatic space. ×135 000

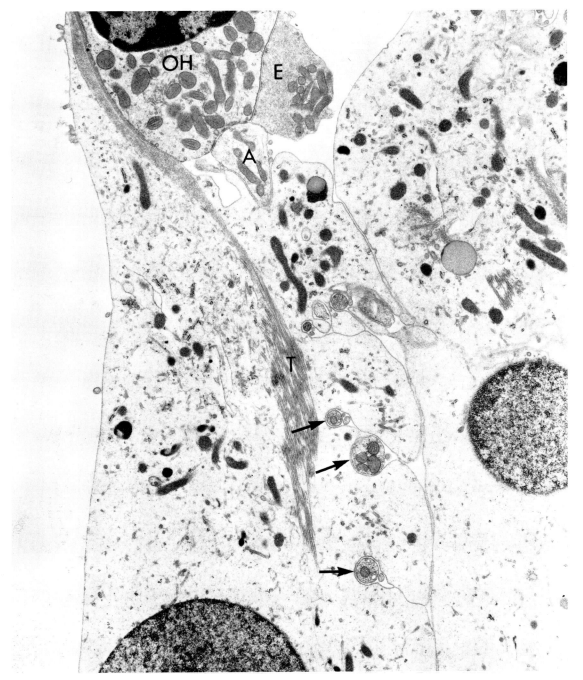

Figure 5.24 *Mid-portion of Deiters' cells (M. fascicularis). Deiters'
cells, or outer phalangeal cells, in addition to forming the reticular
lamina, support the bases of the outer hair cells (OH) by semi-cup-
shaped cytoplasmic extensions which are reinforced by dense ground
substance and tonofibrils (T). Another main feature is the cytoplasm
surrounding nerve fibres (arrows) to provide mechanical support and
possibly nutrition. The cells contain a small number of ribosomes,
rough endoplasmic reticulum, tubules and pigment granules. The
nucleus is round and larger than that of the outer hair cell and
chromatin is diffusely distributed. Note afferent (A) and efferent (E)
nerve endings on the outer hair cells. ×11 500*

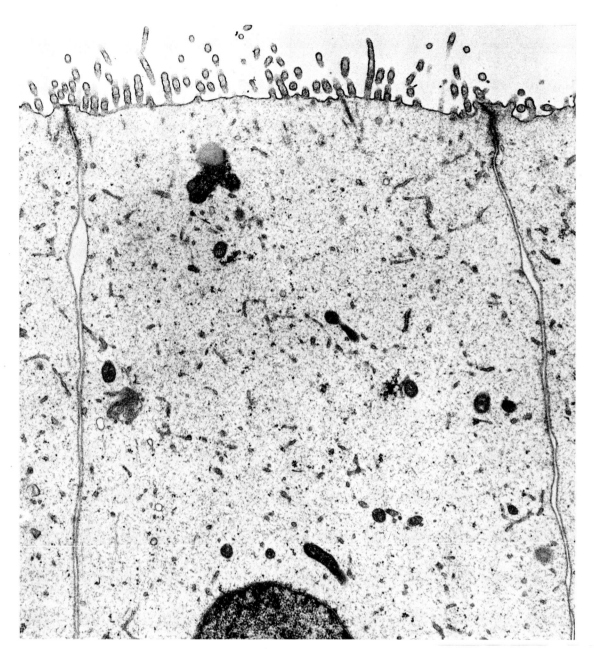

Figure 5.25 *Hensen's cell (M. fascicularis). The cell is very tall with its nucleus located centrally or close to the cell apex. The cytoplasm is filled almost entirely with ground substance and the distribution of mitochondria, lipofuscin and other organelles is rather small. One of the prominent features is the presence of numerous tall microvilli from which dense roots extend into the cytoplasm. The presence of microvilli suggests that these cells may be engaged, to some extent, in fluid absorption. Hensen's cells, Claudius' cells, and inner sulcus cells are similar in their cytoplasmic characteristics, though their sizes and shapes vary. ×19 900*

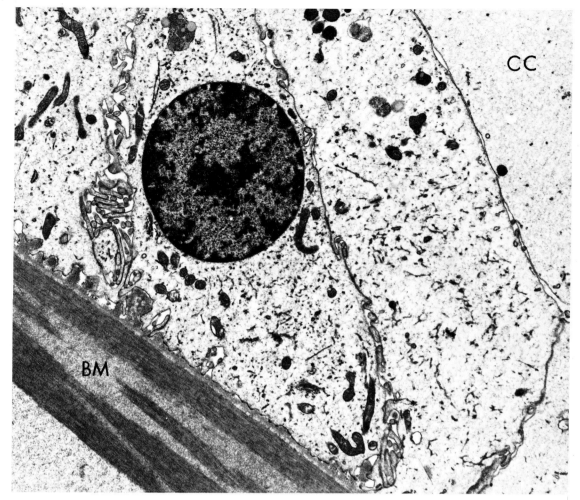

Figure 5.26 *Böttcher cell (M. fascicularis). Böttcher cells are located on the pars pectinata of the basilar membrane (BM) in the basal turn. These cells are found in clusters under the Claudius cells and their cell apices do not extend to the endolymphatic surface. Their nuclei are round-to-oval in shape and show an irregular chromatin condensation. The cytoplasm does not show any unique features and is filled with scattered mitochondria, lipofuscin, tubules and ribosomes. These cells interdigitate frequently among themselves. The basal portion towards the basilar membrane is irregular, electron dense at the plasmalemma, and supported by a thick basal lamina. Plasma membrane infoldings towards the Claudius cells (CC) are not extensive. Their function is not known. ×9600*

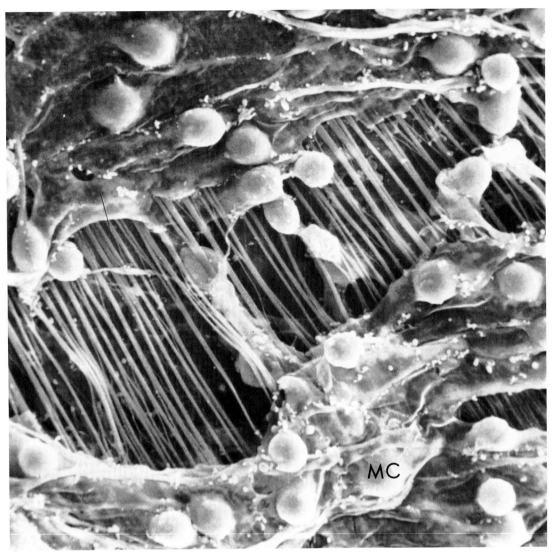

MC

Figure 5.27 *Basilar membrane of the chinchilla (courtesy of Dr I. Hunter-Duvar). This scanning electron micrograph demonstrates the pectinate arrangement of the basilar membrane fibrils in a radial direction, as seen through the gaps in the mesothelial cells (MC) or tympanic covering cells. The long axis of the mesothelial cells lies in the spiral direction, that is, at right angles to the basilar membrane fibrils. In the pars pectinata, a zone between the outer pillar cells and the basilar crest of the spiral ligament, the basilar membrane fibrils are separated into an upper semi-continuous layer and a lower layer with a loose arrangement of bundles like piano strings. At the pars tecta, a zone between the inner and outer pillar cells, the fibrils take the form of a single continuous and compact layer. A capillary, vas spiralis, may or may not be present, depending on the species*

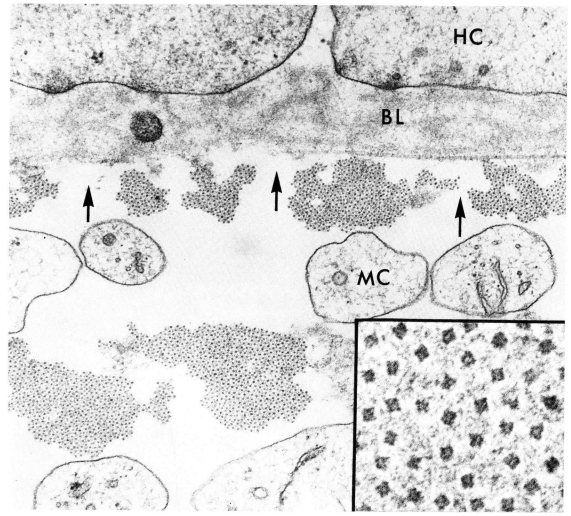

Figure 5.28 *Pars pectinata of the basilar membrane (M. fascicularis). The basilar membrane is composed of fibrils, homogeneous ground substance and mesothelial cells (MC, cells of the tympanic covering layer). In this basilar membrane, taken from an upper cochlear turn, the fibrils near the basal lamina (BL) of the Hensen's cells (HC) do not form a continuous layer. Gaps (arrows) are common, thus suggesting that a tracer substance injected into the perilymphatic space of the scala tympani can enter the organ of Corti readily at these locations. ×46 300. The inset shows a high-magnification view of fibrils embedded in the homogeneous ground substance. The cross-section of the fibrils shows a rectangular profile with a size of 140 Å (14 nm). The mesothelial cells form almost a single cell layer at the basal end but toward the apical cochlear turns they increase in number and are loosely stacked (Cabezudo, 1978). ×255 200*

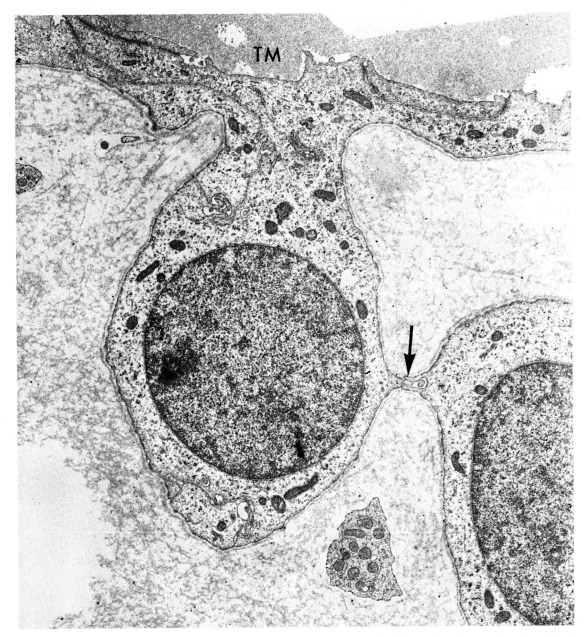

TM

Figure 5.29 *Interdental cell (S. sciureus). The phalangeal portions of the interdental cells cover the endolymphatic surface of the limbus spiralis and serve as an anchor for attachment of the tectorial membrane (TM). Below the phalangeal portions are narrow necks which may sometimes contain deep, duct-like invaginations extending from the apical surfaces. The cell bodies are oval in shape, embedded in a connective-tissue matrix, and in close contact with each other (arrow). The nuclei are large and are located in the central portions of the cell bodies. Mitochondria are sparse and are scattered throughout, and the Golgi network is located at the supranuclear zone. The function of these cells is not well known; however, it is thought to be related to the formation and maintenance of the tectorial membrane (Lim, 1970). ×6300*

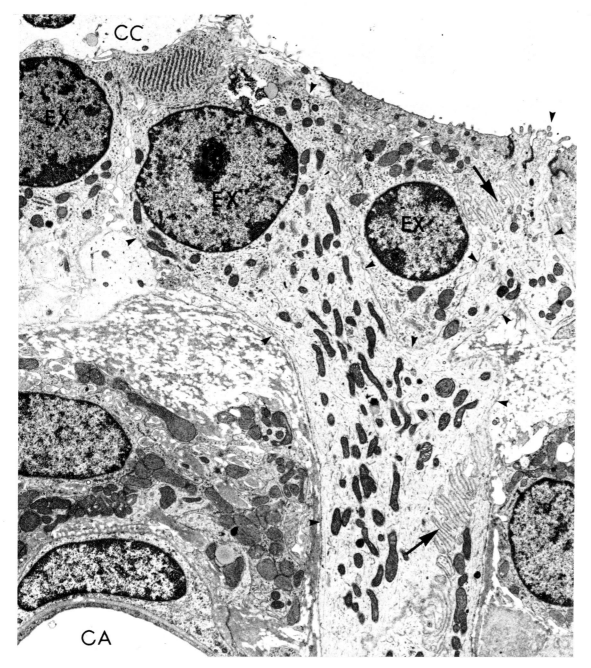

Figure 5.30 *External sulcus cells (M. fascicularis). The external sulcus cells (EX) or root cells are very prominent in the basal turn and decrease in number towards the cochlear apex. In the basal turn, they may be covered entirely by Claudius cells (CC), but more often they are exposed to the endolymphatic surface in other turns. Their shape is irregular (as outlined by arrowheads) and the cell sends a long cytoplasmic extension into the spiral ligament and even behind the spiral prominence. These cells form compact aggregates and frequently interdigitate (arrows) among themselves. As they extend deeper, they carry the basal lamina around them and are often closely associated with capillaries (CA). Their nuclei are often located at the luminal side; however, it is not uncommon to find their nuclei within root portions deep in the spiral ligament. The cytoplasm is rich with ribosomes in rosette form around the nucleus and the Golgi networks are located in the subnuclear zone. Their cell processes contain numerous elongated mitochondria, tubular fibrils, and some lipofuscin granules. Their function is not clearly understood, although fluid absorption and phagocytosis are suggested (Duvall, 1969). ×8300*

References

Ades, H. W. and Engström, H. (1972) Inner ear studies. *Acta Otolaryngologica*, Supplement **301**, 1.

Cabezudo, L. M. (1978) The ultrastructure of the basilar membrane in the cat. *Acta Otolaryngologica*, **86**, 160.

Corti, A. (1851) Recherches sur l'organe de l'ouie des mammifères. *Zeitschrift für Wissenschaftliche Zoologie*, **3**, 109.

Duvall, A. J. (1969) The ultrastructure of the external sulcus in the guinea pig cochlear duct. *Laryngoscope*, **79**, 1.

Engström, H. and Ades, H. W. (1973) The ultrastructure of the organ of Corti. *The Ultrastructure of Sensory Organs*, ed. I. Friedmann, Ch. 2, pp. 83–151. Amsterdam: North-Holland.

Flock, Å. (1965) Electron microscopic and electrophysiological studies on the lateral line canal organ. *Acta Otolaryngologica*, Supplement **199**, 1.

Flock, Å., Cheung, H. C., Flock, B. and Utter, G. (1981) Three sets of actin filaments in sensory cells of the inner ear. Identification and functional orientation determined by gel electrophoresis, immunofluorescence and electron microscopy. *Journal of Neurocytology*, **10**, 133.

Gulley, R. L. and Reese, T. S. (1977) Freeze-fracture studies on the synapses in the organ of Corti. *Journal of Comparative Neurology*, **171**, 517.

Hunter-Duvar, I. M. (1978) Electron microscopic assessment of the cochlea. Some techniques and results. *Acta Otolaryngologica*, Supplement **351**, 1.

Iurato, S. (1967) Organ of Corti. *Submicroscopic Structure of the Inner Ear*, ed. S. Iurato, p. 80. London: Pergamon Press.

Iurato, S., Franke, K., Luciano, L., Wermbter, G., Pannese, E. and Reale, E. (1976) Intercellular junctions in the organ of Corti as revealed by freeze fracturing. *Acta Otolaryngologica*, **82**, 57.

Kiang, N. Y. S., Rho, J. M., Northrop, C., Liberman, M. C. and Ryugo, D. K. (1982) Hair cell innervation by spiral ganglion cells in adult cats. *Science*, **217**, 175.

Kimura, R. S. (1966) Hairs of the cochlear sensory cells and their attachment to the tectorial membrane. *Acta Otolaryngologica*, **61**, 55.

Kimura, R. S. (1975) The ultrastructure of the organ of Corti. *International Review of Cytology*, **42**, 173.

Kimura, R. S. (1978) Differences in innervation of cochlear inner and outer hair cells. *Abstracts of the Midwinter Research Meeting, Jan. 30–Feb. 1*, ed. D. Lim, p. 40. Columbus, Ohio: Association for Research in Otolaryngology.

Lawrence, M. and Burgio, P. A. (1980) The attachment of the tectorial membrane revealed by scanning electron microscope. *Annals of Otology, Rhinology and Laryngology*, **89**, 325.

Lim, D. J. (1970) Morphology and function of the interdental cell. *Journal of Laryngology and Otology*, **84**, 1241.

Lim, D. J. (1972) Fine morphology of the tectorial membrane. Its relationship to the organ of Corti. *Archives of Otolaryngology*, **96**, 199.

Lim, D. J. (1980) Cochlear anatomy related to cochlear micromechanics. A review. *Journal of the Acoustical Society of America*, **67**, 1686.

Nadol, J. B., Jr. (1981) Reciprocal synapses at the base of outer hair cells in the organ of Corti of man. *Annals of Otology, Rhinology and Laryngology*, **90**, 12.

Smith, C. A. (1968) Ultrastructure of the organ of Corti. *Advances in Science*, **24**, 419.

Spoendlin, H. (1970) Auditory, vestibular, olfactory and gustatory organs. In *Ultrastructure of the Peripheral Nervous System and Sense Organs*, eds. J. Babel, A. Bischoff and H. Spoendlin, pp. 177–261. St. Louis: C. V. Mosby.

Spoendlin, H. (1979) Neural connections of the outer hair cell system. *Acta Otolaryngologica*, **87**, 381–387.

Warr, W. B. (1978) The olivocochlear bundle: its origins and terminations in the cat. *Evoked Electrical Activity in the Auditory Nervous System*, eds. R. F. Naunton and C. Fernandez, pp. 43–65. New York: Academic Press.

Wersäll, J., Flock, Å. and Lundquist, P-G. (1965) Structural basis for directional sensitivity in cochlear and vestibular sensory receptors. *Cold Spring Harbor Symposia in Quantitative Biology*, **30**, 115.

6
Primary Neurons and Synapses

H. Spoendlin

Two systems of primary neurons connect the cochlea with the brain-stem:

1. The afferent neurons with the bipolar ganglion cells, which form the spiral ganglion in Rosenthal's canal in the modiolus.
2. The efferent olivocochlear neurons, known as the olivochochlear bundle, originating in the homo- and contralateral superior olivary complex, as originally described by Rasmussen (1942).

A third class of neurons consists of the adrenergic autonomic innervation originating in the cervical sympathetic trunk (*Figure 6.1*).

The number of afferent cochlear neurons varies considerably in different species, with about 30 000 in man, 50 000 in the cat (Schuknecht, 1962; Spoendlin, 1969) and about 250 000 in whales (Hall, 1966). In the cochlear nerve all fibres have a very similar diameter with a

unimodal distribution from 4 to 6 μm and a myelin sheath of about 50 lamellae, which is the structural basis of a uniform conduction velocity, probably a very important basic functional feature (*Figure 6.31*). Only very few unmyelinated fibres are found within the trunk of the cochlear nerve, in contrast to the vestibular nerve where many unmyelinated fibres are present and where the calibres of the myelinated fibres vary considerably, between 2 and 15 μm (Spoendlin, 1972). In contrast to the cochlear nerve trunk, there are many unmyelinated fibres of varying calibre among the peripheral axons of the cochlear neurons in the osseous spiral lamina where afferent, efferent and adrenergic fibres intermingle (*Figure 6.28*) (Paradiesgarten and Spoendlin, 1976). All fibres lose their myelin sheath before they enter the organ of Corti through the habenular openings (*Figure 6.20*).

The material presented in this chapter is based essentially on findings in cats unless otherwise stated.

The efferent innervation consists of more than 1000 neurons of different types, originating in the homo- and contralateral superior olivary complex. They reach the periphery together with the vestibular nerve to cross

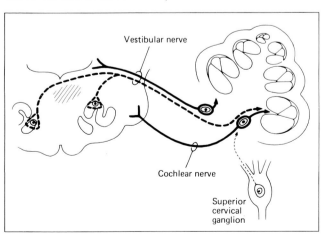

Figure 6.1 *Schematic representation of the three innervation components of the cochlea:*
1. *The bipolar cochlear sensory neurons of the cochlear nerve (full heavy lines), which connect the organ of Corti with the cochlear nucleus in the brain-stem.*
2. *The olivocochlear efferent neurons (interrupted heavy lines), which originate in homo- and contralateral ganglion cells in the superior olivary complex.*
3. *The adrenergic innervation from the superior cervical ganglion (interrupted thin lines)*

133

over to the cochlear nerve through the anastomosis of Oort in the internal acoustic meatus (*see Figure 6.1*).

In order to study the distribution of the afferent and efferent nerve fibres within the organ of Corti, the two systems must be distinguished. The efferents can be selectively eliminated by selective trans-section of the olivocochlear fibres at the level of the vestibular nerve. After such lesions the efferent neurons degenerate very promptly within a few hours and disappear completely within a few days, including practically all inner spiral fibres (*Figures 6.10–6.12*), the upper tunnel radial fibres (*Figures 6.6*) and all the numerous large vesiculated nerve endings at the base of the outer hair cells, showing clearly that all of them belong to the efferent innervation system (Spoendlin, 1966, 1969) (*Figures 6.2* and *6.33*). The great majority of all nerve endings at the base of the outer hair cells are of this large vesiculated efferent type, which is certainly surprising considering the paucity of small afferent nerve endings associated with the outer hair cells (*Figures 6.3* and *6.9*). The number of efferent terminals at the base of the outer hair cells is greatest in the basal turn and decreases gradually towards the cochlear apex, where they are restricted to the first row of outer hair cells. Using the Maillet stain the efferents can be stained quasi-selectively and can be clearly demonstrated in surface preparations (*Figure 6.4*).

According to recent studies on the identification of efferent neurons by the uptake of horse-radish peroxidase from the perilymph (Warr, 1978; Spoendlin, 1981) and by the injection of radioactive amino acids into the superior olivary complex (Warr, 1978), the great majority of the efferent fibres to the outer hair cells originate in large cells in the contralateral accessory olivary nuclei; whereas the great majority of the efferent fibres at the level of the inner hair cells originate in small cells in the homolateral main superior olivary nucleus (*Figures 6.15–6.17*).

There is evidence that the efferent fibres for the outer and inner hair cells represent two different types of neurons. The efferents of the outer hair cells are usually large fibres with diameters up to 1.5 µm and they degenerate very promptly after trans-section, whereas

(*continued on page 138*)

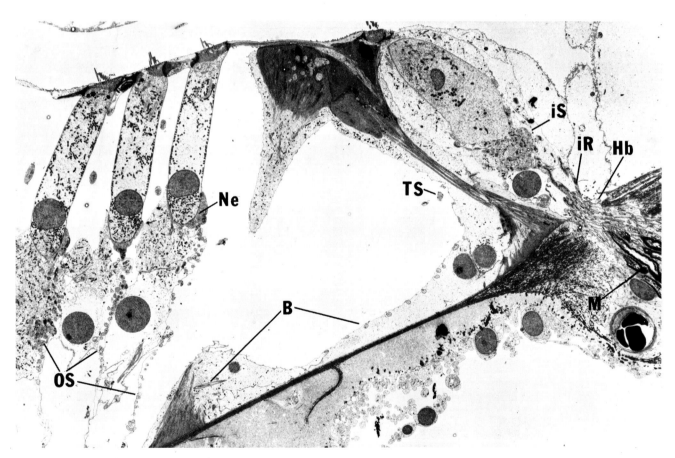

Figure 6.2 *Survey of the organ of Corti of the upper basal turn of a cat. All nerve fibres lose their myelin sheath (M) before they enter the organ of Corti through the habenula perforata (Hb). Within the organ of Corti they are distributed in radial and spiral nerve tracts, according to a well-defined distribution pattern. The main nerve tracts are the inner radial fibres (iR), representing the afferent neurons to the inner hair cells, and the inner spiral fibres (iS), which are the efferent fibres of the inner hair cell area together with the tunnel spiral fibres (TS). The basilar fibres (B) constitute the afferent nerve supply of the outer hair cells. They continue as outer spiral fibres (OS) before ending at the base of the outer hair cells, where the efferent and afferent nerve endings (Ne) are found*

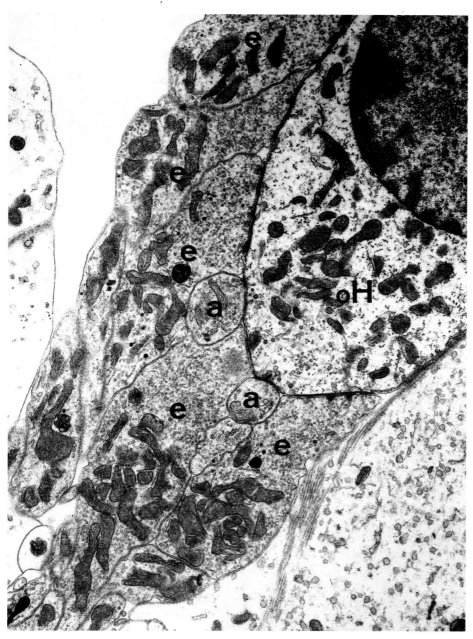

Figure 6.3 *Base of an outer hair cell (oH) of the basal turn of a guinea-pig with a great number of efferent nerve endings (e), which are filled with synaptic vesicles, contain many mitochondria and have a large contact area with the hair cell. In the basal turn they usually outnumber much smaller afferent nerve endings (a), which have a much smaller contact area with the hair cell and only occasionally synaptic complexes in the guinea-pig (but never in the cat)*

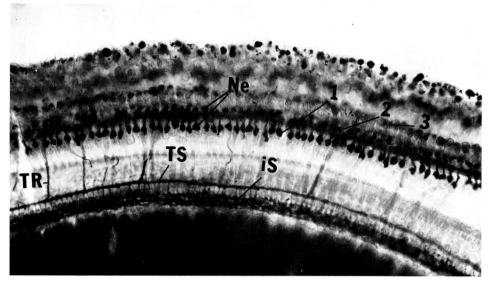

Figure 6.4 *Surface view of the organ of Corti of the second turn of a cat stained according to the method of Maillat, where the efferent nerve fibres and endings are quasi-selectively stained in black. All the efferent elements are clearly seen, such as the inner spiral fibres (iS), the tunnel spiral fibres (TS), the tunnel radial fibres (TR) and the large nerve endings (Ne) at the base of the outer hair cells. The* *efferent innervation is most abundant in the basal turn and decreases gradually towards the apex. There is also a decreasing efferent innervation density from the first to the third row of outer hair cells (1, 2, 3). In the upper cochlear turns there are practically no efferent terminals in the third row of outer hair cells*

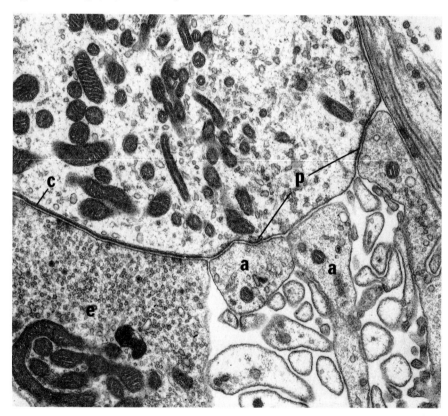

Figure 6.5 *Detail of synaptic contacts between efferent nerve endings (e) and the outer hair cell, and between afferent nerve endings (a) and the outer hair cell in the cat. The postsynaptic membrane on the efferent synapse is paralleled by a very narrow sub-synaptic cysterna (c). The synaptic vesicles in the efferent ending agglomerate at certain areas of the pre-synaptic membrane, which is probably the* *place where the transmitter of the vesicles is released in the synaptic cleft. The differentiation of the afferent synapses is rather poor in the cat. There is only slight thickening of the pre-synaptic membranes (p). The afferent endings (a) contain relatively few cytoplasmic organelles*

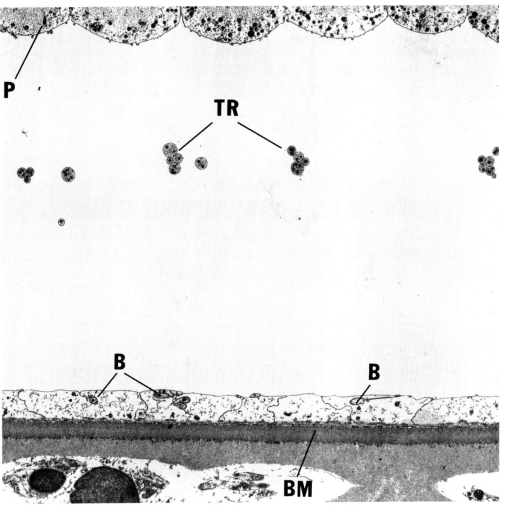

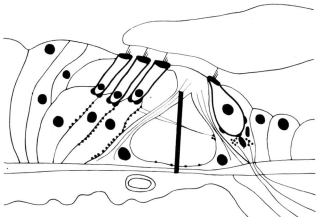

Figure 6.6 *Longitudinal section through the tunnel of Corti in the upper basal turn of a cat. The basilar membrane (BM) is covered by a layer of supporting cells. The basilar fibres (B), which are the only afferent nerve fibres leading to the outer hair cell system, are usually hidden in deep invaginations of these supporting cells. The upper tunnel radial fibres (TR) run entirely free through the tunnel space in fascicles of several fibres of very different calibres. P = Heads of outer pillars. The plane of the section is indicated in the diagram*

the efferents of the inner hair cell region are small and take several days to degenerate. Furthermore, the inner spiral fibres have a tendency to increase from base to apex, whereas the efferent endings at the outer hair cells decrease from base to apex. Finally, there is a basic difference in the synaptic connections of the efferents to the outer and inner hair cells. At the level of the outer hair cells the efferent fibres synapse almost exclusively with the receptor cell as such; at the level of the inner hair cells, they form synaptic contacts almost exclusively with the afferent nerve fibres associated with the inner hair cells. In respect of the afferent synapse these efferent connections are pre-synaptic at the level of the outer hair cells and post-synaptic at the level of the inner hair cells (*Figures 6.3, 6.11, 6.13, 6.14* and *6.26*).

The afferent fibres that remain after elimination of the efferent fibres are the inner radial fibres, the basilar fibres and the outer spiral fibres (*Figure 6.33*). They represent the afferent innervation of the organ of Corti. The few basilar fibres are the only afferent nerve fibres reaching the area of the outer hair cells (*Figures 6.2* and *6.6*). They can be counted as they pass between the base

of the outer pillars in tangential sections through this area (*Figure 6.7*). Evaluated over long distances of the cochlea, there is an average of one basilar fibre penetrating between two outer pillars. The number of afferent fibres going to the outer hair cells is therefore about equal to the number of outer pillars, which amount to approximately 2500 in the cat cochlea. This is an extremely small number compared with the entire population of about 50 000 cochlear neurons in a cat cochlea. It means that only about 5 per cent of all afferent cochlear neurons are associated with the outer hair cell system, which represents more than three-quarters of the receptor cells of the cochlea. More evidence can be provided for this surprising 20:1 ratio of afferent innervation of inner and outer hair cells. If we reconstruct the area of some inner hair cells by means of serial sections after eliminating the efferent innervation, we find that the great majority of all nerve fibres entering the organ of Corti through the habenular openings lead directly, unbranched, to the base of the nearest inner hair cell, and that only about one fibre in 20 turns towards the outer hair cells (*Figures 6.20, 6.21* and *6.35*).

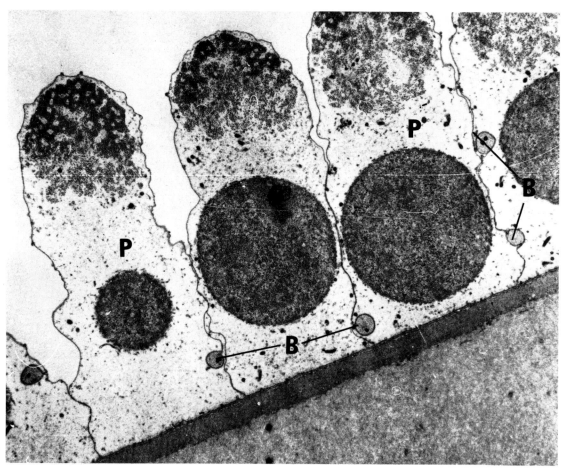

Figure 6.7 *Longitudinal section through the base of the outer pillars (P), with the basilar fibres (B) penetrating between the pillars. In the cat there is an average of one basilar fibre between each pair of pillars*

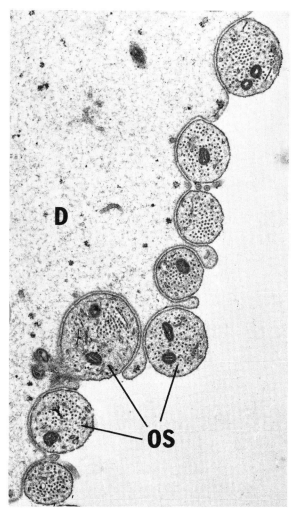

Figure 6.8 *Some outer spiral fibres (OS) running in invaginations or between the Deiters' cells (D). The axoplasm of these afferent nerve fibres contains a great number of neurotubules*

Finally, we can compare the total number of afferent and efferent nerve fibres entering the organ of Corti at the level of the habenula with the number of all fibres crossing the tunnel towards the outer hair cells in corresponding tangential section in normal animals (*see Figures 6.6* and *6.22*). The numbers show that only about 15 per cent of all fibres entering the organ of Corti cross the tunnel towards the outer hair cells. Since two-thirds of the tunnel-crossing fibres are the efferent upper tunnel radial fibres, the basilar fibres represent only about 5 per cent of the total afferent neuron population, the same small percentage as we found on the basis of other evaluations.

There is little doubt that the outer hair cell system is associated only with a small minority of afferent neurons whereas the great majority of all afferent neurons are associated with the inner hair cell system. After the demonstration of this surprising situation in the cat, similar ratios between neurons associated with the outer and inner hair cells have been found in the

guinea-pig (Spoendlin, 1969; Morrison *et al.*, 1975) and in man (Nomura, 1976).

Within the habenula, the fibres associated with the outer hair cells cannot be distinguished from the other fibres but they are usually situated in the most distal portion of the habenular opening (*Figures 6.21* and *6.25*). In contrast to the fibres for the inner hair cells they take an independent spiral course (*Figures 6.23* and *6.34*), immediately after the habenula, for about five pillars before penetrating between the inner pillars to cross the base of the tunnel as basilar fibres; reaching the area of the outer hair cells, they form the outer spiral fibres (*see Figures 6.2* and *6.8*) which gradually climb up towards the base of the outer hair cells. In its terminal portion each fibre gives off collaterals, which end as afferent nerve endings at the bases of the outer hair cells. Each fibre sends collaterals to about 10 outer hair cells and each outer hair cell receives collaterals from several outer spiral fibres, according to the principle of multiple innervation (Spoendlin, 1970, 1973) (*Figure 6.34*).

The average length of the basalward spiral course of the outer spiral fibres can be estimated from the number of basilar fibres reaching the outer hair cell area (1 fibre/pillar cell) and the number of outer spiral fibres (about 100 at any one place). It amounts to about 0.6 mm (Spoendlin, 1968), as also measured directly in stained silver preparations by Smith (1975).

There is no morphological evidence at any level for direct functional interaction between the afferent fibres from inner and outer hair cells. At no place are there any direct contacts or even synapses between the axons of the two fibre systems. Within the habenular openings all the fibres are individually surrounded by processes of a special habenular satellite cell, which seems to take the role of individual Schwann cells proximal to the habenula (*see Figures 6.20* and *6.24*).

The calibre of the fibres varies considerably along their course through the organ of Corti. Especially where the fibres pass mechanically important supporting structures, such as the basilar membrane at the habenular region or the pillars, their diameter is considerably reduced and the axoplasm is rather empty (*Figures 6.24* and *6.36*). The ultrastructural organization of the axoplasm varies to a certain extent between afferent fibres to the outer and to the inner hair cells. The fibres for the outer hair cells contain mainly neurocanaliculi in their axoplasm whereas the fibres for the inner hair cells contain predominantly neurofibrils (*see Figure 6.8*). The endings at the inner hair cells form usually synaptic complexes with synaptic bars of varying sizes (Smith and Sjöstrand, 1961) (*Figures 6.18* and *6.19*). The endings of the outer hair cells, on the other hand, have no synaptic ribbons in the cat and only relatively small ones in the guinea-pig (*see Figures 6.3* and *6.5*) (Rodriguez, 1967; Spoendlin, 1970; Dunn and Morest, 1975).

(*continued on page 155*)

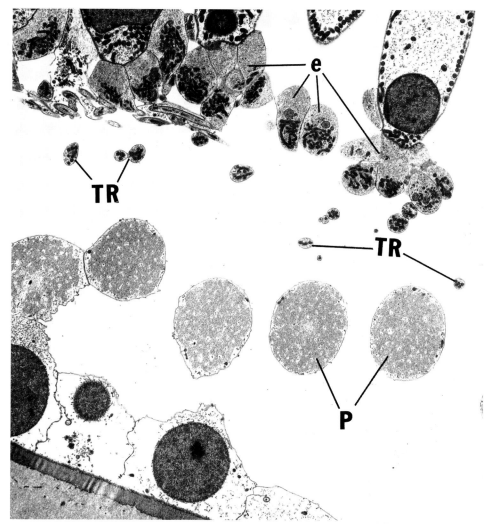

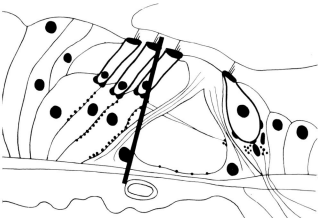

Figure 6.9 *Longitudinal section through the first row of outer hair cells and the outer pillars (P), as indicated in the diagram. The tunnel radial fibres (TR) ramify and form large efferent endings (e) at the base of the outer hair cells*

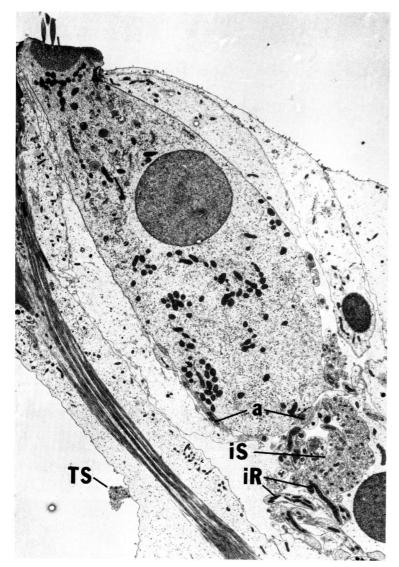

Figure 6.10 *Survey of an inner hair cell with associated nerve fibres, such as the afferent inner radial fibres (iR), which intermingle with the efferent inner spiral fibres (iS). The inner radial fibres end at the base of the inner hair cells, usually 20 fibres per hair cell, and each fibre has one nerve ending (a). The tunnel spiral bundle (TS) belongs most probably to the system of inner spiral fibres*

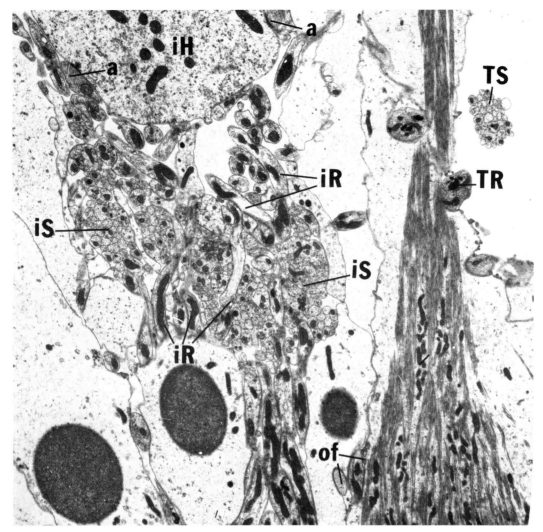

Figure 6.11 *Area below the base of the inner hair cells (iH) with the nerve plexus of intermingled afferent inner radial fibres (iR) and efferent inner spiral fibres (iS). The inner spiral fibres consist of several hundred very small fibres with diameters around 0.2 μm and varicose enlargements containing many synaptic vesicles and forming synaptic contacts with the afferent inner radial fibres. The afferent nerve endings (a) at the inner hair cells usually have synaptic complexes. Adjacent to the inner pillar one finds the afferent fibres (of) of the outer hair cell system, which after a short spiral course will penetrate between the inner pillars to form the basilar fibres. The large efferent tunnel radial fibres (TR) penetrate between the inner pillars at a higher level before crossing the tunnel. The tunnel spiral fibres (TS) resemble closely the inner spiral fibres*

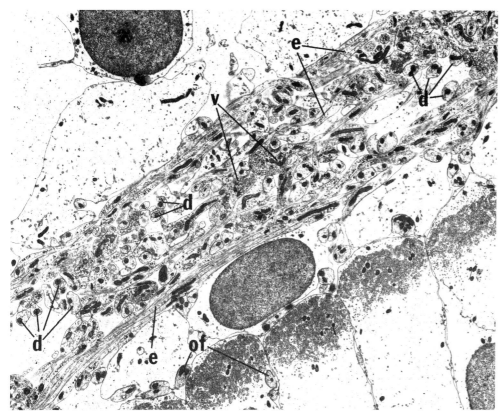

Figure 6.12 *Horizontal section through the inner spiral bundle below the base of the inner hair cells at low magnification. It is clearly seen how the afferent dendrites (d) intermingle with the efferent (e) inner spiral fibres with their varicose enlargements (v) containing numerous synaptic vesicles. The afferent fibres of the outer hair cell system (of) appear between the inner pillars*

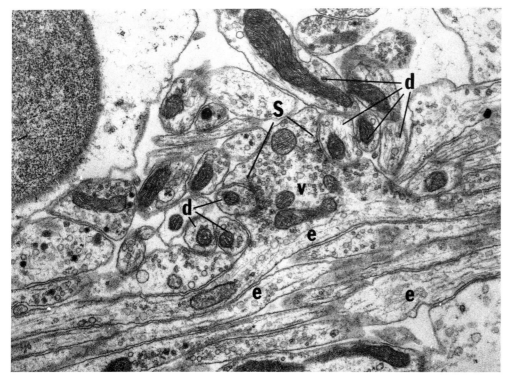

Figure 6.13 *Detail of inner spiral plexus with efferent inner spiral fibres (e) with varicose enlargement (v), which makes 'en passant' synaptic contacts with the afferent dendrites (d)*

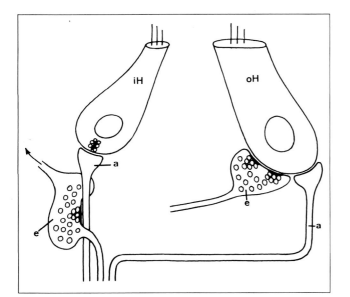

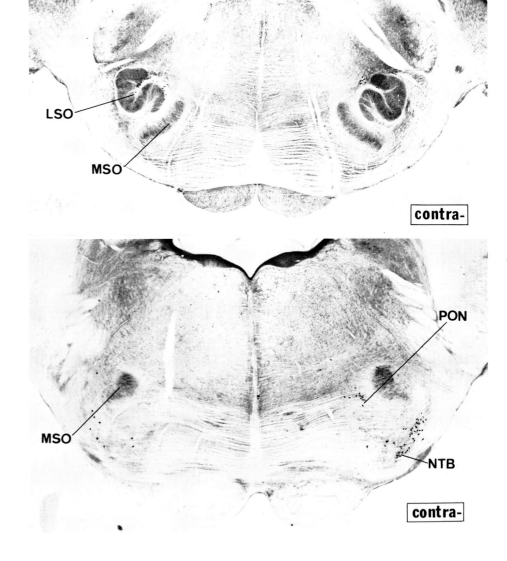

Figure 6.14 *Schematic representation of the relationship between the efferent and afferent nerve fibres and the hair cells. In the inner hair cell system (iH) the efferents (e) have almost exclusive synaptic contacts with the afferent fibres (a), but not with the inner hair cells. In the outer hair cell system the efferents (e) have almost exclusive synaptic contacts with the hair cells (oH) and not with the afferent fibres (a). In the cat this pattern is the rule, whereas in rodents there are some exceptions*

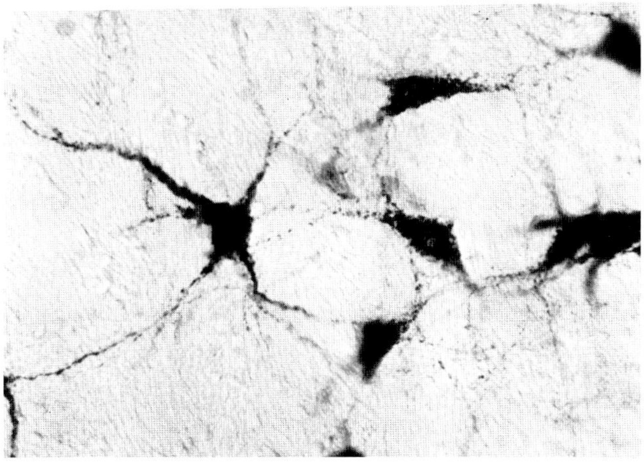

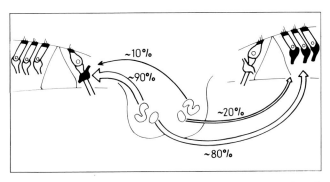

Figure 6.16 *Detail of some large multipolar efferent neurons in the nucleus of the trapezoid body of the contralateral side containing the horse-radish peroxidase reaction product*

◀ **Figure 6.15** *Sections through the brain-stem of a cat 24 hours after horse-radish peroxidase injection into the right cochlea. Histochemical reaction with TMB. Section A is slightly more caudal than section B. LSO = Lateral superior olivary nucleus; MSO = medial superior olivary nucleus; PON = periolivary nucleus; NTB = nucleus of trapezoid body; Contra = contralateral. Horse-radish peroxidase is taken up by the efferent endings in the cochlea and transported retrogradely to the ganglion cells of origin of the efferents. There is one group of numerous small ganglion cells, mainly in the hilus of the lateral superior olivary nucleus of the homolateral side. Another group of larger ganglion cells is found in the nucleus of the trapezoid body and in the periolivary nucleus, mainly of the contralateral side. The first group of small ganglion cells in the lateral superior olivary nucleus constitutes the efferent neurons of the inner hair cell system and the second group of larger ganglion cells in the nucleus of the trapezoid body and the periolivary nucleus constitute the efferent neurons for the outer hair cell system (Compare with schematic representation in Figure 6.17.)*

Figure 6.17 *Schematic representation of the homo- and contralateral efferent nerve supply from the superior olivary nucleus and accessory olivary nucleus*

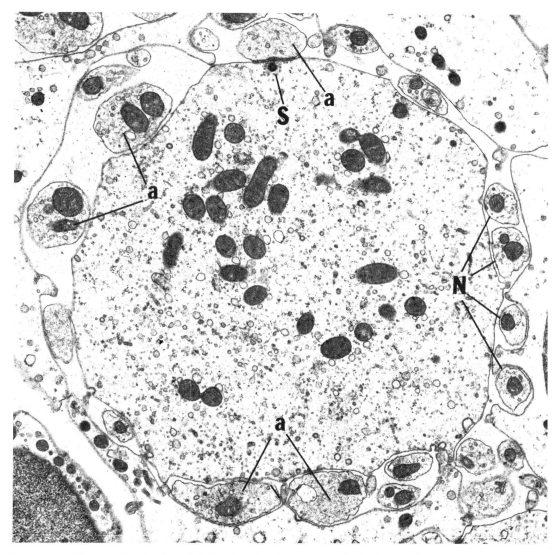

Figure 6.18 *Horizontal section through the basal portion of an inner hair cell. The afferent nerve fibres (N) surround the base of the hair cell like a basket and end at different levels with single nerve endings (a) having usually one synaptic complex (S) with the hair cell. There are about 20 afferent neurons associated with each inner hair cell*

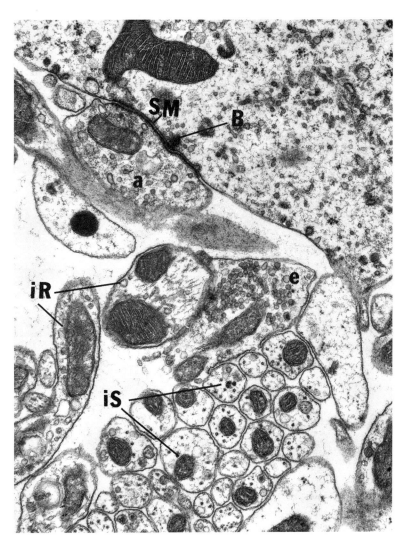

Figure 6.19 *Detail of the basal portion of an inner hair cell with an afferent nerve terminal (a) making a synapse with the hair cell. The synaptic membranes show very clear thickenings (SM), especially the pre-synaptic membrane. There is usually an accessory synaptic structure in the form of a synaptic bar (B) surrounded by synaptic vesicles. Below some efferent inner spiral fibres (iS) of varying calibre are seen with a varicose enlargement (e) making synaptic contacts with the afferent inner radial fibres (iR)*

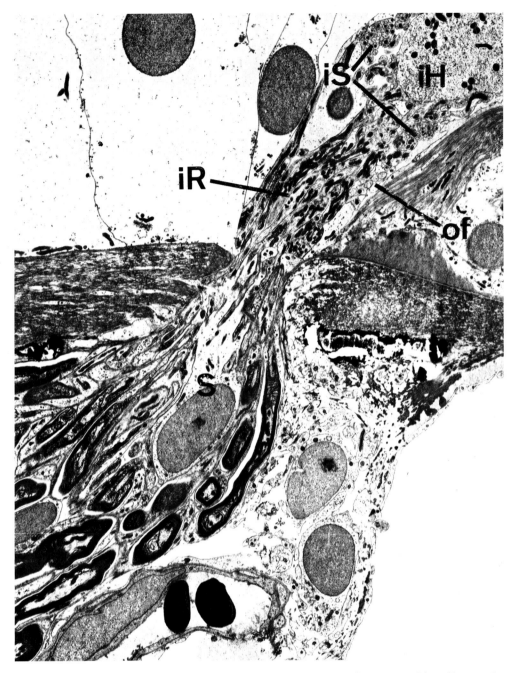

Figure 6.20 *Low-magnification picture of the habenular area. All nerve fibres lose their myelin sheaths before entering the organ of Corti through the habenular openings. Within the habenular openings the nerve fibres are surrounded by the processes of a special satellite cell (S). Within the organ of Corti, above the habenular opening, the nerve fibres run without any sheathing in tight bundles.*

The majority of these fibres are the inner radial fibres (iR), which are the afferent nerve supply to the inner hair cells and lead unbranched (radially) directly to the closest inner hair cell. They intermingle with the efferent inner spiral fibres (iS) before terminating at the base of the inner hair cell (iH). The few smaller afferent fibres for the outer hair cell system (of) are adjacent to the inner surface of the inner pillars

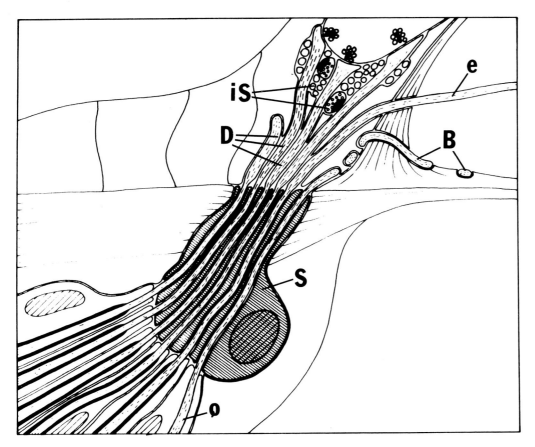

Figure 6.21 *Schematic representation of the area of the habenula perforata with the special satellite cell (S), which surrounds the nerve fibres in the habenula. Afferent dendrites (D) form the great majority of all fibres. They lead to the inner hair cells and originate from myelinated nerve fibres. They cross the efferent nerve bundles of the inner spiral fibres before terminating at the base of the inner hair cells.*

The afferent fibres for the outer hair cells (o) are unmyelinated, run in the most distal part of the habenular openings and continue as basilar fibres (B). The efferent fibres for the outer hair cell system (e) run without any synaptic contacts directly (radially) through the tunnel to the base of the outer hair cells

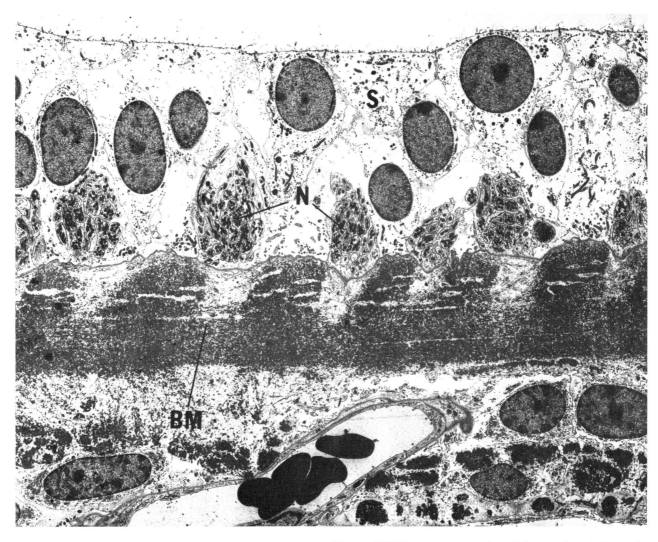

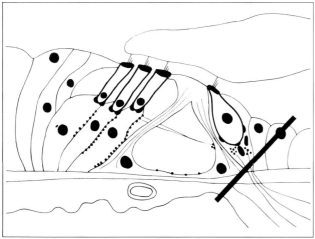

Figure 6.22 *Transverse section through the area above the habenula with nerve bundles (N) of relatively large unmyelinated nerve fibres surrounded by supporting cells of the inner hair cell area (S). Plane of section as indicated in the diagram*

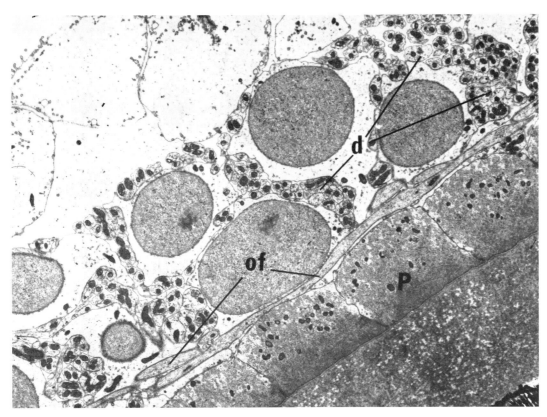

Figure 6.23 *Same plane of section as in Figure 6.22 at a somewhat higher level (in a cat, in which the efferents have been eliminated). The afferent dendrites (d) for the inner hair cells run between the supporting cells. The afferent fibres for the outer hair cell system (of) take a spiral course at the inner surface of the inner pillars (P) before they penetrate between the pillars to cross the tunnel as basilar fibres .*

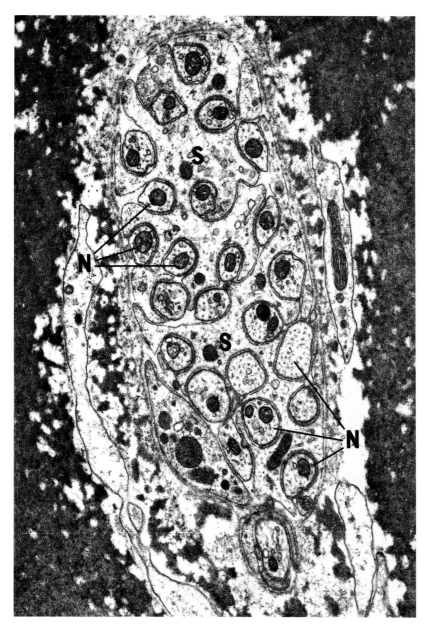

Figure 6.24. *Transverse section through one habenular opening showing the nerve fibres (N) surrounded by extensions of the satellite cell (S). In this narrow space within the habenular opening the nerve fibres are small and contain only very few mitochondria*

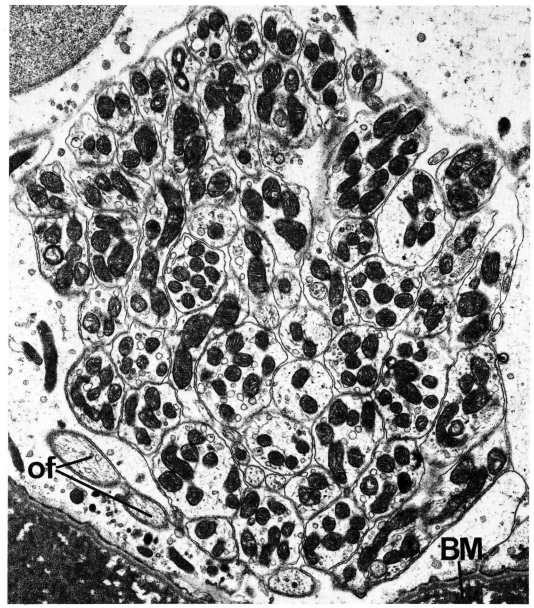

Figure 6.25 *Transverse section through a nerve bundle immediately above the habenular opening in the basilar membrane (BM) showing the relatively large tightly packed nerve fibres without any sheathing and containing a great number of mitochondria. Only the afferent fibres for the outer hair cell system (of), having a relatively smaller calibre, separate from the bundle to take an independent spiral course. There is no evidence of synaptic or ephaptic contacts between the fibres*

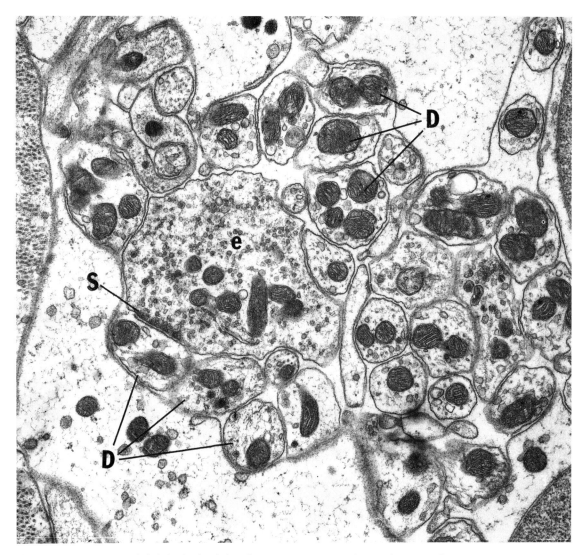

Figure 6.26 *Section at a slightly higher level than that in Figure 6.25 showing the afferent dendrites of the inner hair cell system (D) in contact with a varicose enlargement of an efferent fibre (e) with a clear condensation of synaptic vesicles (S) at a site of contact with an afferent dendrite, indicating a synapse*

In all mammalian species studied so far – namely the cat, the guinea-pig, the rat, the rabbit and the chinchilla – two distinct types of ganglion cells have been found in the normal spiral ganglion (Kellerhals *et al.*, 1967; Spoendlin, 1971); a majority (90–95 per cent) of large bipolar type I cells and 5–10 per cent of smaller type II cells with a number of distinctive features. The most common feature of the type II cell in all species is its smaller size and a lighter, more filamentous cytoplasm, with fewer ribosomes or Nissl substance (*Figures 6.29 and 6.30*). The absence of a myelin sheath around the type II cells is a frequent – but not constant – finding. In the human, there are just as many myelinated as unmyelinated small cells (Ota and Kimura, 1980), and in the cat some type II cells may have a thin myelin sheath. The excentric position and lobulation of the nucleus are typical in the cat, but lacking in rodents and humans. Although the type II cells are clearly distinguishable in all species, their morphological features are most pronounced in the cat, in which they are also best studied. Type II neurons are predominantly pseudomonopolar and their axons are unmyelinated as far as they can be followed (*Figure 6.30*). The two types of ganglion cells are already present in the new-born kitten (Spoendlin, 1981).

The observation that type II axons are small in calibre and remain unmyelinated throughout their entire course is very important in respect of their function but difficult to prove, since it is practically impossible to follow them over longer distances in normal animals (Spoendlin, 1979). To a limited extent this is possible with electron microscopic (EM) serial sections or with light-microscopic techniques following retrograde degeneration of the type I cells using interference contrast in extra-thick sections or with methylene blue staining. In areas of the cochleas in which, long after trans-section of the VIIIth nerve, only type II cells remain, we have found almost exclusively unmyelinated nerve fibres in the osseous spiral lamina, thus indicating that the type II axons remain unmyelinated throughout their peripheral course. In the same areas the afferent innervation of the outer hair cells remained intact.

To follow the central axon of type II neurons is more difficult because EM serial sections are possible only over limited distances, and in light-microscopic preparations the unmyelinated axons become extremely thin at a certain distance from the perikarya and are usually lost at the entrance of the modiolus. Being so small at this level they will most probably remain very small and unmyelinated in their further centralward course within the cochlear nerve. Unmyelinated fibres on the other hand are very rare within the normal cochlear nerve (*see Figure 6.31*), so that possibly the type II neurons are not effectively connected to the central nervous system (Spoendlin, 1979).

When small amounts of horse-radish peroxidase (HRP) are injected into the cochlear nerve in the internal acoustic meatus, many neurons take up the tracer, which is then transported along the axon down to the nerve endings, where it can be demonstrated by histochemical techniques.

Using this technique we found HRP-containing fibres only at the level of the inner hair cells (*Figure 6.27*), but not in the afferent fibres associated with the outer hair cells.

In EM preparations the amorphous fine granular reaction product of the DAB procedure was easily identified within many afferent radial fibres to the inner hair cells and their nerve endings, but not in the afferent nerve fibres and nerve endings of the outer hair cell system.

This suggests that the type II axons are either not present in the cochlear nerve, or that for some reason they cannot readily take up HRP, possibly because they are too small and metabolically inactive, even though other investigators have described some HRP-containing fibres at the level of the outer hair cells after application of HRP to the cochlear nerve.

After section of the cochlear nerve in the cat almost all type I neurons (which belong to the inner hair cell system) degenerate and disappear, and only the type II neurons (which represent the afferent innervation of the outer hair cell system) remain unchanged in normal numbers. In electrocochleographic recordings in such animals, no compound action potentials can be found despite the remaining normal type II neurons (Spoendlin, 1977).

This and the obvious difference in HRP uptake between the afferent neurons of the outer and inner hair cells supports the view that the outer hair cells and the inner hair cells, together with the associated neurons, represent two different systems; and that the neurons of the outer hair cell system are incompletely and inefficiently connected to the central nervous system. The extreme paucity of small afferent neurons in the outer hair cell system constitutes a very poor information transmission system. In the small unmyelinated axons, saltatory action-potential conduction is not possible, so that presumably the propagation of nerve potentials is very slow, with considerable decrement. All this suggests that the outer hair cell system, which represents the great majority of all cochlear hair cells, has its main functional role at the level of the receptor cells, possibly by electrically monitoring the inner hair cell system, rather than by transmitting direct neural information to the central nervous system.

In order to demonstrate the adrenergic innervation, the histochemical method of Falck *et al.* (1962) can be used in which, with an appropriate filter combination, all adrenergic nerve fibres appear with a very specific green fluorescence. Using this technique in normal animals and after various lesions (such as extirpation of the superior cervical or stellate ganglion or by section of the tympanic plexus), we found in the cat two different independent adrenergic innervations in the inner ear (*Figure 6.37*). The one is strictly perivascular, originating in the stellate ganglion and reaching the inner ear through the perivascular plexuses of the vertebral, basilar and inferior anterior cerebellar and labyrinthine arteries. These perivascular plexuses can be followed as far as the greater arteriolar branches in the modiolus, but they are never found in the blood vessels of the spiral ligament or stria vascularis; the second type of adrenergic innervation is independent of the blood vessels. It originates in the superior cervical ganglion and reaches the inner ear via the tympanic plexus. This component is most pronounced in the osseous spiral lamina, where it forms a fairly rich terminal plexus just below the habenula, especially pronounced in the apical turns of the cochlea (*Figure 6.32*). This blood-vessel-independent adrenergic innervation, however, does not enter the organ of Corti through the habenular openings and therefore does not interfere directly with the receptor process in the organ of Corti (Lichtensteiger and Spoendlin, 1967; Paradiesgarten and Spoendlin, 1976). The two independent types of adrenergic innervation are less distinct in other species, such as the rabbit (Densert, 1974) and the guinea-pig (Terrayama, Holz and Beck, 1966) but they have been confirmed in the cat (Ross, 1969).

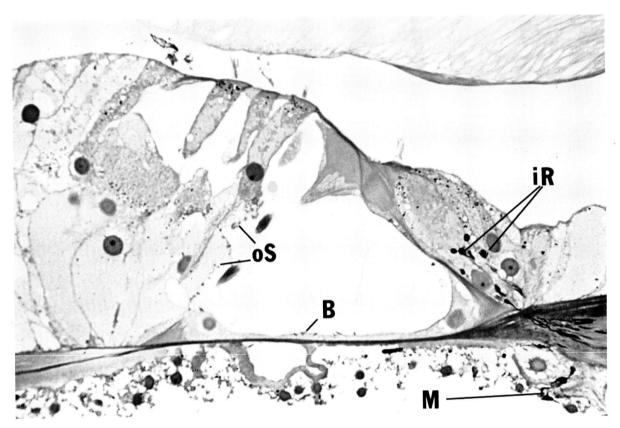

Figure 6.27 *Light-microscopic section through the organ of Corti of a cat 24 hours after the injection of 20 μl of horse-radish peroxidase (HRP) into the cochlear nerve in the internal acoustic meatus. HRP-positive fibres are found only in the area of the inner radial fibres (iR) with their nerve endings at the inner hair cells, but not in the basilar fibres (B) or the outer spiral fibres (oS), which constitute the afferent nerve supply to the outer hair cell system. M indicates a macrophage of the tympanic lamina with some phagocytosed HRP reaction product*

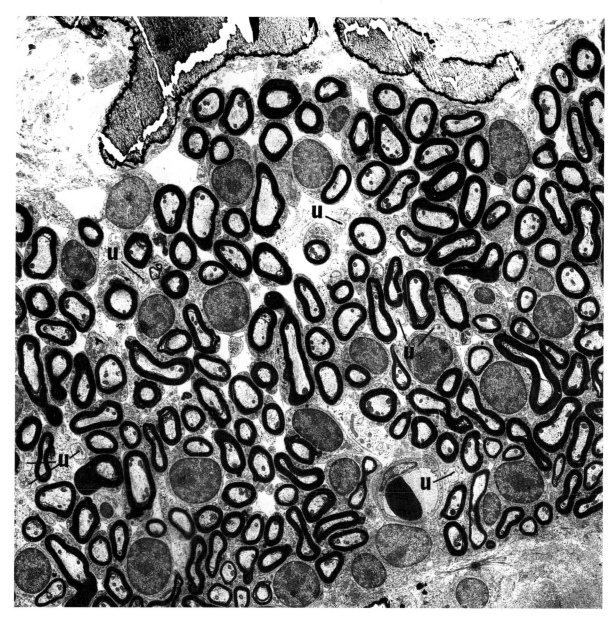

Figure 6.28 *Transverse section through a portion of the osseous spiral lamina with many myelinated nerve fibres of varying calibre and a fair number of unmyelinated nerve fibres (u)*

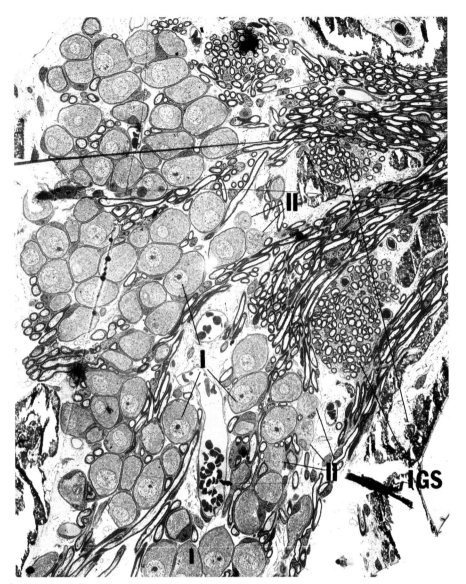

Figure 6.29 *Low-magnification survey of the spiral ganglion of the upper basal turn in a cat, showing a great majority of large myelinated type I ganglion cells (I) and a few smaller unmyelinated type II ganglion cells (II), which lie predominantly at the peripheral portion of the spiral ganglion and which constitute the afferent nerve supply of the outer hair cell system*

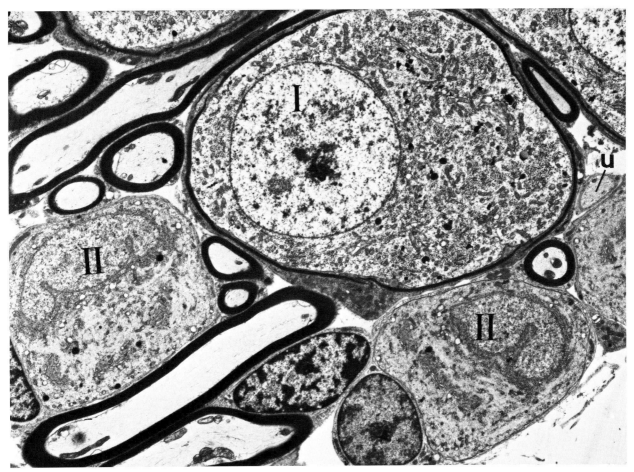

Figure 6.30 *Detail of type I (I) and type II (II) ganglion cells of a cat spiral ganglion. The type I cells are large, myelinated, with a round nucleus and a pronounced nucleolus. The cytoplasm is rich in ribosomes and mitochondria. The type II cells are smaller and unmyelinated. The nucleus is usually excentric and lobulated and the cytoplasm is poor in ribosomes and mitochondria but contains a great number of filaments. The axons of the type II ganglion cells (u) are unmyelinated. Only an occasional type II cell has a thin myelin sheath*

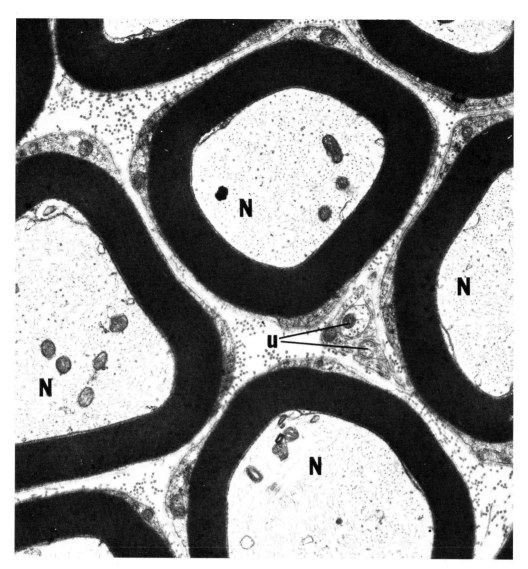

Figure 6.31 *Portion of the cochlear nerve in the internal acoustic meatus with the large axons of the cochlear neurons (N), of fairly uniform diameter and with a fairly uniform myelin sheath thickness. Between these fibres very small unmyelinated (u) fibres may occasionally be found. Whether they belong to the autonomic innervation or to the type II neurons cannot definitely be decided*

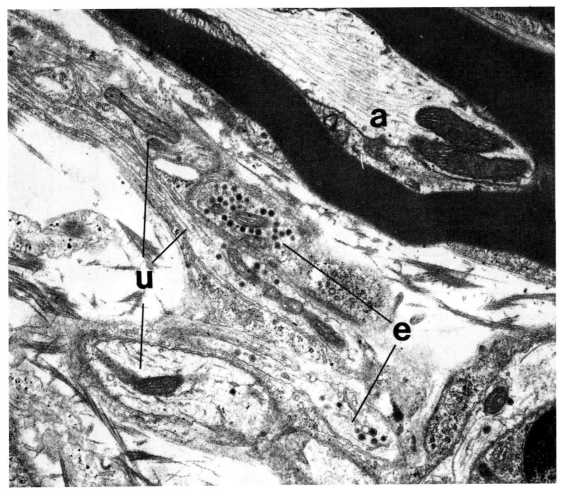

Figure 6.32 *Detail of nerve fibres below the habenula perforata with a myelinated afferent cochlear neuron (a) and some adrenergic unmyelinated nerve fibres (u) with enlargements (e) containing* dense-core vesicles, which probably represent noradrenalin stores of the adrenergic fibre system

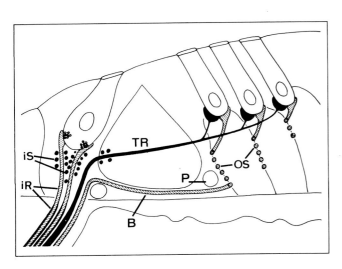

Figure 6.33 *Radial innervation scheme of the organ of Corti with efferent nerve fibres (full black lines) consisting of the inner spiral fibres (iS), the tunnel radial fibres (TR) and the large nerve endings at the base of the outer hair cells and the afferent nerve fibres (hatched lines), consisting of inner radial fibres (iR), basilar fibres (B) and outer spiral fibres (oS), with small nerve endings at the outer hair cells*

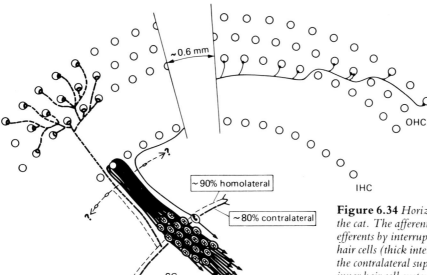

Figure 6.34 *Horizontal innervation scheme of the organ of Corti of the cat. The afferent nerve fibres are represented by full lines, the efferents by interrupted lines. The efferent innervation of the outer hair cells (thick interrupted lines) originates (up to 80 per cent) from the contralateral superior olivary complex and the efferents of the inner hair cell system (thin interrupted lines) originate (up to 90 per cent) from the homolateral superior olivary complex*

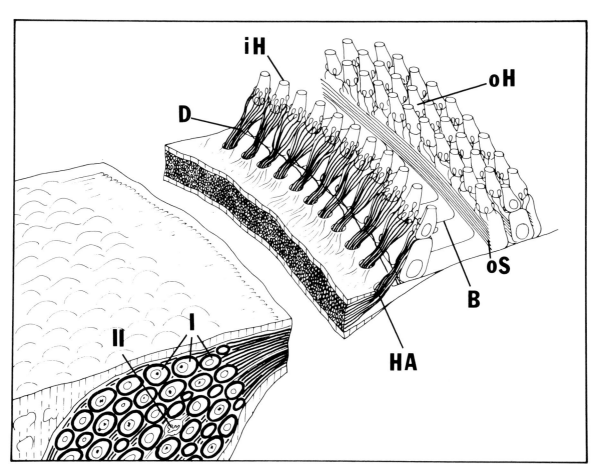

Figure 6.35 *Schematic representation of the afferent innervation of the organ of Corti with a great majority of type I cochlear neurons. The type I neurons are directly associated with the inner hair cells (iH) with radial unbranched dendrites (D), which lead from the habenular opening (HA) directly to the closest inner hair cell, about 20 neurons to one inner hair cell. The afferent nerve supply of the outer hair cell system (oH), on the other hand, originates in the few type II ganglion cells (II) and crosses the tunnel as basilar fibres (B), to form the outer spiral fibres (oS), where they gradually climb up to the base of the outer hair cells and finally give off collaterals with nerve endings to about 10 outer hair cells*

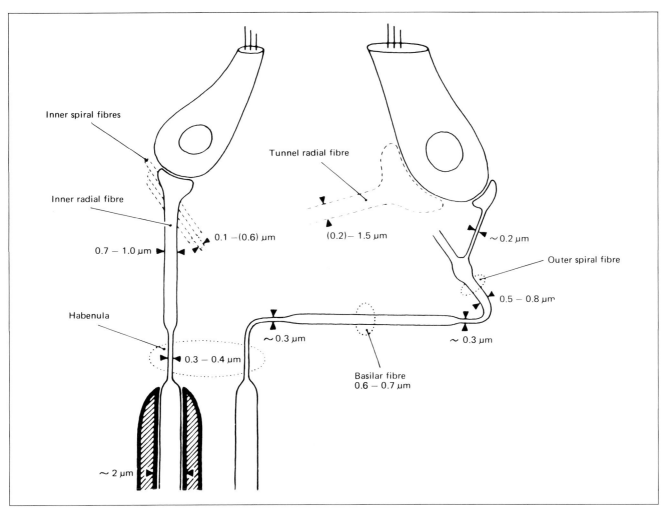

Inner spiral fibres

Tunnel radial fibre

Inner radial fibre

0.1 — (0.6) μm

(0.2) — 1.5 μm

~ 0.2 μm

0.7 — 1.0 μm

Outer spiral fibre

0.5 — 0.8 μm

Habenula

~ 0.3 μm

~ 0.3 μm

0.3 — 0.4 μm

Basilar fibre
0.6 — 0.7 μm

~ 2 μm

Figure 6.36 *Average diameters of the nerve fibres in the organ of Corti in the cat*

Figure 6.37 *Schematic representation of the two types of adrenergic innervation of the cochlea. One system is independent of the blood vessels and originates in the superior cervical ganglion, whereas the other is perivascular and originates in the stellate ganglion. It forms a continuous perivascular plexus around the vertebral, basilar, inferior anterior cerebellar and labyrinthine arteries*

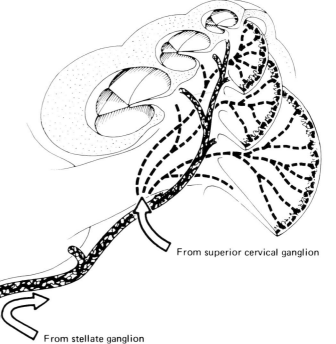

From superior cervical ganglion

From stellate ganglion

References

Densert, O. (1974) Adrenergic innervation in the rabbit cochlea. *Acta Otolaryngologica (Stockholm)*, **78**, 1–12

Dunn, R. A. and Morest, D. K. (1975) Receptor synapses without synaptic ribbons in the cochlea of the cat. *Proceedings of the National Academy of Science of the USA*, **72**, 3599–3603.

Falck, B., Hillarp, N. S., Thieme, G. and Dorp, A. (1962) Fluorescence of catecholamines and related compounds condensed with formaldehyde. *Journal of Histochemistry and Cytochemistry*, **10**, 348.

Hall, J. G. (1966) Hearing and primary auditory centres of the whales. *Acta Otolaryngologica (Stockholm)*, Supplement **224**, 244–250.

Kellerhals, B., Engström, H. and Ades, H. W. (1967) Die morphologie des ganglion spirale cochleae. *Acta Otolaryngologica (Stockholm)*, Supplement **226**, 1–78.

Lichtensteiger, W. and Spoendlin, H. (1967) Adrenergic pathways to the cochlea of the cat. *Life Sciences*, **6**, 1639–1645.

Morrison *et al.* (1975) Quantitative analysis of the afferent innervation of the organ of Corti in guinea pig. *Acta Otolaryngologica (Stockholm)*, **79**, 1–23.

Nomura, Y. (1976) Nerve fibers in the human organ of Corti. *Acta Otolaryngologica*, **82**, 317–324

Ota, C. Y. and Kimura, R. S. (1980) Ultrastructural study of the human spiral ganglion. *Acta Otolaryngologica (Stockholm)*, **89**, 53.

Paradiesgarten, A. and Spoendlin, H. (1976) The unmyelinated nerve fibres of the cochlea. *Acta Otolaryngologica (Stockholm)*, **82**, 157–164.

Rasmussen, G. L. (1942) An efferent cochlear bundle. *Anatomical Record*, **82**, 441.

Rodriguez, E. E. L. (1967) An electron microscopic study on the cochlear innervation. I. The receptoneural junctions at the outer hair cells. *Zeitschrift für Zellforschung und mikroskopische Anatomie*, **78**, 30–46.

Ross, M. D. (1969) The general visceral efferent component of the eighth cranial nerve. *Journal of Comparative Neurology*, **135**, 453–478.

Schuknecht, H. (1962) Neuroanatomical correlates of auditory sensitivity and pitch discrimination in the cat. In *Neural Mechanisms of the Auditory and Vestibular Systems*, eds. Rasmussen and Windle, Springield, Il.: Thomas.

Smith, C. A. (1975) Innervation of the cochlea of the guinea pig by use of the Golgi stain. *Annals of Otology, Rhinology and Laryngology*, **84**, 443

Smith, C. A. and Sjöstrand, F. (1961) A synaptic structure in the hair cell of the guinea pig cochlea. *Journal of Ultrastructural Research*, **5**, 184.

Spoendlin, H. (1966) The organization of the cochlear receptor. *Advances in Otorhinolaryngology*, **13**, 1–231.

Spoendlin, H. (1968) Ultrastructure and peripheral innervation pattern of the receptor in relation to the first coding of the acoustic message. In *Hearing Mechanisms in Vertebrates*, eds. A. V. S. De Reuck and J. Knight, pp. 89–119. London: Churchill.

Spoendlin, H. (1969) Innervation patterns on the organ of Corti of the cat. *Acta Otolaryngologica (Stockholm)*, **67**, 239–254.

Spoendlin, H. (1970) Structural basis of peripheral frequency analysis. In *Frequency Analysis and Periodicity Detection in Hearing*, eds. R. Plomp and F. G. Smoorenburg, pp. 2–36. Leiden: Sijthoff.

Spoendlin, H. (1971) Degeneration behaviour of the cochlear nerve. *Archiv für klinische und experimentelle Ohren-, Nasen-, und Kehlkopfheilkunde*, **200**, 275–291.

Spoendlin, H. (1972) Innervation densities of the cochlear. *Acta Otolaryngologica (Stockholm)*, **73**, 233–243.

Spoendlin, H. (1973) The innervation of the cochlear receptor. In *Basic Mechanisms in Hearing*, ed. A. B. Möller, pp. 185–234. New York: Academic Press.

Spoendlin, H. (1979) Neural connections of the outer hair cell system. *Acta Otolaryngologica (Stockholm)*, **87**, 130.

Spoendlin, H. (1981) Differentiation of cochlear afferent neurons. *Acta Otolaryngologica*, **91**, 451–456.

Spoendlin, H. and Baumgartner, H. (1977) Electrocochleography and cochlear pathology. *Acta Otolaryngologica (Stockholm)*, **83**, 130–135.

Spoendlin, H. and Lichtensteiger, W. (1966) The adrenergic innervation of the labyrinth. *Acta Otolaryngologica (Stockholm)*, **61**, 423–434.

Spoendlin, H. and Lichtensteiger, W. (1967) The sympathetic nerve supply to the inner ear. *Archiv für klinische und experimentelle Ohren-, Nasen-, und Kehlkopfheilkunde*, **189**, 346.

Terayama, Y., Holz, E. and Beck, Ch. (1966) Adrenergic innervation of the cochlea. *Annals of Otology, Rhinology and Laryngology*, **75**, 1–18.

Warr, W. B. (1978) The olivocochlear bundle: Its origins and terminations in the cat. In *Evoked Electrical Activity in the Nervous System*, eds. R. F. Naunton and C. Fernandez. New York: Academic Press.

7
The Spiral Ganglion

R. Romand   M. R. Romand

Introduction

The spiral ganglion has been considered to be a ganglion of simple structure, populated mainly by bipolar cells, whose peripheral extensions made their way towards the organ of Corti, while the axon advances towards the cochlear nucleus, thus forming a link between the auditory receptors and the central nervous system. Subsequent observations by electron microscopy have shown that the spiral ganglion is a more complex structure than originally thought and is made up of various components; moreover, the functional significance of some of its components continues to present a challenge to researchers.

The spiral ganglion is populated mainly by the perikaryon of bipolar cells, whose processes are located at opposite sides of the cell body, as was recognized at a very early date (Retzius, 1895; Cajal, 1909, Held, 1926; Lorente de Nó, 1937). The ultrastructural features of this type of cell are known thanks mainly to detailed descriptions of various different species such as the goldfish (Rosenbluth and Palay, 1961), the rat (Rosenbluth, 1962; Merck et al., 1977), the guinea-pig (Kellerhals et al., 1967; Thomsen, 1967; Trevisi, Testa and Riva, 1977), the rabbit (Suzuki, Watanabe and Osada, 1963), the cat (Spoendlin, 1972, 1974; Adamo and Daigneault, 1973), and man (Ylikoski, Collan and Palva, 1978; Ota and Kimura, 1980).

It has long been recognized that the perikaryon of these cells may be surrounded by a myelin sheath in the majority of species studied. This peculiarity is rare, though not exceptional (see Rosenbluth and Palay, 1961). In addition to these cells, which are easily recognizable under the light or electron microscope, there are other types of ganglion cells, fewer in number, identifiable either as multipolar cells (Retzius, 1895; Ross and Burkel, 1973) or by their cytoplasmic content (Rosenbluth, 1962; Kellerhals, Engström and Ades, 1967); other distinguishing features include the shape of the nucleus and the myelin sheath (Spoendlin, 1972, 1974). There is no doubt about the existence of several types of cell, as has been established in other sensory ganglia (Noden, 1980), but the criteria that facilitate their identification vary for the spiral ganglion from species to species. In fact, it is difficult to make any generalization for all species of vertebrates – since man, for example, has been shown to be different from others (Kimura, Ota and Takahashi, 1979; Ota and Kimura, 1980).

The most important sources of information concerning the ultrastructure of different cell populations in the spiral ganglion were studies by Kellerhals, Engström and Ades (1967) on the guinea-pig, and by Spoendlin (1972, 1974) on the cat. The latter described three populations of ganglion cells and we have used most of his criteria in this chapter. Most of the results given here concern the adult cat, starting with a few observations made with the light microscope. This is followed by observations on the ultrastructure of the various components of the spiral ganglion: the different types of ganglion cells, with emphasis on the myelination of certain cells wherever this may occur; and the different types of fibre that may be visible between the perikarya of the ganglion cells or in the intraganglionic spiral bundle, in connection with their satellite cells where applicable. In addition to these observations, examples are given of spiral ganglion cells in man, in order to demonstrate the main features that distinguish them from similar cells found in most of the small mammals.

Acknowledgement

This work was supported by grants from INSERM (ASR 6-10) and DGRST, 79.71073.

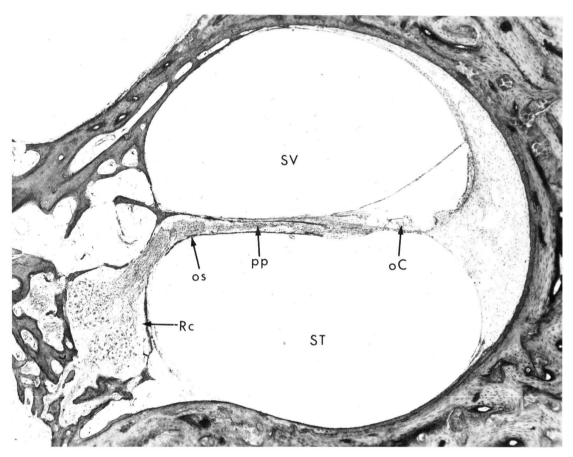

Figure 7.1 *Mid-modiolar section of a human cochlear duct. From a 9-year-old-boy. The spiral ganglion cells, which are mainly bipolar, are clustered in Rosenthal's canal (Rc). The central processes (CP) join the auditory nerve, while the peripheral processes (PP) take a direct radial course to the organ of Corti (OC). SV = Scala vestibuli; ST = scala tympani; OS = osseous spiral lamina. Haematoxylin–Eosin stain. Reproduced by courtesy of H. F. Schuknecht*

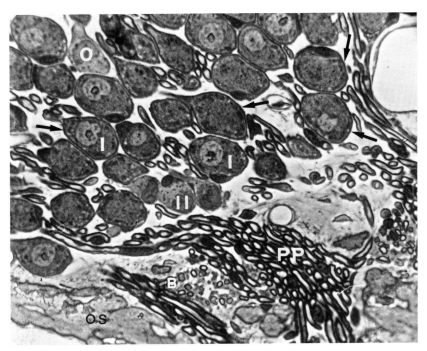

Figure 7.2 *Various constituents of the spiral ganglion. Phase-contrast photomicrograph of the lateral portion of the spiral ganglion from the second turn of the cochlea (this and subsequent pictures are taken from the cat unless otherwise noted)*

(a) Type I ganglion cells (I): in this picture, these cells can be distinguished by a dark cytoplasm and a round nucleus. The perikaryon is surrounded by a dark line, which in most cases represents the myelin sheath (arrow).

(b) Type II ganglion cells (II): these are very often found in clusters of two or three cells in the lateral portion of the spiral ganglion near the root of the osseous spiral lamina (OS) and close to the intraganglionic spiral bundle (B). Some other cells are individually scattered in the ganglion (as demonstrated by serial sections) like one at the top of the photomicrograph (O). In general the perikarya

of these cells can be easily differentiated in a semi-thin section stained with Toluidine blue from type I cells by their smaller size, lighter cytoplasm, and eccentric nuclei. Under light-microscopic observation they appear to be unmyelinated.

(c) Peripheral processes of spiral ganglion cells (PP) projecting towards the organ of Corti and in particular towards the inner hair cells. Notice the heavy myelination of these fibres.

(d) Intraganglionic spiral bundles (B): these bundles of fibres are localized as a rule in the lateral part of the spiral ganglion. They are composed of two main populations of fibres, myelinated ones which can be easily recognized by their black ovals and unmyelinated ones which are found between the former but are difficult to differentiate at this magnification. Semi-thin section. Toluidine blue stain. Original magnification ×90 reduced to 81%

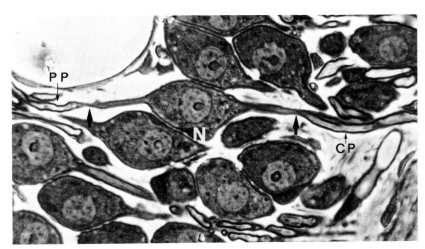

Figure 7.3 *Phase-contrast photomicrograph of a longitudinal section through a type I ganglion cell. The central process (CP) that represents the axon of the cell and builds up the auditory nerve is well myelinated past the first node of Ranvier (arrow). The peripheral process (PP) that projects towards the cochlea represents the dendrite of the ganglion cell. The perikaryon is surrounded by a thin myelin*

sheath forming a slight black line around the cell body. In this micrograph, the central process is larger than the peripheral process and this is a common feature of most type I ganglion cells. N = Nucleus of Schwann cell. Semi-thin section. Toluidine blue stain. Original magnification ×300 reduced to 81%

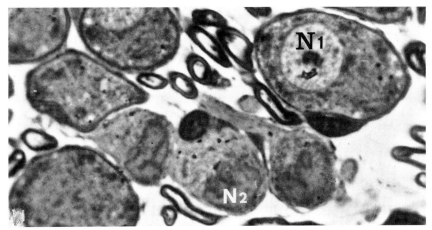

Figure 7.4 *Phase-contrast micrographs of type I and type II ganglion cells. A cluster of three type II ganglion cells is found close to the intraganglionic spiral bundle. Note the lobular shape of the eccentric nucleus (N2) compared with the round nucleus of the type I cell (N1). The cytoplasm of type II cells stains lightly with Toluidine blue in contrast to type I cells which stain more darkly. Second turn, semi-thin section. Toluidine blue stain. Original magnification ×550 reduced to 87%*

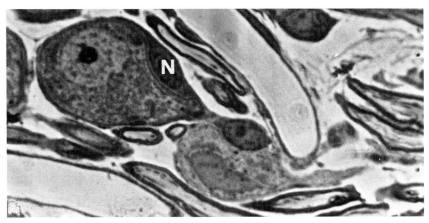

Figure 7.5 *The two types of ganglion cells with their satellite cells in both. The nucleus of the Schwann cell (N) is generally located in the depression where the perikaryon gives rise to a process. Some sections of spiral ganglion cells may show two or even three Schwann cells. Second turn, semi-thin section. Toluidine blue stain. Original magnification ×550 reduced to 87%*

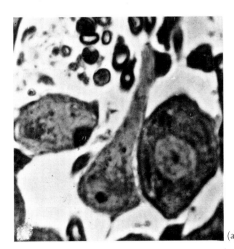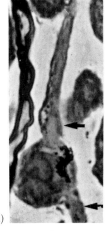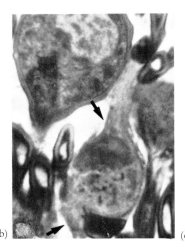

(a)   (b)   (c)

Figure 7.6 *Type II ganglion cells of various shapes. (b) and (c) show two processes which originate in the perikarya. In a type I cell, the two processes are usually found at opposite sides of the cell body (see Figure 7.3); in type II cells they tend to emerge from the same side of the cell body (arrows). S = Schwann cell nucleus. Semi-thin sections, Toluidine blue stain. (a), Original magnification ×450 reduced to 87%; (b), Original magnification ×450 reduced to 87%; (c), Original magnification ×550 reduced to 87%*

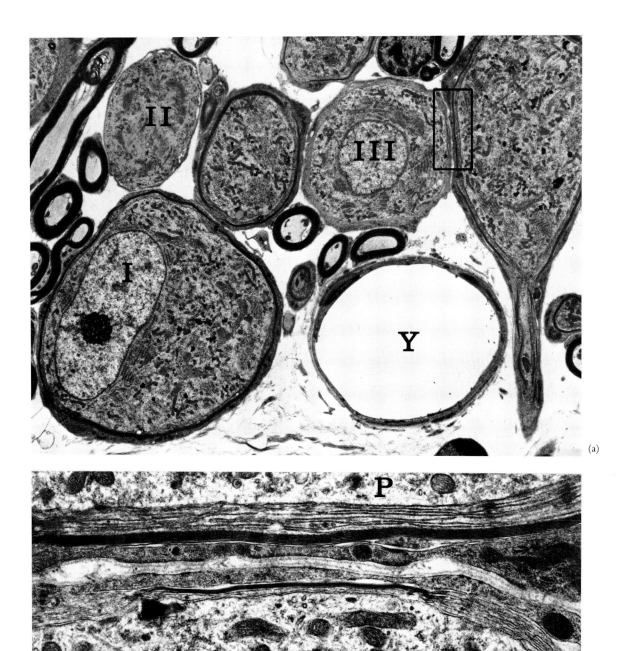

(a)

(b)

Figure 7.7 *Cross-section of spiral ganglion cells. (a) On the left a ganglion cell (I) with a myelin sheath shows a granular cytoplasm; this cell resembles a type I ganglion cell according to Spoendlin's classification (1974). In the cat this type of cell represents the main population (90–95 per cent). They degenerate after section of the cochlear nerve in the internal acoustic meatus. This type of cell (type I) is large and myelinated, and its cytoplasm contains many ribosomes and few filaments. Close to this cell, a smaller perikaryon (II) is visible without a myelin sheath. This cell corresponds to a type II spiral ganglion cell that does not degenerate after section of the cochlear nerve. It usually has a lobulated nucleus, and its cytoplasm contains only a few ribosomes but a large number of neurofilaments.*

Another cell's perikaryon (III) above the capillary (Y) seems to be, in structural appearance, half-way between the two previous ones. The body of this cell has a cytoplasm that is filled with numerous organelles, like a type I cell, but lacks the thick myelin sheath. However, as may be seen in (b) – a higher magnification of the boxed area in (a) – a very small area of the perikaryon shows a compact myelin composed of a few myelin lamellae. This sheath may be compared with a normal myelin sheath from a nearby cell (P). The two populations of type II and III ganglion cells represent approximately 8 per cent of the overall ganglion cell population. First turn of the cochlea. (a) ×3800; (b) ×26 000

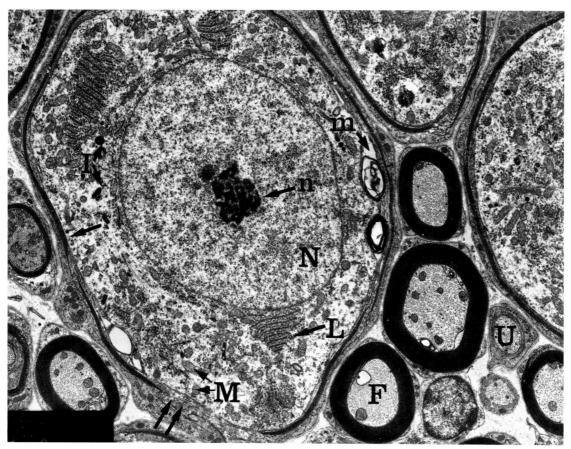

Figure 7.8 *Electron micrograph of a cross-section of a type I ganglion cell. Note the round nucleus (N), with light chromatin and a well-differentiated nucleolus (n). An accumulation of mitochondria (M), endoplasmic reticulum, Nissl bodies (L) and inclusion bodies (I) are clearly visible in the ganglion cell's cytoplasm at this magnification. The perikaryon is surrounded by a myelin sheath whose structure changes with its site; it may appear either compact (single arrow) or loose (double arrows). Myelin-like bodies (m) can be seen between the myelin sheath and the cytoplasm. Two types of fibres are present between the perikarya throughout the spiral ganglion; large and well-myelinated ones (F); and thinner unmyelinated fibres (U). The latter are surrounded by glial processes. Basal part of the first turn of the cochlea. ×6400*

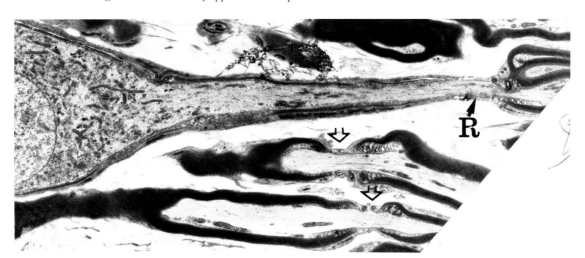

Figure 7.9 *Longitudinal section through the axon hillock and the initial segment of a type I ganglion cell. Note the difference in thickness of the myelin sheath between the cell body and the axon, which changes at the first node of Ranvier (R). Open arrows point to the node of Ranvier of two fibres. Hook of cochlea. ×2000*

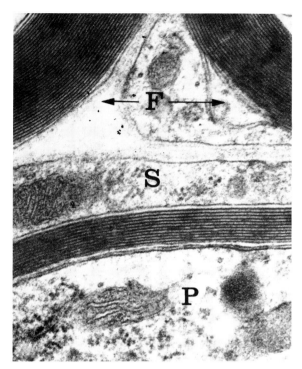

Figure 7.10 *Electron micrograph of myelin sheaths of fibres and cells, comparing the myelin sheath from two fibres (F) and from a ganglion cell (P). Note the large number of myelin lamellae in both fibres in comparison with the cell. Major dense lines and minor dense lines can be differentiated in both structures, fibres and cell. S = Schwann cell process. Second turn of cochlea. Original magnification ×62 000 reduced to 91%*

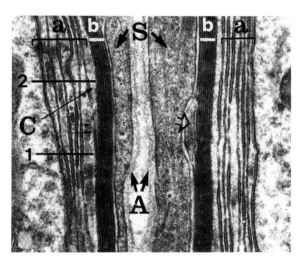

Figure 7.11 *Myelin sheaths of two adjacent ganglion cells. These two myelin sheaths are composed of loose (a) and compact (b) myelin. The lamellae adjacent to the cytoplasm on the left of the picture represent loose myelin as they contain Schwann cell cytoplasm (C). The number of lamellae in the loose myelin area changes at two different levels of the myelin sheath. There are five lamellae at level 1 and six at level 2; this difference is caused by a change in the compact myelin, where the innermost major dense line (double arrows) splits into the two basic membranes separated at level 2 by Schwann cell cytoplasm (C). At the same time the number of major dense lines*

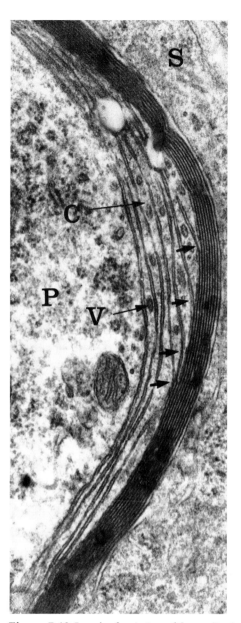

Figure 7.12 *Sample of variations of the myelin sheath around the perikaryon (P). Each large arrow points to the separation of Schwann cell lamellae from the compact myelin sheath. The thickness of the included Schwann cell cytoplasm (C) varies between each lamella and also with distance. Numerous ovoid structures that may be microtubules are visible in the cytoplasm (V). S = Schwann cell process. Second turn of the cochlea. Original magnification ×53 700 reduced to 91%*

decreases by one between levels 1 and 2. The same type of myelin sheath is visible on the neighbouring cell on the right, with seven loose lamellae and eight major dense lines in the compact myelin portion of the sheath. On the outermost compact lamellae at the top of the figure it appears as a major dense line and changes for a short distance to a loose lamella (open arrow) then returns to its initial shape. S = Schwann cell processes separated by the intercellular space; A = basal lamina. Basal part of the first turn of the cochlea. Original magnification ×41 000 reduced to 91%

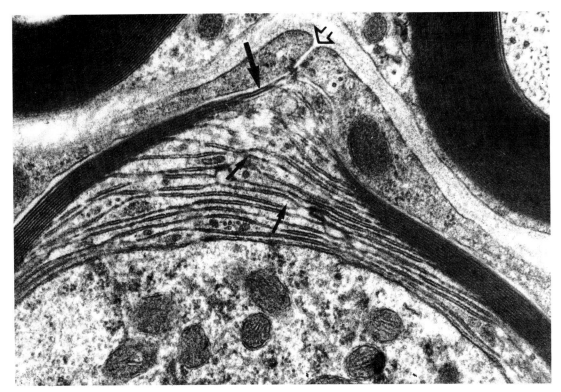

Figure 7.13 *Electron micrograph of a complex figure of myelin at a pole of a spiral ganglion perikaryon showing the transition of compact myelin lamellae into loose lamellae, many of the latter ending at this level (arrows). At the upper part of the compact myelin sheath from the left side, the outermost compact lamellae (large arrow) become a large Schwann cell process at the right side of the perikaryon. An external mesaxon is present at the top (open arrow). Second turn of cochlea. Original magnification ×47 700 reduced to 97%*

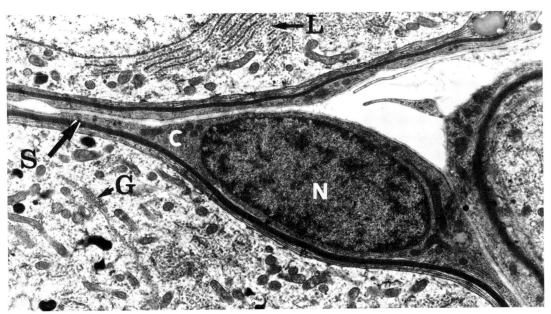

Figure 7.14 *Electron micrograph of a Schwann cell. The ovoid nucleus (N) contains some scattered dense chromatin material, but some of the chromatin material is clumped and lines up at the inner lamina of the nuclear envelope. The dark perinuclear cytoplasm (C) contains several mitochondria, granular endoplasmic reticulum and many clumps of ribosomes that are responsible for the deeper colour of the cytoplasm. Glial processes (S) are given off from the cell body to ensheath the perikaryon of the ganglion cell. Compare the lighter cytoplasm of two myelinated spiral ganglion cells that also contain many cytoplasmic organelles such as mitochondria, Golgi complex (G), Nissl bodies (L) and clusters of ribosomes. First turn of cochlea. Original magnification ×82 000 reduced to 97%*

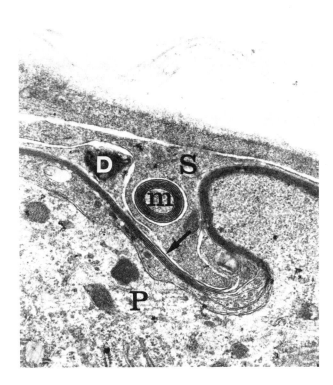

◀ **Figure 7.15** *Redundant myelin from type I ganglion cell. Invagination of the myelin sheath into the cytoplasm of the cell body (P) followed by a Schwann cell process (S). This contains a round myelinated body (m), with the same number of lamellae as the myelin sheath. The material inside the myelinated body resembles the Schwann cell cytoplasm. This myelinated body may represent some redundant myelin. The arrow points to a Schwann cell process changing into a compact myelin lamella. This Schwann cell process seems to contain degenerated myelin (D). First turn of cochlea.* ×21 200

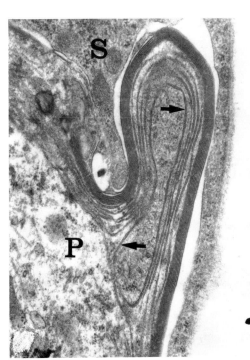

Figure 7.16 *Another form of myelin sheath that protrudes outside the cell body (P); although no myelinated body is visible, a limiting membrane (arrow) surrounds a dense material which may represent an inclusion of Schwann cell cytoplasm. This complicated figure of myelin is surrounded by a Schwann cell process (S). Second turn of cochlea.* ×25 900

Figure 7.17 *Electron micrograph of a myelin sheath specialization. Stack of desmosome-like structures (arrow) limited to the loose myelin lamellar section. They are characterized by an aggregation of osmiophilic material on the cytoplasmic side of the lamellae that become wider at this location. First turn of cochlea.* ×50 000

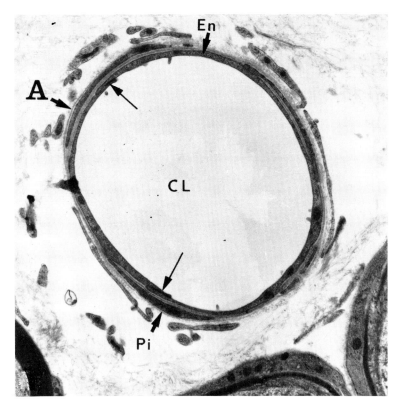

Figure 7.18 *A capillary from the spiral ganglion composed of an uninterrupted endothelium (En) and some pericytes (Pi). A = Basal membrane; CL = capillary lumen; arrows point to endothelial cell junctions. First turn of cochlea. ×9700*

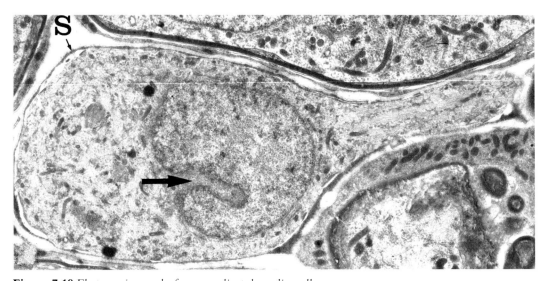

Figure 7.19 *Electron micrograph of an unmyelinated ganglion cell. This cell is surrounded by only one or two thin dark Schwann cell processes (S). Only one neuronal process is visible in this figure. Myelin sheaths from neighbouring cells are identified. Note infolding of the nuclear membrane into the nucleus (large arrow). Basal part of the first turn of the cochlea. ×6700*

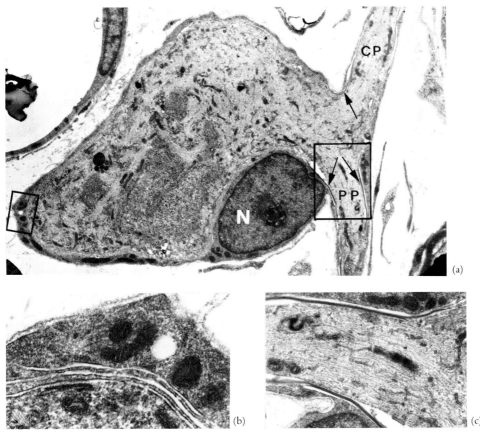

Figure 7.20(a) In another unmyelinated cell, the section passes through two processes, a peripheral process (PP) and a centrally oriented one (CP). Note the position of the nucleus of the glial cell (N) in the portion of the cell that gives rise to the two neuronal processes. These appear to arise from a common trunk and subsequently split into two components. Around the perikaryon, the myelin sheath is composed of a few loose lamellae as seen at a higher magnification of the boxed area, shown in (b). The presence of compact myelin lamellae (arrows) at the origin of the neuronal processes can be observed in (c). First turn of cochlea. (a), Original magnification ×6500 (b), ×35 600; (c), ×18 000 all reduced to 84%

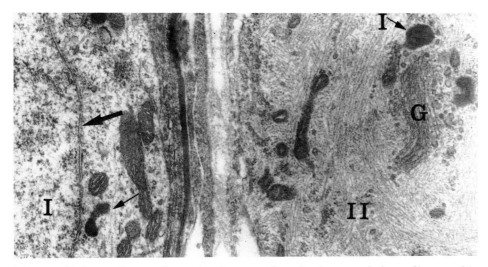

Figure 7.21 Contrast between the cytoplasmic content of type I and type II ganglion cells. The cytoplasm of type I cells contains many organelles such as mitochondria, clumps of ribosomes that are mainly responsible for the granular appearance of the cytoplasm. Few neurofilaments are visible (thin arrow). On the opposite side of the photograph the cytoplasm of a type II cell is characterized by densely packed neurofilaments of the cytoplasm; only a few mitochondria and ribosomes are present. G = Golgi complex with vesicles and lamellae; I = dark inclusion body near the Golgi complex. The large arrow points to the nuclear membrane. First turn of cochlea. Original magnification ×29 000 reduced to 84%

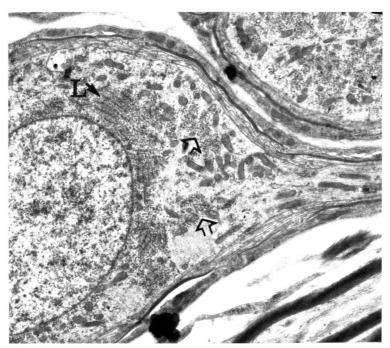

Figure 7.22 *Axon hillock of a myelinated cell. Note the Nissl bodies (L), largely confined to around the nucleus. Large clusters of ribosomes (open arrow) are also present in the cytoplasm, but they are not found in the axon hillock where mitochondria and filamentous structures (that are revealed to be neurofilaments and neurotubules at* *higher magnification) are found in greater numbers, becoming increasingly oriented with the axis of the axon at the initial segment. First turn of the cochlea. Original magnification ×7800 reduced to 90%*

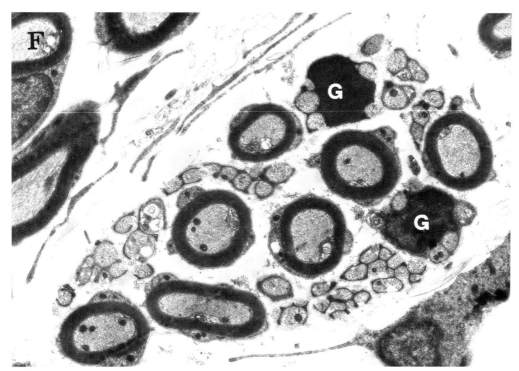

Figure 7.23 *Cross-section through part of the intraganglionic spiral bundle. Two types of fibres can be easily distinguished by their myelin sheaths and their diameter. One internode of myelinated fibres is surrounded by only one Schwann cell whereas a glial cell envelops several unmyelinated fibres (G). Unmyelinated fibres outnumber* *myelinated ones, but the ratio between the two types of fibres is not constant throughout the cochlea. F: This myelinated fibre and the neighbouring ones do not belong to the intraganglionic spinal bundle. Basal part of the first turn of the cochlea. Original magnification ×11 400 reduced to 90%*

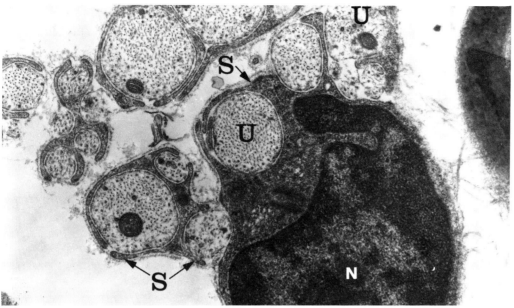

Figure 7.24 *Cross-section through a glial cell and unmyelinated fibres. Cytoplasmic processes of a glial cell ensheathing completely or in part several unmyelinated fibres (U) of various sizes.*

N = Nucleus of the glial cell. S = Glial cell process. Second turn of cochlea. Original magnification ×24 600 reduced to 91%

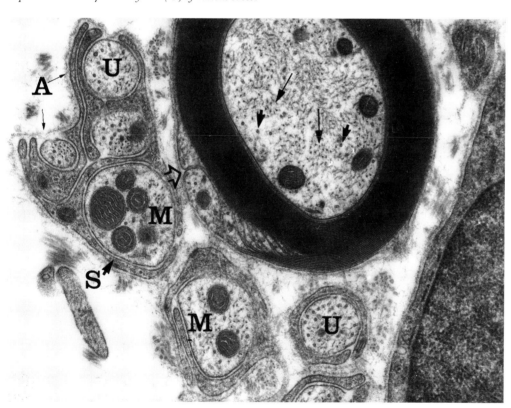

Figure 7.25 *Fibres from the intraganglionic spiral bundle. In this section three types of fibres can be differentiated:*

1. A large fibre whose axoplasm contains several clusters of neurofilaments (thin arrows), few neurotubules (larger arrows) and some mitochondria. The axoplasm is surrounded by a thick myelin sheath possessing about 30 myelin lamellae.

2. Unmyelinated fibres which display numerous mitochondria (M) in their axoplasm along with neurofilaments and neurotubules.

These fibres are ensheathed by a single layer of glial cell process (S).

3. Unmyelinated fibres that have fewer or no mitochondria (U). Unmyelinated fibres always have a smaller axoplasm than myelinated fibres. All fibres are, however, surrounded by a basal lamina (A). External mesaxon (open arrow). First turn of cochlea. Original magnification ×36 800 reduced to 91%

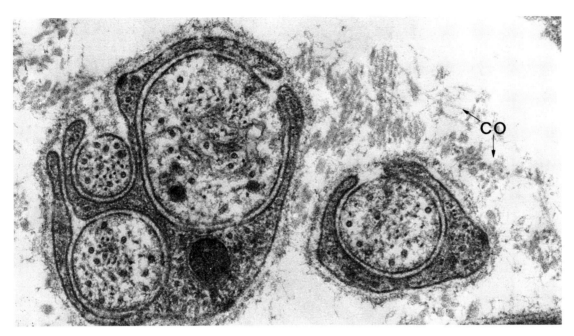

Figure 7.26 *Higher magnification of some unmyelinated fibres. Note the great variation in size between fibres. Neurofilaments and numerous neurotubules are clearly visible. In unmyelinated fibres the number of neurotubules tends to be higher per unit area than in most myelinated ones. CO = Fibrils of collagen. Basal part of the first turn of the cochlea. ×61 500*

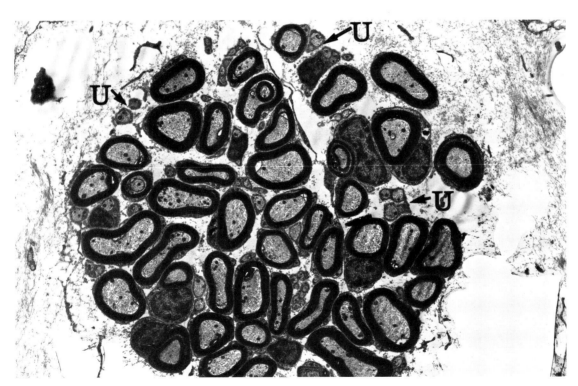

Figure 7.27 *Cross-section of preganglionic fibres obtained by a tangential section through the osseous spiral lamina close to the spiral ganglion. Two populations of fibres are present, one myelinated, represented by the larger axoplasm and the other unmyelinated (U) the latter being smaller. It is worth pointing out that the ratio of myelinated to unmyelinated fibres is very often the inverse of the one in this section. Most of the myelinated fibres probably represent the dendrites of type I spiral ganglion cells, which will subsequently project to the inner hair cell. Numerous unmyelinated fibres arise from the intraganglionic spiral bundle that is made up of several components, others come from the cochlear nerve and in a few cases they may come directly from the spiral ganglion cells. First turn of cochlea. ×4900*

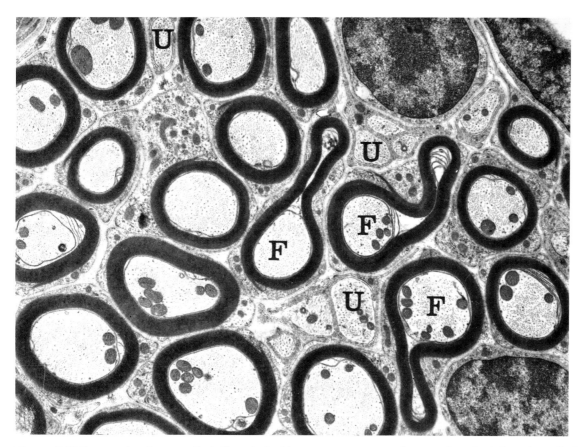

Figure 7.28 *Cross-section of postganglionic fibres obtained just at the exit of Rosenthal's canal. Large axoplasms are surrounded by a thick myelin sheath, whose three fibres (F) display some redundancy. Unmyelinated fibres are also present at this level (U), but in a lower ratio than at the preganglionic level. Most of the myelinated fibres represent the axon of type I spiral ganglion cells that will subsequently reach the cochlear nucleus via the cochlear nerve. The exact origin of unmyelinated fibres is not well known. It is thought that some of them may be adrenergic. First turn of cochlea. ×12 700*

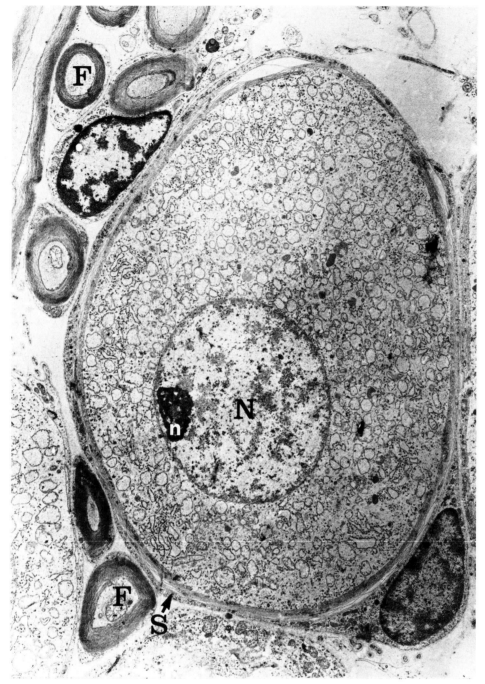

Figure 7.29 *Large ganglion cells of the human spiral ganglion that represent the main population (94 per cent). This type of cell can be either unmyelinated or myelinated. However, in contrast to small mammals (guinea-pig, rat, cat) in the human the majority of the ganglion cells are not myelinated. The cell shown in this figure is myelinated with a cytoplasm filled with ribosomes, rough endoplasmic reticulum and mitochondria represented in this picture by small vacuoles, due to post-mortem changes. The round nucleus (N)* *contains dispersed chromatin substance. However, the eccentric nucleolar substance is more compact. Possibly due to post-mortem changes, the myelin sheath of this cell appears to be composed of semi-compact and loose myelin surrounded by processes of a Schwann cell (S) whose nucleus is present at the bottom right of the perikaryon. This cell is surrounded by many myelinated fibres (F). ×26 000 (From Ota and Kimura 1980, reproduced by kind permission of authors and publishers.)*

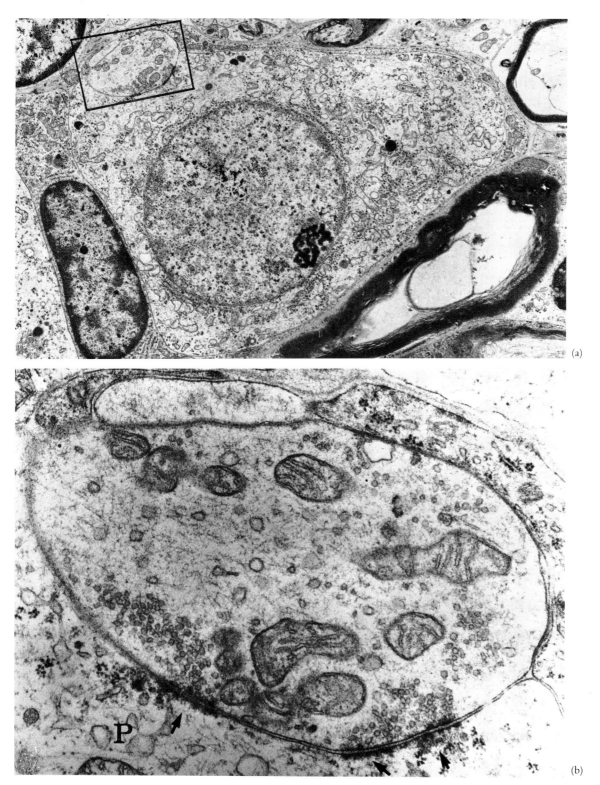

(a)

(b)

Figure 7.30(a) *An unmyelinated ganglion cell that represents the smallest cellular population in the human cochlea. This type of cell includes many neurofilaments and a relatively small number of rough endoplasmic reticulum, few ribosomes, and mitochondria that are oedematous due to post-mortem changes. One axosomatic synapse shown in the boxed area becomes apparent at a higher magnification (b). The enlargement of the nerve terminal shows several mitochon-* *dria, neurofilaments and synaptic vesicles aggregated close to the pre-synaptic membrane. The post-synaptic side, on the ganglion cell body (P), shows thickening at three different synaptic contacts (arrows). (a), ×7000; (b), ×38 000 (From Kimura, Ota and Takahashi 1979, reproduced by kind permission of authors and publishers.)*

181

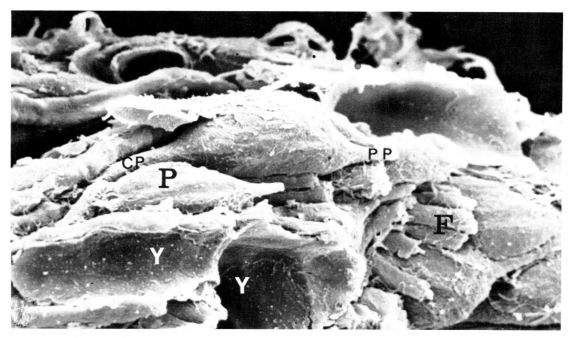

Figure 7.31 *Scanning electron micrograph of spiral ganglion cells. The two processes of the bipolar cell at the centre of the figure are located approximately at opposite sides of the cells. The peripheral process (PP) is smaller than the central one (CP). Adjacent fibres (F) and ganglion cells (P) are oriented in the same direction as blood vessels (Y).* ×2300

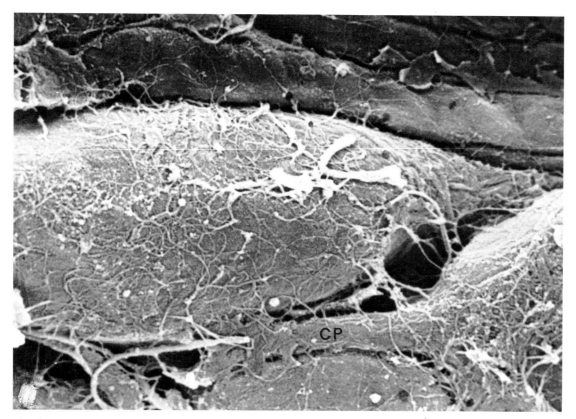

Figure 7.32 *Perikaryal surface of a bipolar ganglion cell. An extensive network of collagen filaments is clearly visible on the surface of the perikaryon and the processes. CP = Central process of a ganglion cell.* ×6100

References

Adamo, N. J. and Daigneault, E. A. (1973) Ultrastructure features of neurones and nerve fibres in the spiral ganglia of cats. *Journal of Neurocytology*, **2**, 91–103.

Cajal, S. R. (1909) *Histologie du Système Nerveux de l'Homme et des Vertébrés*. Paris: Maloine.

Held, H. (1926) Die Cochlea der Säuger und der Vögel, ihre Entwicklung und ihr Bau. In *Handbuch der Normalen und Pathologischen Physiologie*, vol. II. *Receptionorgane*, ed. A. Bethe. Berlin: Julius Springer.

Kellerhals, B., Engström, H. and Ades, H. W. (1967) Die Morphologie des Ganglion spirale cochleae. *Acta Otolaryngologica (Stockholm)*, Supplement **226**, 1–78.

Kimura, R. S., Ota, C. Y. and Takahashi, T. (1979) Nerve fiber synapses on spiral ganglion cells in the human cochlea. *Annals of Otology, Rhinology and Laryngology*, Supplement **62, 88**, 1–17.

Lorente de Nó, R. (1937) The neural mechanism of hearing. I: Anatomy and Physiology. b: The sensory endings of the cochlea. *Laryngoscope*, **47**, 373–377.

Merck, W., Riede, U. N., Löhle, E. and Cürten, I. (1977) Vergleichende ultrastrukturen-morphometrische Studie des Ganglion spirale cochleae der Ratte und des Meerschweinchens. *Archives of Oto-Rhino-Laryngology*, **217**, 441–449.

Noden, D. M. (1980) Somatotopic and functional organization of the avian trigeminal ganglion: An HRP analysis in the hatchling chick. *Journal of Comparative Neurology*, **190**, 405–428.

Ota,C. Y. and Kimura, R. S. (1980) Ultrastructural study of the human spiral ganglion. *Acta Otolaryngologica (Stockholm)*, **89**, 53 –62.

Retzius, G. (1895) *Biologische Untersuchungen*, vol. 6. Leipzig: Vogel.

Rosenbluth, J. (1962) The fine structure of acoustic ganglia in the rat. *Journal of Cell Biology*, **12**, 329–359.

Rosenbluth, J. and Palay, S. L. (1961) The fine structure of nerve cell bodies and their myelin sheaths in the eighth nerve ganglion of the goldfish. *Journal of Biophysical and Biochemical Cytology*, **9**, 853–877.

Ross, M. D. and Burkel, W. (1973) Multipolar neurons in the spiral ganglion of the rat. *Acta Otolaryngologica (Stockholm)*, **76**, 381–394.

Spoendlin, H. (1972) Innervation densities of the cochlea. *Acta Otolaryngologica (Stockholm)*, **73**, 235–248.

Spoendlin, H. (1974) Neuroanatomy of the cochlea. In: *Psychophysical Models and Physiological Facts in Hearing*, eds. E. Zwicker and E. Terhardt, pp. 1–33. Berlin: Springer.

Suzuki, Y., Watanabe, A. and Osada, M. (1963) Cytological and electron microscopic studies on the spiral ganglion cells of the adult guinea pigs and rabbits. *Archivum Histologicum Japonicum*, **24**, 9–33.

Thomsen, E. (1967) The ultrastructure of the spiral ganglion in the guinea pig. *Acta Otolaryngologica (Stockholm)*, Supplement, **224**, 442–448.

Trevisi, M., Testa, F. and Riva, A. (1977) Fine structure of neurons of cochlear ganglion of the guinea pig after prolonged sound stimulations of the inner ear. *Journal of Submicroscopic Cytology*, **9**, 157–172.

Ylikoski, J., Collan, Y. and Palva,T. (1978) Ultrastructural features of spiral ganglion cells. *Archives of Oto-Rhino-Laryngology*, **104**, 84–88.

8
The Stria Vascularis

Matti Anniko   Dan Bagger-Sjöbäck

Introduction

The cytoarchitecture of the stria vascularis with its intra-epithelial network of blood vessels has been known for over a century. The stria vascularis is presumed to be involved in the formation of endolymph (Guild, 1927; Anniko and Nordemar, 1980), the supply of endolymphatic oxygen (Misrahy et al., 1958; Vosteen, 1960), and the maintenance of the endocochlear potential (EP) (Davis et al., 1958; Tasaki and Spyropoulos, 1959).

The tissue architecture, composed of three layers of cells (marginal, intermediate, and basal cells) and blood vessels, is essentially the same in all mammals (Figure 8.1). Under the light microscope, the strial cell layers are rather ill defined but are roughly distinguished by their location and cytoplasmic staining characteristics. The cells of the basal layer connect the stria vascularis to the spiral ligament and constitute both an anatomical and functional barrier in the compartmentalization of the scala media (Jahnke, 1975a; Reale et al., 1975).

The marginal and intermediate cells are in close contact with the cells of Reissner's membrane at its insertion at the stria vascularis. There is a rather gradual transition between the cells of the stria vascularis and the cells of Reissner's membrane. The anatomical border towards the spiral prominence is distinctly defined. The intra-epithelial blood vessels lie parallel to one another in the direction of the cochlear turns. The location of the blood vessels in relation to the strial cells can vary slightly in different regions of the stria vascularis (Axelsson, 1974).

Embryonic Development of the Stria Vascularis

The embryonic development has been studied in detail in only a few species (guinea-pig, mouse, rabbit and rat) and ultrastructural analyses are scarce (Weibel, 1957; Schmidt and Fernandez, 1963; Kikuchi and Hilding, 1966; Sher, 1971; Anniko and Nordemar, 1980; Anggård, 1965). Correlations have been made between the structural maturation of the stria vascularis and the maturation of the specific composition of endolymph (Anniko and Wroblewski, 1981); between the electrochemistry of the cochlear fluids and the development of the endocochlear potential (Bosher and Warren, 1971); and between the structure and the maturation of cochlear potentials (Alford and Ruben, 1963; Pujol and Hilding, 1973).

When the anatomical location of the stria vascularis becomes evident, the epithelial lining comprises one layer of cuboidal or columnar cells which are separated from the underlying tissue by a basal lamina. In mice this occurs on the 16th–17th gestational days (the gestational age is 20–21 days). Thereafter, the staining characteristics of these cells change to a dense appearance both in light and electron microscopic preparations. According to Ruben (1967), the cells of the stria vascularis pass through their terminal mitotic phase considerably later than cochlear or vestibular hair cells (mouse). It can be assumed that Ruben's observations are valid for mammalian species other than mice because the developmental sequences are similar between species.

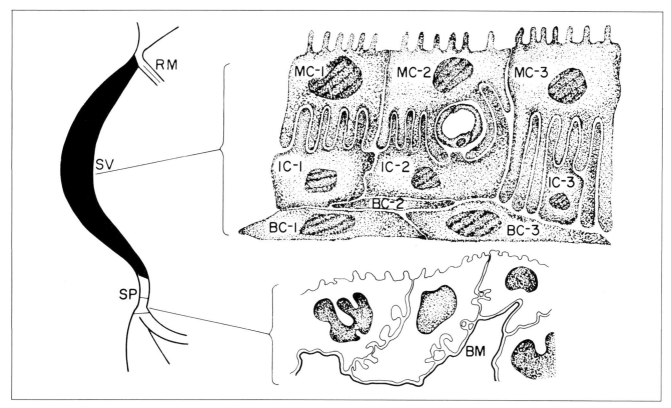

Figure 8.1 *The principal morphology of the three cell types comprising the stria vascularis (SV): marginal cells (MC), intermediate cells (IC) and basal cells (BC). SP = Spiral prominence; RM = Reissner's membrane. Marginal cells: The surface is covered with a thick layer of microvilli. Adjacent apical parts of MC are connected with each other by 5–7 sealing elements of the tight junction and are thus closer to each other than the remaining parts of the cell bodies. The nucleus is located close to the region where the basal portion of the cell is split into an extensive network of cytoplasmic processes containing numerous mitochondria. The digitations from the marginal cells often meet a similar but apically directed network of cytoplasmic processes from intermediate cells (MC-1/*

IC-1). The cytoplasmic processes from MC surround blood vessels in toto or only in part (MC-2) and can also reach the basal cells (MC-3). Intermediate cells: These cells have their main network of cytoplasmic processes directed apicalwards towards those from MC (IC-1). The IC processes can also constitute 'vascular feet' to blood vessels, i.e. short cytoplasmic projections reaching the blood vessel from below (IC-2). The IC are smaller than MC and are basalwards in contact with the BC layer (IC-1/BC-1; IC-1/BC-2; etc.). Basal cells: These cells are spindle shaped and slightly flattened. A number of flattened processes extend laterally from each cell and penetrate between neighbouring BC (BC-2) giving the impression of several layers of BC in cross-sections of the stria vascularis

The intermediate and basal cells are formed after the marginal cells. The marginal cells are of ectodermal origin whereas the intermediate and basal cells are of mesenchymal origin. Progress of the maturation occurs from basal to apical coil. In mice, there can be a 1–3 day difference in morphological maturation between the basal and apical turns during the most active period of differentiation, i.e. the first postnatal week (Anniko and Nordemar, 1980). In other species where the gestational period is longer, this difference may be even more pronounced. Analysis of the development of the specific composition of endolymph has shown that a gradient of potassium – decreasing from basal to apical turns – occurs during morphological maturation of the stria vascularis (Anniko and Wroblewski, 1981).

The onset of maturation of the stria vascularis occurs as an ingrowth of the marginal cells towards the underlying mesenchyme. Thus, the marginal-cell extensions come into contact with cells that have just passed their mitosis and with blood vessels under development

(*Figures 8.2–8.6*). The basal lamina below the marginal cells becomes fragmented and gradually disappears. Remnants of the basal lamina can occasionally be identified in the morphologically newly mature stria vascularis. As the marginal cells start their ingrowth, a fine network of cytoplasmic projections starts to develop. Cells of the future intermediate cell layer form a similar type of network and meet that from the marginal cells (*Figure 8.6(c)*).

During early strial maturation when the intermediate cells are formed, the immature blood vessels are in close contact with the developing intermediate cells which surround the major part of the blood vessel. Later, when the marginal-cell extensions grow toward the blood vessel, these cytoplasmic projections separate the blood-vessel endothelium and the intermediate cells (*Figures 8.7 and 8.8*). The blood vessels become surrounded on three sides by the marginal-cell extensions (*see Figure 8.1*). During this period the stria vascularis develops its border towards the spiral ligament (*Figure*

185

8.9). Cells differentiating into basal cells become orientated with their cell nuclei parallel to the endolymphatic surface of the stria vascularis. Occasionally, cells from the marginal-cell layer are released into the endolymphatic space (*Figure 8.9*).

During the next phase of maturation, the marginal- and intermediate-cell extensions intermingle and move into close contact with one another. The nucleus of the marginal cell moves basally in the cell close to the area from which the cytoplasmic projections extend. The stria vascularis is initially more cellular than at the adult stage but rapidly reaches maturation (*Figure 8.10*).

(*continued on page 192*)

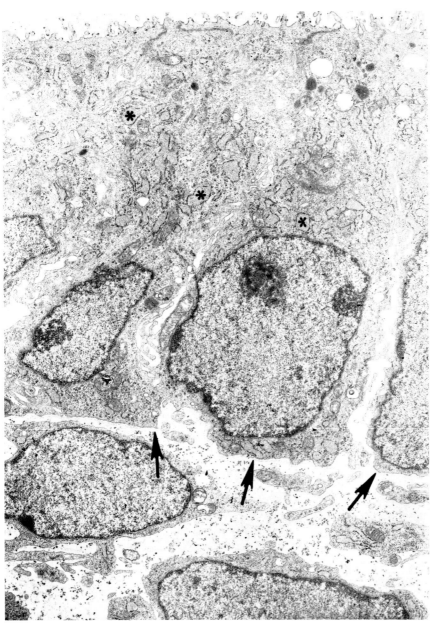

Figure 8.2 *Epithelial cells at the anatomical location of the future stria vascularis. Apical coil of the cochlea in the newborn mouse. The nuclei of the columnar cells are located basally and contain one or two nucleoli. A distinct basal lamina is found below the epithelial cell layer (arrows). There is no network of cytoplasmic processes between or below the cells. The surface is covered with numerous microvilli.*

The cytoplasm contains several morphological features believed to be criteria of high metabolic activity: distended rough endoplasmic reticulum (asterisks) and clusters of ribosomes (polysomes). The apical parts of the cells are in closer contact with each other than are the basal parts. There are few mitochondria. The cells below the basal lamina are fibroblast-like. Electron micrograph. ×3100

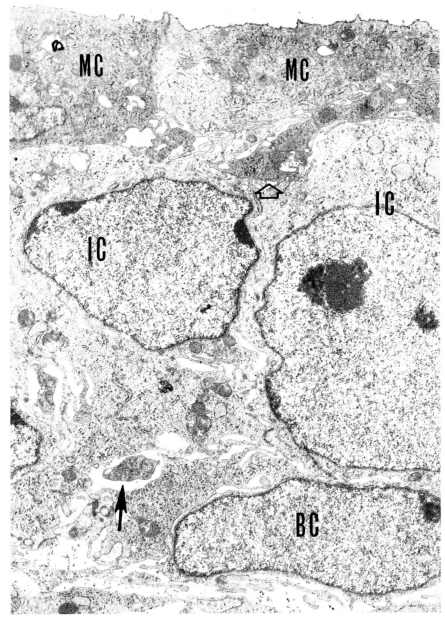

Figure 8.3 *Stria vascularis from the middle part of the cochlea of the 2-day-old mouse. The marginal cells (MC) have an electron-dense cytoplasm and have started the ingrowth towards the underlying cells (unfilled arrows). The future intermediate cells (IC) have large nuclei which contain one or several nucleoli. The extensive network of MC and IC protoplasmic projections has not yet formed. However, some MC extensions can be found near the future basal cells (BC) whose nuclei run parallel with the endolymphatic surface. Electron micrograph. ×3900*

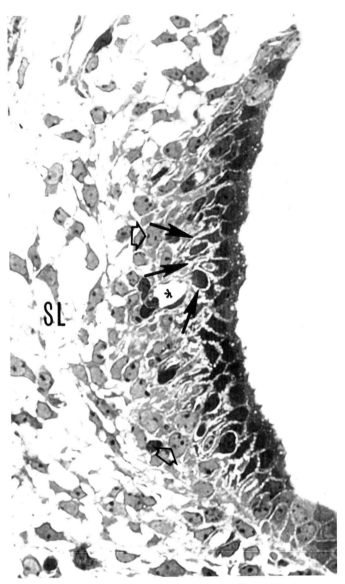

Figure 8.4 *Stria vascularis from the basal coil of the cochlea of the newborn mouse. The marginal cells stain intensely, have some vacuoles in the cytoplasm, and show outgrowth of cytoplasmic extensions towards underlying cells. The marginal-cell extensions (arrows) have not yet reached the blood vessel (asterisk). The intermediate cells are difficult to distinguish. The future basal cells are close to the spiral ligament (SL) and are numerous (unfilled arrows). The border between the stria vascularis and the SL is difficult to delineate. Light micrograph. ×160*

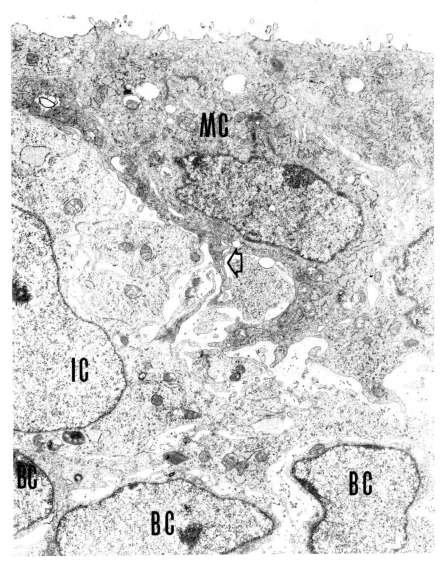

Figure 8.5 *The same cochlea as in Figure 8.4 but the section is taken from the basal coil. The marginal cells (MC) have a large number of microvilli on the surface. The cytoplasm contains several vesicles and abundant rough endoplasmic reticulum. The nucleus is located basally as found in the mature MC. Several large protoplasmic projections are advancing basalwards but have not penetrated the basal lamina. Smaller, slender projections have also started to form (unfilled arrow). The intermediate cells (IC) have large nuclei and have started to form extensions. The basal cells (BC) are identified by their location and the orientation of the nucleus in parallel with the endolymphatic surface. Electron micrograph. ×3800*

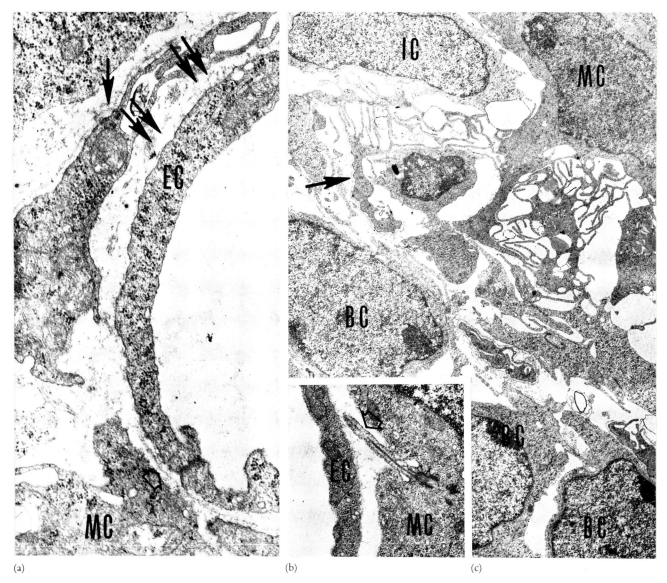

(a) (b) (c)

Figure 8.6(a) and (b) Stria vascularis from the basal coil of the
cochlea of the 1-day-old mouse. Note formation of strial blood vessels.
(a) The marginal cell (MC) extensions still have a basal lamina
towards the inner part of the stria vascularis (arrow). An MC
protoplasmic extension starts to surround the endothelial cell (EC).
The EC has started its secretion of a basal lamina but this is only
partially developed (unfilled arrow). At adjacent regions of the EC a
foamy structure is present close to the EC (double arrows). The EC
cytoplasm is filled with free ribosomes. Electron micrograph. ×8800.
(b) The EC has, like the MC, a very electron-dense cytoplasm. No

basal lamina has yet been formed around the EC but a foam-like
material is adjacent to it. The MC has a kinocilium directed towards
the EC (arrow). Electron micrograph. ×6900. (c) Stria vascularis of
the basal coil of the 3-day-old mouse cochlea. MC and intermediate
cells (IC) have a network of cytoplasmic projections but these are not
fully developed and they are not in close contact with each other. A
basal lamina can still occur adjacent to MC extensions (arrow). The
basal cells (BC) have not yet attained their spindle shape. Electron
micrograph. ×3900

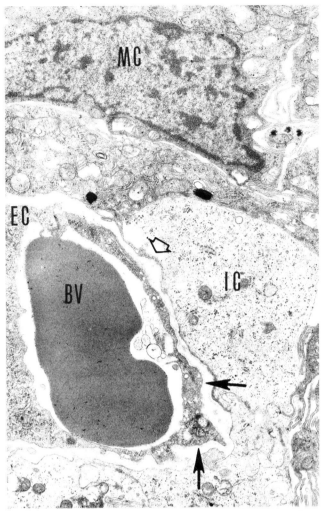

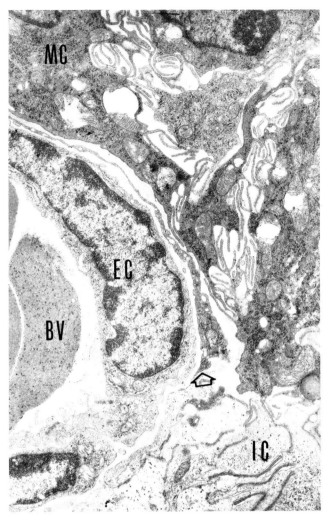

Figure 8.7 *Stria vascularis of the 1-day-old mouse. Basal coil. Ā marginal cell (MC) is close to the upper part of a strial blood vessel (BV). The endothelial cells (EC) have an immature morphology. The EC stain as electron dense as the MC and are filled with a large number of rough endoplasmic reticulum free ribosomes. The intermediate cell (IC) is in close contact with the BV both below it and at one side. A basal lamina has started to form around the EC (arrows). A small cytoplasmic projection from the MC has started its ingrowth between the EC and IC (unfilled arrow). At this stage of embryonic development the MC cytoplasm is filled with rough endoplasmic reticulum and free ribosomes. A few pigments occur in the MC. Electron micrograph. Original magnification ×6400 reduced to 89%*

Figure 8.8 *Stria vascularis of the 3-day-old mouse. Basal coil. The marginal cells (MC) have a rich network of cytoplasmic projections which have started to surround the endothelial cell (EC) of the blood vessel (BV) (unfilled arrows). The EC cytoplasm is no longer as electron dense as at previous developmental states (compare with Figure 8.7). IC = Intermediate cell. Electron micrograph. Original magnification ×6800 reduced to 84%*

191

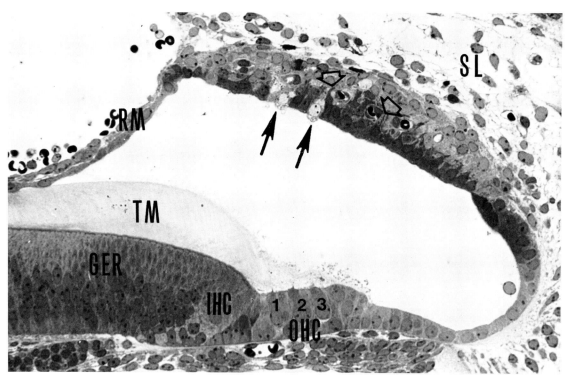

Figure 8.9 *Section through the endolymphatic portion of the 2-day-old mouse cochlea. Basal coil. The stria vascularis is more cellular than in the mature animal. A possible way to reduce the number of cells is to expel some strial cells into the endolymphatic space (arrows). At this stage of postnatal development marginal-cell projections have reached the strial blood vessels (unfilled arrows).*

The separation of the stria vascularis from the spiral ligament (SL) is distinct. The organ of Corti is immature but three outer hair cells (OHC 1-3) and one inner hair cell (IHC) are identified. GER = Greater epithelial ridge; TM = tectorial membrane; RM = Reissner's membrane. Light micrograph. ×90

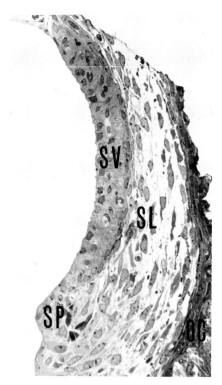

◄**Figure 8.10** *Stria vascularis (SV) of the basal coil of the 6-day-old mouse cochlea. Mature conditions. SP = Spiral prominence; SL = spiral ligament; OC = otic capsule. Light micrograph. ×70*

During the development of marginal- and intermediate-cell extensions, the marginal cells show many morphological characteristics of high metabolic and secretory activity (*Figure 8.11*). The cytoplasm is filled with distended rough endoplasmic reticulum (RER), coated vesicles and polysomes (*Figure 8.12*), vesicles with granulated material (*Figure 8.13(a)*), and occasionally inclusion bodies surrounded by a limiting membrane (*Figure 8.13(b)*). The surface of the marginal cell is covered with a large number of microvilli. The marginal cell has a coated cell membrane and exocytotic vesicles occur (*Figures 8.12* and *8.13(c)*). Shortly after the formation of the marginal- and intermediate-cell interdigitations, many degenerating mitochondria with a myelin-figure-like substructure occur in marginal- and intermediate-cell extensions (*Figure 8.14(a)* and *(b)*). Other fragments of electron-dense material appear with a myelin-like substructure surrounded by a limiting membrane (*Figure 8.14 (c)* and *(d)*).

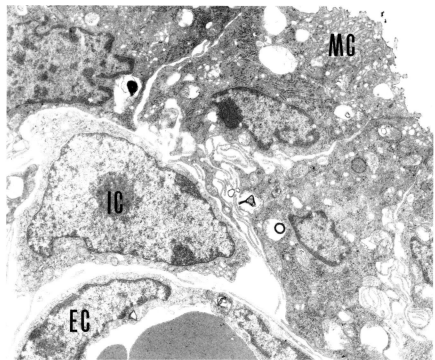

Figure 8.11 *Stria vascularis of the basal coil of the 3-day-old mouse cochlea. The marginal cells (MC) have a rough surface with many microvilli and contain a large number of vesicles and rough endoplasmic reticulum. The nucleus is located basally in the cell. The cytoplasmic projections are few as are those of the intermediate cells (IC). The strial blood vessel has a basal lamina around the endothelial cell (EC). Electron micrograph. ×1800*

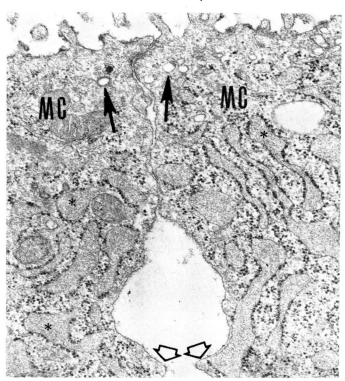

Figure 8.12 *Marginal cells (MC) from the stria vascularis of the 1-day-old mouse, basal coil. The surface is coated and covered with microvilli. The cytoplasm is filled with extended rough endoplasmic reticulum (asterisks). In the apical part of the cell small vesicles occur (arrows). The two MC are close to each other in the apical part, thus indicating a tight junction. Further down, the cell membranes again come closer together (unfilled arrows), at a location where in the adult MC desmosomes ofter occur. Electron micrograph. ×9000 (Compare with Figures 8.11 and 8.14.)*

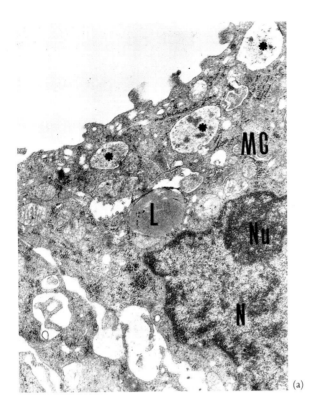

(a)

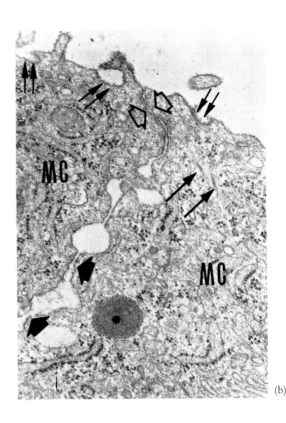

(b)

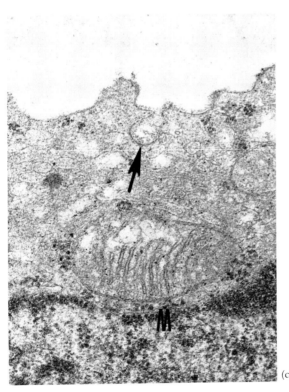

(c)

Figure 8.13(a) *Marginal cell (MC) from the stria vascularis of the 2-day-old mouse, basal coil. The cytoplasm is very electron dense and contains large vesicles filled with granular material (asterisks). A lipoid-like inclusion (L) is also identified. The network of cytoplasmic protrusions is sparse. A large nucleolus (Nu) is present in the cell nucleus (N). Electron micrograph. ×12 000. (b) Upper part of MCs from the 1-day-old mouse, basal coil. An inclusion body surrounded by a limiting membrane is present in the cell cytoplasm which otherwise contains distended endoplasmic reticulum (asterisks), microvesicles, and microtubules or microfilaments (arrows). The cell surface is coated and several invaginations occur, indicating the formation of exo- or endocytotic vesicles (double arrows). The area between the two MC adjacent to the endolymphatic surface is sealed by closed apposition of the two cell membranes (tight junctions) (unfilled arrows). Further down, the two MC separate but come closer together at intervals. This can possibly indicate the early formation of desmosomes or gap junctions (filled arrows). Electron micrograph. ×9200. (c) MC from the 6-day-old stria vascularis of the basal coil (mouse). The cell surface is coated. A pinocytotic/exocytotic vesicle is indicated with an arrow. The cytoplasm contains considerably smaller amounts of rough endoplasmic reticulum or free ribosomes compared with earlier developmental stages. M = Mitochondrion. Electron micrograph. ×28 000*

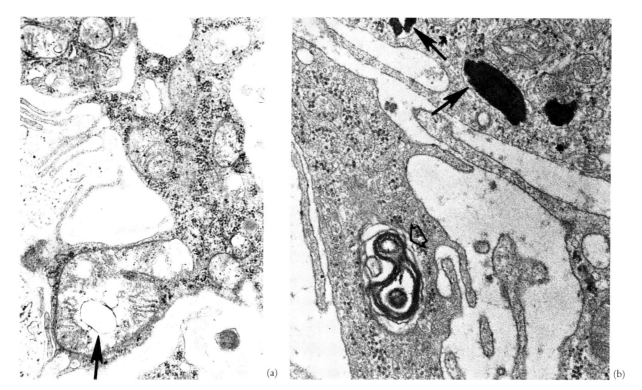

(a)

(b)

Figure 8.14 *Marginal cell (MC) extensions from the basal coil (mouse). Reconstruction of cells during maturation. (a), 3-day-old; (b–d), 1-day-old. Electron micrographs. (a) A degenerating large mitochondrion shows disintegration of its internal structure. A few vesicles have formed within the mitochondrion (arrow). ×28 000. (b) Myelin figure in an MC extension (unfilled arrow). A few electron-dense inclusions still have a surrounding membrane (arrows). ×18 800. (c) A limiting membrane may surround several small inclusions. This may indicate the formation of melanin pigments. ×16 000. (d) Some inclusions appear to display a myelin-like substructure. ×18 000*

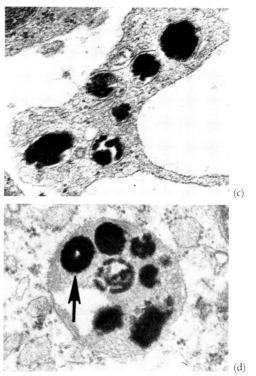

(c)

(d)

The analysis of the sequential development of the melanin pigments in marginal and especially intermediate cells is difficult. Hilding and Ginzberg (1977) claimed that, in the rat, melanocytes penetrate the basal membrane early during development and transfer melanin granules to intermediate cells.

A prerequisite for the maintenance of a high potassium concentration of endolymph is that the cells of the scala media can efficiently seal the scala media. During the maturation of the stria vascularis there is a gradual increase in the thickness of the tight junctional complexes both with regard to the number of sealing strands and the depth of the junction. In the mouse, when the marginal cells can first be identified (16th gestational day), the tight junctions consist of 1–3 strands of sealing elements with large inter- and intracellular variations. Gap junctions are few. A considerable maturation occurs during the first postnatal days before the onset of ionic maturation of the endolymph. The sealing strands become arranged in a parallel fashion. The number of sealing elements has increased to 5–7, thus reaching an intermediate to tight type of junction (according to the classification of Claude and Goodenough, 1973) in the adult animal (*Figures 8.15* and *8.16*) (Jahnke, 1975b; Anniko and Bagger-Sjöbäck, 1982). All mature strial cells are coupled with an extensive network of gap junctions. Tight junctions in the adult stria vascularis are also found in the basal cell layer sealing the stria vascularis from the spiral ligament and against the possible inflow of perilymph (Reale *et al.*, 1975).

Mature Stria Vascularis

The ultrastructure of the mature stria vascularis has been studied in a large number of species showing the same features of structural organization (Engström, Sjöstrand and Spoendlin, 1955; Smith, 1957; Hinojosa and Rodri-guez-Echandia, 1966; Spoendlin, 1967).

Recently, Santi and Muchow (1979) published mor-phometric data on the stria vascularis of the chinchilla. According to their measurements, the stria vascularis is approximately 1.4 per cent longer than the organ of Corti. The strial area increases gradually from the cochlear apex towards the base, reaching a maximum at approximately 80 per cent of its total length. Thereafter,

Figure 8.16 *Freeze-fractured specimens. Marginal cells from the mouse. Electron micrographs. (a) Sixteenth gestational day. The anatomical location of the future stria vascularis can be identified at this stage of embryonic development. The epithelium consists of one layer of columnar or cuboidal cells. The tight junction comprises an irregular network of a few sealing elements which do not completely close the tight junction towards the endolymphatic space (arrow). Original magnification × 18 000 reduced to 91%. (b) Nineteenth gestational day. The intermediate or basal cells have not yet started to form (see Anniko and Nordemar, 1980). The epithelium consists of one layer of columnar cells. The network of sealing elements in the tight junction is still irregular but now extends basalwards (arrows). The number of sealing elements has increased to 3–5 parallel strands. The junction has the morphological feature of a leaky to intermediate type according to the classification of Claude and Goodenough (1973). Gap junctions are indicated by unfilled arrows. Original magnification × 14 000 reduced to 91%. (c) Adult stria vascularis (3 months old). The tight junction has 8–10 parallel strands of sealing elements. The lower part of the tight junction consists of a rather irregular network. However, some intercellular variation occurs with regard to the number of sealing strands. In general there are only 5–7 sealing elements. This mature configuration of the tight junction in the stria vascularis is present already in the 6-day-old animal, i.e. immediately prior to the development of the high potassium concentration of endolymph (Anniko and Wroblewski, 1981). Original magnification × 14 000 reduced to 91%*

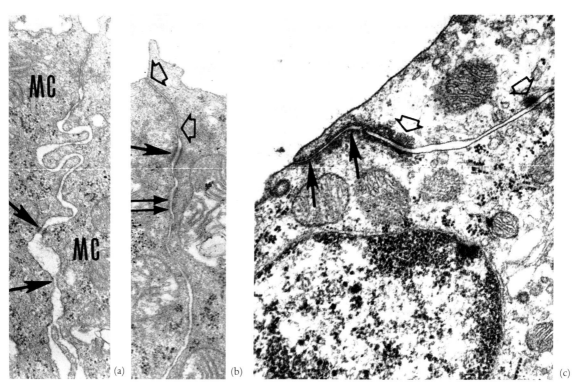

Figure 8.15 *Junctions between marginal cells (MC) from the basal coil, stria vascularis (mouse). (a) 1 day old; (b), 6 days old; (c), 3 months old; all are electron micrographs. (a) The apical parts of the two MC are close together (tight junctions). Below this region there is a large variation in the intercellular distance. However, at two sites (arrows) the cells are close together indicating the formation of junctional complexes. × 11 000. (b) Immediately below the tight junctional area (unfilled arrows) a desmosome (arrow) is disting-uished. This is followed by a region of even closer proximity to the cells and thereafter a new desmosome (double arrow). Further down, the two cell membranes are wider apart but parallel with each other. × 11 000. (c) The mature conditions are comparable with those observed in the 6-day-old animal. The tight junction is indicated with arrows; the desmosomes with unfilled arrows. Compare with (b). × 14 800*

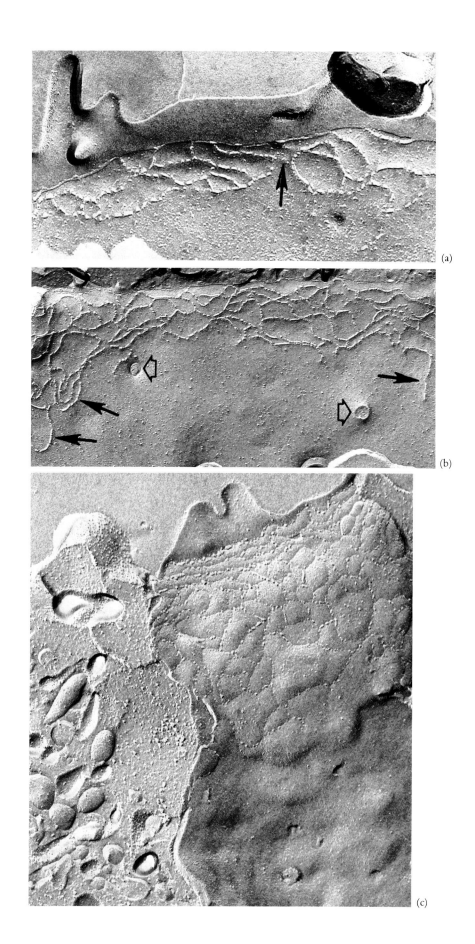

(a)

(b)

(c)

the area of the stria vascularis decreases towards the base, except for the last 5 per cent of its length, where it again increases. Strial width and thickness follow the same approximate function although the strial thickness is more variable. The estimated mean total volume of the stria vascularis is 0.15 mm³. These data can be compared with data on the organ of Corti (Bohne and Carr, 1979). The cross-sectional area of the organ of Corti averaged 0.02 mm² near the apex and gradually decreased to a value of 0.0005 mm² near the basal end. Thus, a small-volume organ of Corti lies adjacent to a large-volume stria vascularis and vice versa.

Little information is available on the morphology of the *spiral ligament* with its attachment to the stria vascularis and the otic capsule, respectively. An ultrastructural analysis of the human spiral ligament was recently performed by Morera, del Sasso and Iurato (1980). Three different types of fibrocytes occur and make numerous connections with one another. All three types of cells have a specific morphology which may reflect differences in mechanical function in different parts of the spiral ligament. There are no significant ultrastructural differences between the human ligament and the ligament of other mammalian species.

Scanning Electron Microscopy

The *surface* of the *marginal cells* and cells comprising the *spiral prominence* are covered with numerous microvilli: grooves of different depths and opening widths are often encountered showing morphological evidence of a high endo- or exocytotic activity of these cells. Most marginal cells have a hexagonal or almost hexagonal surface structure. Cells of the spiral prominence are elongated and spindle shaped. In between, there is a gradual transition between these two cell types (*Figures 8.17–8.20*) (Lim, 1969; Anniko, 1976).

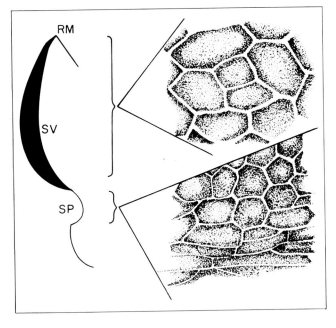

Transmission Electron Microscopy

The *marginal cells* have an electron-dense cytoplasm after staining. Depending on species and the location of the stria vascularis in the cochlea, the marginal cells can constitute the major part of the whole stria vascularis (*Figures 8.21* and *8.22*). The nucleus is spherical or slightly ellipsoidal. The cytoplasm is homogeneous with a large number of free ribosomes, a Golgi apparatus with clusters of associated small vesicles, and coated vesicles (*Figures 8.23* and *8.24*). The rough endoplasmic reticulum is sparse and located mainly in the apical third of the cell. Mitochondria occur particularly in the long infoldings toward the intermediate and basal cells. The cytoplasmic projections can constitute up to two-thirds of the total length of the marginal cell (*see Figures 8.16–8.25*). The network of infoldings between the marginal and intermediate cells is often so complex that it is difficult to decide the origin of these cytoplasmic processes; the differences in staining properties are therefore of great help. As compared with intermediate-cell extensions, mitochondria are found predominantly in marginal-cell extensions (*Figure 8.26*).

The body of the *intermediate cells* is smaller than that of marginal cells. Some radiating cytoplasmic processes are directed upward between marginal cells but never reach the endolymphatic surface. Most cytoplasmic extensions intermingle with those coming from marginal cells. A few are in contact with basal cells (*Figure 8.27*). Cytoplasmic processes from the intermediate cells extend also towards the blood vessels and may contact the endothelial cells from the direction of the basal cell layer, thus forming 'vascular feet' (*see Figure 8.1*).

The *basal cells* are flattened and spindle shaped with processes that penetrate between neighbouring basal cells. Basal cells have relatively few organelles or inclusions (*see Figures 8.21, 8.25, 8.28* and *8.29*). The staining properties of the cytoplasm are similar to those of the intermediate cells.

The strial *blood vessels* (*see Figure 8.22*) have a continuous endothelial lining with an underlying basal lamina. The endothelial cells have a flattened nucleus and few cell organelles. Occasionally microvilli and vesicular invaginations occur at the luminal side. The basal membrane sends out a number of prolongations which extend into the extensions of all three types of strial cells. Pericytes occasionally surround the endothelial cells and are enclosed by the basal lamina of the endothelial cells. The ultrastructure of the pericytes is similar to that of endothelial cells but there are fewer vesicles in them.

(*continued on page 206*)

◀ **Figure 8.17** *Schematic drawing of the surface structure of the mature stria vascularis (SV) and the spiral prominence (SP). The cells of the SV are hexagonal like while those of the SP are elongated and spindle shaped. A gradual transition occurs between the two cell types*

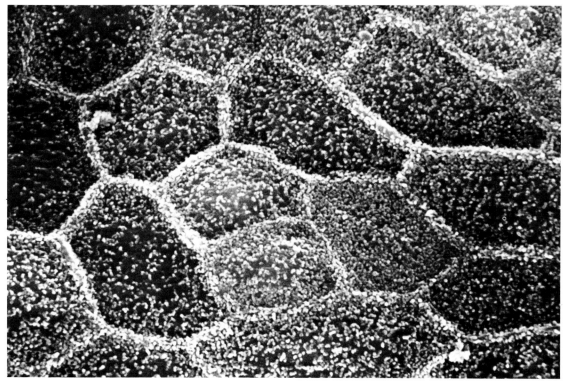

Figure 8.18 *Cells of the mature stria vascularis with a hexagonal or hexagonal-like surface. Guinea-pig. All cell surfaces are covered with a thick layer of microvilli. Scanning electron micrograph. Original magnification ×4200 reduced to 92%*

Figure 8.19 *The transition zone between the cells of the stria vascularis (SV) and the spiral prominence (SP). Mature guinea-pig. The hexagonal shape of cells gradually changes on the SP and becomes more irregular. Several cells have a structure of elongated hexagons. The cells of the SV have more microvilli than those of the SP. Scanning electron micrograph. Original magnification ×4100 reduced to 92%*

Figure 8.20 *Cells of the lower part of the spiral prominence. Adult guinea-pig. The cells are spindle shaped and have fewer microvilli than the cells of the stria vascularis. Compare with Figure 4.18. Original magnification ×5500 reduced to 92%*

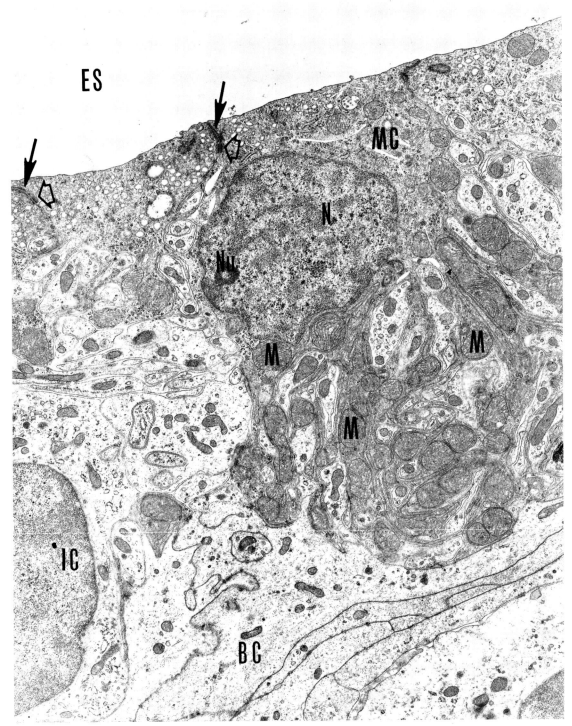

Figure 8.21 *Mature stria vascularis of the guinea-pig. Basal coil. MC = Marginal cell; IC= intermediate cell; BC= basal cell; ES = endolymphatic space. The MC has a rounded nucleus (N) with one prominent nucleolus (Nu). The cytoplasm is very electron dense after staining and contains a large number of coated vesicles in the supranuclear region. The amount of rough endoplasmic reticulum is small. The cytoplasmic extensions reach the BC and are filled with a large number of mitochondria (M). Apically, the MC are connected to each other with tight junctions (arrows) and desmosomes (unfilled arrows). The IC have a less electron-dense cytoplasm after staining. The nucleus occupies a large part of the cell. Cytoplasmic organelles are few as compared with MC. The cytoplasmic projections of IC meet those of MC. The staining properties of BC are similar to the IC. Cytoplasmic organelles are few. There is no basal lamina below the BC. The BC appear flattened. The penetration of cytoplasmic processes laterally (i.e. in parallel with the surface of the endolymphatic space) between neighbouring BC gives the appearance of several layers of cells. Electron micrograph.* ×8900

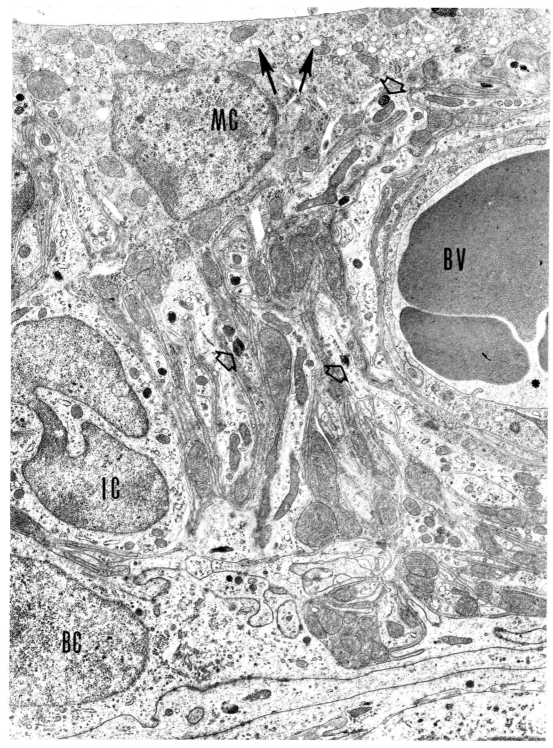

Figure 8.22 *Mature stria vascularis of the guinea-pig. Basal coil. MC = Marginal cell; IC = intermediate cell; BC = basal cell; BV = blood vessel. The electron-dense cytoplasm of MC contains coated vesicles in the supranuclear region but few free ribosomes and only a small amount of rough endoplasmic reticulum. The MC extensions contain large mitochondria and surround the BV and reach the BC.*

Most electron-dense pigments (melanin pigments) are located in the IC (unfilled arrows). The IC nucleus is lobulated in contrast to the BC nucleus which is slightly elongated. The lumen of the BV is filled with red blood corpuscles and electron optically granulated material (asterisk) (plasma?). Transmission electron micrograph. ×11 000

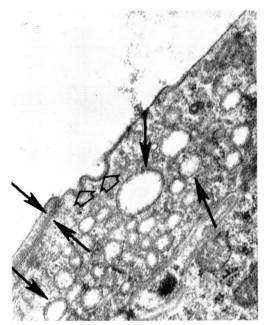

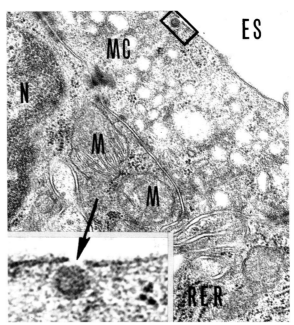

Figure 8.23 *Mature stria vascularis of the guinea-pig. 2nd coil. The apical part of the marginal cell cytoplasm is filled with coated vesicles of varying sizes (arrows). The cell membrane is also coated. Two invaginations (unfilled arrows) of the surface membrane indicate exo- or endocytosis. The tight junction between the adjacent cells is indicated by the very close apposition of the cell membranes (double arrows). Electron micrograph. Original magnification ×26 000 reduced to 91%*

Figure 8.24 *Similar conditions as in Figure 8.23. In the apical part of the marginal cell (MC) an electron-dense particle can be observed close to the endolymphatic space (ES) (enclosed; see inset). M = Mitochondria; N = nucleus; RER = rough endoplasmic reticulum. Electron micrograph. ×27 000. Inset: The coated membrane is interrupted (arrow). Fixation artefact or release of substance? Original magnification ×76 000 reduced to 91%*

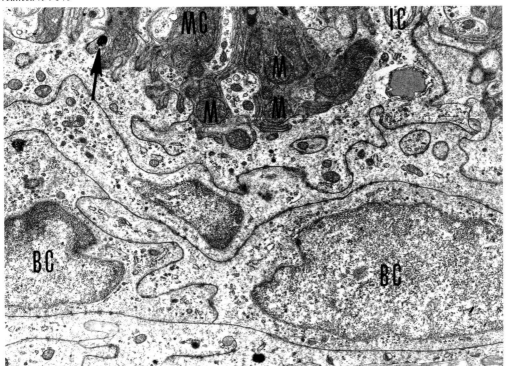

Figure 8.25 *Stria vascularis of the adult guinea-pig. Intermediate coil. The electron-dense cytoplasmic projections of marginal cells (MC) are filled with mitochondria (M) and extend close to the basal cells (BC). The MC and intermediate cell (IC) extensions interdigitate. In this micrograph electron-dense pigments are found only in IC extensions (arrow). The number of cell organelles in BC is low. The large nuclei of the BC have their long axes parallel to the surface of the endolymphatic space. Electron micrograph. Original magnification ×19 000 reduced to 91%*

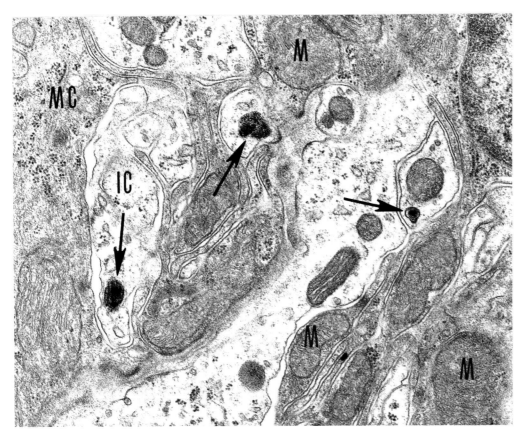

Figure 8.26 *Stria vascularis of the adult guinea-pig. Cytoplasmic projections of marginal cells (MC) and intermediate cells (IC) intermingle. The origin of the two types of extensions is distinguished by the different staining properties. The MC projections are very electron dense and contain a large number of mitochondria (M). The cristae mitochondriales are well distinguished in many of them. Cell organelles are fewer in IC extensions which, however, contain several pigment inclusions (melanin) (arrows). The cell membranes from MC and IC are in close contact with each other. Electron micrograph. ×24 000*

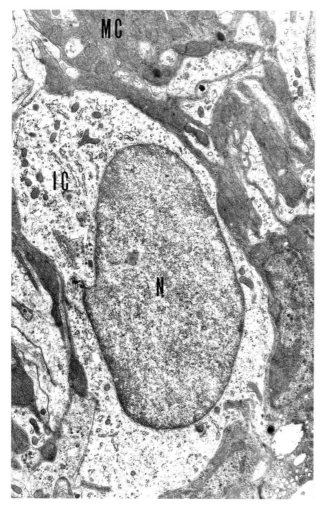

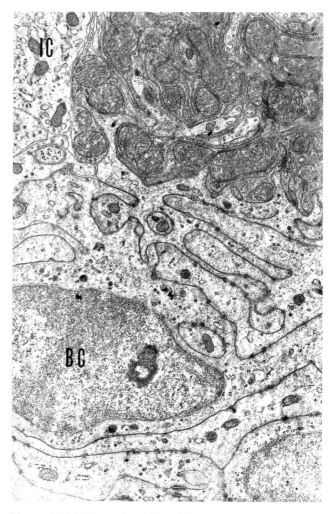

Figure 8.27 *Stria vascularis of the adult guinea-pig. Basal coil. The intermediate cell (IC) is filled with a large nucleus (N). The cytoplasm stains less densely than that in the marginal cell (MC) and contains rather fewer cell organelles. The interdigitations of MC and IC are in close contact. Electron micrograph. Original magnification ×9800 reduced to 86%*

Figure 8.28 *Stria vascularis of the adult guinea-pig. Intermediate coil. The electron-dense marginal cell extensions (filled with numerous mitochondria) penetrate so deeply into the stria vascularis that they may be in contact both with intermediate and with basal cells (BC). IC = Intermediate cell. Electron micrograph*

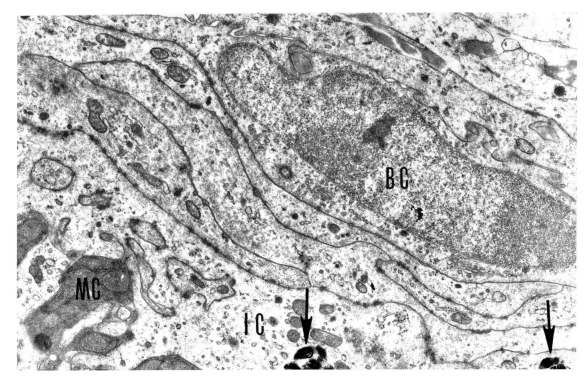

Figure 8.29 *Stria vascularis of the basal coil. Adult guinea-pig. The basal cells (BC) have a nucleus that fills the major part of their volume. The lateral projections of BC cytoplasm overlap, thus giving the impression that the BC layer consists of several layers of* *extensions although it comprises only one layer of cells. IC = Intermediate cell; MC = marginal cell. Melanin pigments are indicated by arrows. Electron micrograph. ×12 000*

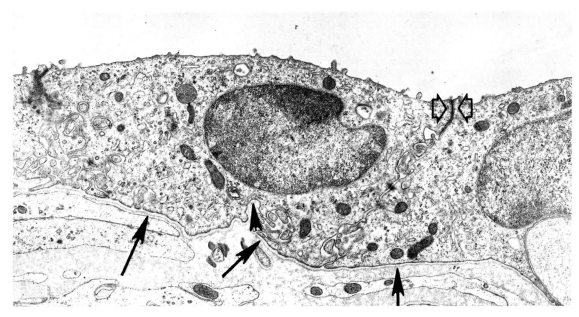

Figure 8.30 *Epithelial cells lining the spiral prominence. Adult guinea-pig. Basal coil. These cells are flattened and have a basal lamina towards underlying structures (arrows). The cells have a large* *nucleus but few cell organelles in the cytoplasm. The cells are close to the endolymphatic space and tightly connected by tight junctions (unfilled arrow). Electron micrograph. ×4800*

The *cells of the spiral prominence* are cuboidal or slightly flattened on the endolymphatic surface and are covered with a thick layer of microvilli (*see Figures 8.20 and 8.30–8.32*). The cells have an electron-dense cytoplasm comparable with that of marginal cells. Beneath the cells of the spiral prominence there is a distinct basal lamina.

The function of the spiral prominence and its adjacent structures is only partially understood. Metabolic studies have shown a high activity comparable with that of the stria vascularis. After perilymphatic horse-radish peroxidase perfusion, tracer-marked vesicles were observed in all cells of the spiral prominence and the root cells but not in the cells of the stria vascularis initially (Mees, 1981). The author implies that the spiral prominence with its blood vessels represents one of the pathways for the absorption of perilymph. However, further studies are warranted before this interpretation can be accepted.

Acknowledgements

This work was supported by grants from the Swedish Medical Research Council, Project 12X-720. A part of this study was performed at the Department of Otolaryngology, Kresge Hearing Research Institute, The University of Michigan, Ann Arbor, MI, USA, and supported by the Foundation Tysta Skolan (Matti Anniko).

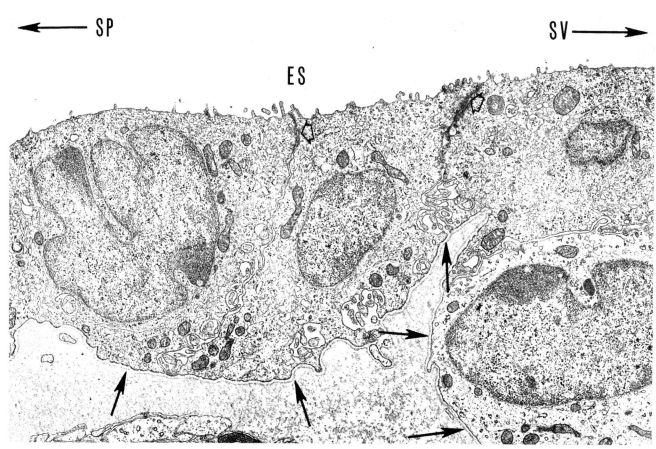

Figure 8.31 *The transition between the spiral prominence (SP) and the stria vascularis (SV). Basal coil. Guinea-pig. The cuboidal cells have an electron-dense cytoplasm and large nuclei. The cell surface is covered with a few short microvilli. Tight junctions (unfilled arrows)* *bring the apical parts of the cells close together at the endolymphatic space (ES). A basal lamina is found beneath the cells (arrows). Electron micrograph. Original magnification ×7200 reduced to 82%*

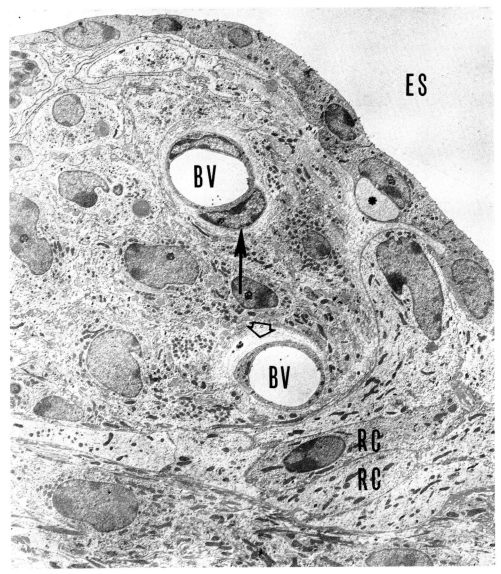

Figure 8.32 *Spiral prominence of the adult guinea-pig. Basal coil. The epithelial cell lining towards the endolymphatic space (ES) consists of cells with an electron-dense cytoplasm similar to that of marginal cells. One epithelial cell contains a large vesicle (asterisk).* *The blood vessels (BV) in this region are sometimes surrounded by pericytes (arrow) and can reveal an 'empty' perivascular space (unfilled arrow). RC = Root cells. ×1800*

References

Alford, B. R. and Ruben, R. J. (1963) Physiological, behavioral and anatomical correlates on the development of hearing in the mouse. *Annals of Otology, Rhinology and Laryngology*, **72**, 327–347.

Anggård, L. (1965) An electrophysiological study of the development of cochlear functions in the rabbit. *Acta Otolaryngologica (Stockholm)*, Supplement **203**, 1–64.

Anniko, M. (1976) The surface structure of the stria vascularis in the guinea pig cochlea. Normal morphology and atoxyl-induced pathological changes. *Acta Otolaryngologica (Stockholm)*, 82, 325–336.

Anniko, M. and Nordemar, H. (1980) Embryogenesis of the inner ear. IV. Post natal maturation of the secretory epithelia of the inner ear in correlation with the elemental composition in the endolymphatic space. *Archives of Otolaryngology*, **229**, 281–288.

Anniko, M. and Wroblewski,R. (1981) Elemental composition of the developing inner ear. *Annals of Otology, Rhinology and Laryngology*, **90**, 25–32.

Anniko, M. and Bagger-Sjöbäck, D. (1982) Maturation of junctional complexes during embryonic and early postnatal development of inner ear secretory epithelia. *American Journal of Otolaryngology*, **3**, 242–253.

Axelsson, A. (1974) The blood supply of the inner ear of mammals. In *Handbook of Sensory Physiology* vol. V/1, eds. W. D. Keidel and W. D. Neff, pp. 213–260. Berlin: Springer-Verlag.

Bohne, B. and Carr, C. (1979) Location of structurally similar areas in chinchilla cochleas of different lengths. *Journal of the Acoustical Society of America*, **66**, 411–414.

Bosher, S. K. and Warren, R. L. (1971) A study of the electrochemistry and osmotic relationships of the cochlear fluids in the neonatal rat at the time of the development of the endocochlear potential. *Journal of Physiology (London)*, **212**, 739–761.

Claude, P. and Goodenough, D. A. (1973) Fracture faces of zonulae occludentes from 'tight' and 'leaky' epithelia. *Journal of Cell Biology*, **58**, 390–400.

Davis, H., Deatherage, B. H., Rosenblut,B., Fernández, C., Kimura, R. and Smith, C. A. (1958) Modification of cochlear potentials produced by streptomycin poisoning and by extensive venous obstruction. *Laryngoscope*, **68**, 596–627.

Engström, H., Sjöstrand, F. S. and Spoendlin, H. (1955) Feinstruktur der Stria vascularis beim Meerschweinchen. *Practica Oto-rhino-laryngologica*, **17**, 69–79.

Guild, S. (1927) Circulation of the endolymph. *American Journal of Anatomy*, **39**, 57–81.

Hilding, D. H. and Ginzberg, R. D. (1977) Pigmentation of the stria vascularis. The contribution of neural crest melanocytes. *Acta Otolaryngologica (Stockholm)*, **84**, 27–37.

Hinojosa, R. and Rodriguez-Echandia, E. L. (1966) The fine structure of the stria vascularis of the cat inner ear. *American Journal of Anatomy*, **118**, 631–664.

Jahnke, K. (1975a) The fine structure of freeze-fractured intercellular junctions in the guinea pig inner ear. *Acta Otolaryngologica (Stockholm)*, Supplement **336**, 1–40.

Jahnke, K. (1975b) Die Feinstruktur gefriergeätzter Zellmembranhaftstellen der Stria vascularis. *Anatomie und Embryologie*, **147**, 189–201.

Kikuchi, K. and Hilding, D. (1966) The development of the stria vascularis in the mouse. *Acta Otolaryngologica (Stockholm)*, **62**, 277–291.

Lim, D. J. (1969) Three dimensional observation of the inner ear with the scanning electron microscope. *Acta Otolaryngologica (Stockholm)*, Supplement **255**, 1–38.

Mees, K. (1981) On the function of the spiral prominence. A HRP study in the guinea pig. *Congress Abstract. XVIIIth Workshop on Inner Ear Biology*, Montpellier, France, September 14–16.

Misrahy, G. A., Hildreth, K. M., Shinabarger, E. W., Clark, L. C. and Rice, E. A. (1958) Endolymphatic oxygen tension in the cochlea of the guinea pig. *Journal of the Acoustical Society of America*, **30**, 247–250.

Morera, C., del Sasso, A. and Iurato, S. (1980) Submicroscopic structure of the spiral ligament in man. *Revue de Laryngologie*, **101**, 73–85.

Pujol, R. and Hilding, D. (1973) Anatomy and physiology of the onset of auditory function. *Acta Otolaryngologica (Stockholm)*, **76**, 1–10.

Reale, E., Luciano, L., Franke, K., Pannese, E., Wermbter, G. and Iurato, S. (1975) Intercellular junctions in the vascular stria and spiral ligament. *Journal of Ultrastructure Research*, **53**, 284–297.

Ruben, R. J. (1967) Development of the inner ear of the mouse: A radioautographic study of terminal mitoses. *Acta Otolaryngologica (Stockholm)*, Supplement **220**, 1–44.

Santi, P. A. and Muchow, S. C. (1979) Morphometry of the chinchilla organ of Corti and stria vascularis. *Journal of Histochemistry and Cytochemistry*, **27**, 1539–1542.

Schmidt, R. S. and Fernández,C. (1963) Development of mammalian endocochlear potential. *Journal of Experimental Zoology*, **153**, 227–236.

Sher, A. E. (1971) The embryonic and postnatal development of the inner ear of the mouse. *Acta Otolaryngologica (Stockholm)*, Supplement **285**, 1–77.

Smith, C. A. (1957) Structure of the stria vascularis and the spiral prominence. *Annals of Otology, Rhinology and Laryngology*, **66**, 521–536.

Spoendlin, H. (1967) Stria vascularis. In *Submicroscopic Structure of the Inner Ear*, ed. S. Iurato, pp. 131–149. New York: Pergamon Press.

Tasaki, I. and Spyropoulos, C. S. (1959) Stria vascularis as source of endocochlear potential. *Journal of Neurophysiology*, **22**, 149–155.

Vosteen, K. H. (1960) The histochemistry of the enzymes of oxygen metabolism in the inner ear. *Laryngoscope*, **70**, 351–362.

Weibel, E. R. (1957) Zur Kenntnis der Differenziersvorgänge im Epithel des Ductus cochlearis. *Acta Anatomica (Basel)*, **29**, 53–90.

The Vestibule

9
Vestibule: Sensory Epithelia

Ivan M. Hunter-Duvar   Raul Hinojosa

The sensory epithelia of the mammalian vestibular system (*Figures 9.1* and *9.2*) consist of the epithelia of the otolithic organs, the saccule and utricle, and the epithelia of the cristae contained in the ampullae of the three semicircular canals. The sensory epithelia of macula sacculi, macula utriculi and the three cristae contain two types of sensory cells with nerve and blood supply and supporting cells. Since the two types of sensory cells are common to saccule, utricle and cristae their description will precede that of the various other structures.

Type I Sensory Cell

The type I sensory or hair cell is best described as flask shaped (*Figures 9.3* and *9.4*). The cell is surrounded to slightly below the upper surface by a chalice or calyx formed from the terminal end of an afferent nerve fibre (*Figures 9.1* and *9.5*). One such ending may surround one or more adjacent type I cells, or one or more type I cells that are not adjacent (*Figures 9.5* and *9.7*).

The upper surface of the cell contains from 20 to 100 stereocilia and one kinocilium (*Figures 9.11* and *9.12*). The stereocilia are continuous with the surface membrane of the cell and may be considered as enlarged, specialized microvilli. They are arranged in step-like fashion according to length, the longest being positioned adjacent to the kinocilium (*Figures 9.10* and *9.11*). The kinocilium is always considerably longer than the longest of the stereocilia and has its upper portion attached to the mesh-like structure that lies between the
· sensory epithelium and the statoconial layer (*Figures 9.10* and *9.15*) described in the following chapter. The stereocilia contain actin filaments; they are narrower at the base and end in rootlets which extend into and sometimes through the cuticular plate (*Figures 9.8* and *9.25*). Blebs containing vesicle-like structures are regularly found on the stereocilia and often give the appearance of making contact with adjacent stereocilia (*Figure 9.13*). The kinocilium, of slightly larger diameter than the stereocilia, is generally positioned on one side of them. It usually has the standard nine pair-plus-two fibrillar structure; however, both this structure and the position described above are subject to exception (*Figures 9.12* and *9.14*).

The cuticular plate that underlies the stereocilia is composed of a dark-staining granular substance. A centriole is normally found in the vicinity of the basal body at the base of the kinocilium. The infracuticular region of the cell is often rich in mitochondria and numerous multivesicular bodies are generally present in this area (*Figures 9.1* and *9.5*).

The large circular nucleus is located in the centre of the bulbous area of the cell (*Figures 9.1* and *9.5*). The cytoplasm surrounding the nucleus generally contains elongated mitochondria, endoplasmic reticulum and Golgi complexes, all in varying numbers. The cytoplasm throughout the cell is rich in ribosomes and contains a large number of vesicles (*Figures 9.1* and *9.5*).

The infranuclear region near the base of the cell usually contains a number of elongated mitochondria. Synaptic bodies of various shapes are found near the cell membrane in this area, usually in the vicinity of an invagination in the cell membrane (*Figure 9.16*). The nerve chalice that surrounds all type I cells is invariably rich in elongated mitochondria throughout. Efferent nerve endings regularly synapse with the afferent chalice (*Figures 9.6* and *9.7*).

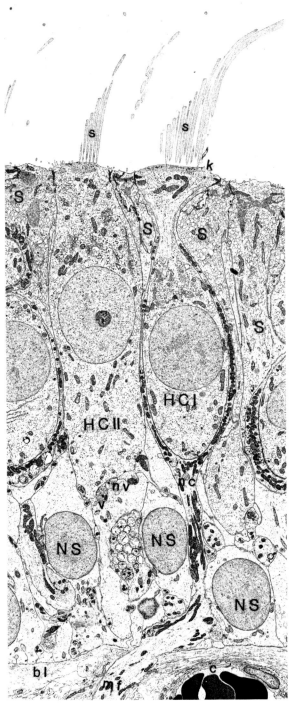

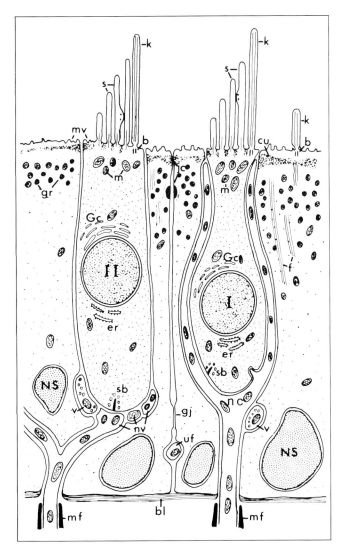

Figure 9.1 *A portion of the macula of the utricle of the cat showing type I (HCI) and type II (HCII) sensory cells separated by supporting cells (S). Type I cells are partially surrounded by a nerve chalice (nc) which is the unmyelinated ending of a large myelinated nerve fibre (mf) seen here crossing the basal lamina (bl). Numerous vesiculated (v) and non-vesiculated (nv) nerve endings make synaptic contact at the infranuclear portion of type II cells. The nuclei of supporting cells (NS) are located at the base of the epithelium. Those of type II cells are at the upper half and those of type I cells are at the centre of the epithelium. Each sensory cell contains a kinocilium (k) and numerous stereocilia (s) protruding from their free surface. c = Capillary containing red blood cells. ×2894*

Figure 9.2 *Schematic drawing illustrating the general structure of the vestibular sensory epithelium. The type I cell (I) is flask shaped and is almost completely surrounded by a nerve chalice (nc). The type II cell (II) is rod shaped and is innervated by small nerve endings of two types, vesiculated (v) and non-vesiculated (nv) boutons. Sensory hairs (k = kinocilium; s = stereocilia) protrude from the free surface of every sensory cell. A kinocilium (k) is occasionally seen protruding from supporting cells. Microvilli (mv) are found on the free surface of sensory and supporting cells. b = Basal body; j = junctional complex; cu = cuticular plate; m = mitochondria; Gc = Golgi complex; er = endoplasmic reticulum; sb = synaptic bar; f = filaments; gr = cytoplasmic vesicles in supporting cells; gj = gap junction between supporting cells; uf = unmyelinated fibres; mf = myelinated fibres; NS = nucleus of supporting cell; bl = basal lamina*

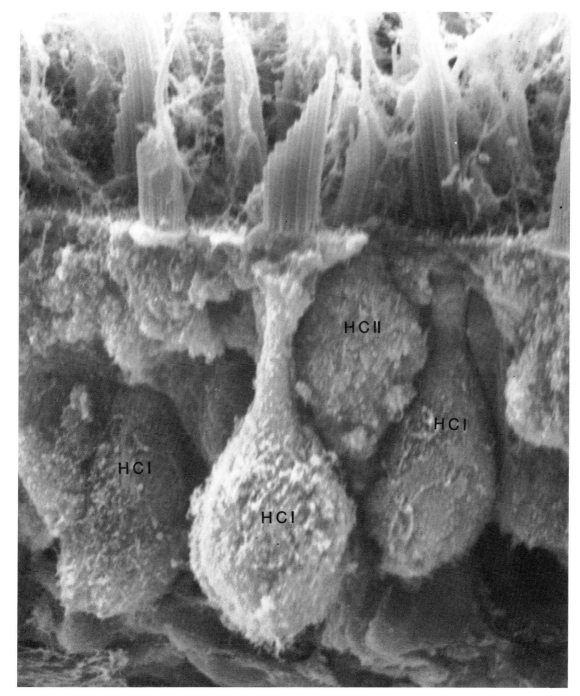

Figure 9.3 *A fracture through the macula sacculi of a chinchilla demonstrates the shape and relationship of type I (HCI) and type II (HCII) sensory cells*

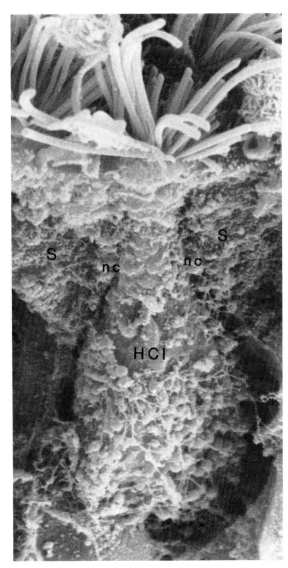

Figure 9.4 *An isolated type I (HCI) cell with the outer membrane of the nerve calyx (nc) stripped away demonstrates the flask shape of the cell and shows the mitochondria and other organelles in the calyx. S = Supporting cell*

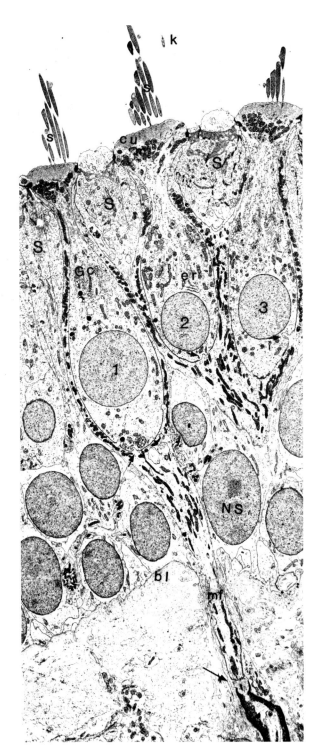

Figure 9.5 *Three type I sensory cells (1, 2, 3) from the crista of a semicircular canal of the cat. Each cell contains a round nucleus located at the widest portion of the cell. Most of the cytoplasmic organelles such as the Golgi complex (Gc), the endoplasmic reticulum (er) and dense and multivesicular bodies are located in the supranuclear area. A group of mitochondria (m) is located immediately below the cuticular plate (cu). The infranuclear area shows numerous invaginations or synaptic zones (arrowhead). Cell 1 is surrounded by a nerve chalice which is the ending of a large myelinated fibre (mf). The myelin sheath ends in a half-node configuration (arrow) a short distance from the basal lamina (bl). Cells 2 and 3 are surrounded by a common chalice. s = Stereocilia; k = kinocilium; S = supporting cells; NS = nucleus of supporting cell. ×2589*

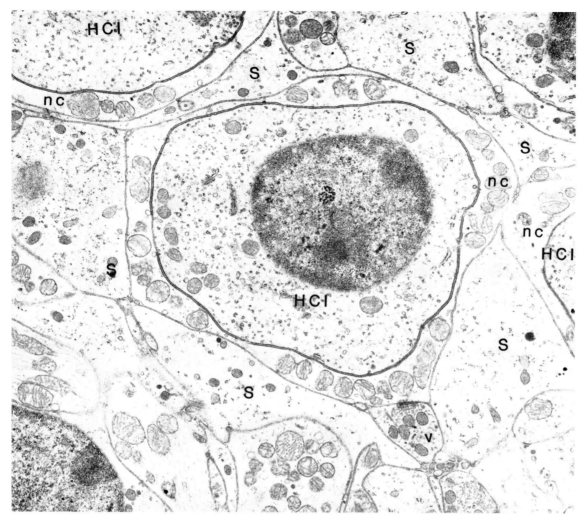

Figure 9.6 *A horizontal section at the nucleus level of a type I cell (HCI) like that shown in Figure 9.4 demonstrates the relationship of the nerve calyx to the cell and to the nerve calyx of an adjoining type I cell. A multivesiculated efferent nerve ending (v) is seen to synapse with the calyx. Supporting cells (S) are seen to interdigitate to completely surround the sensory cells. Macula utriculi of rat. ×12 300*

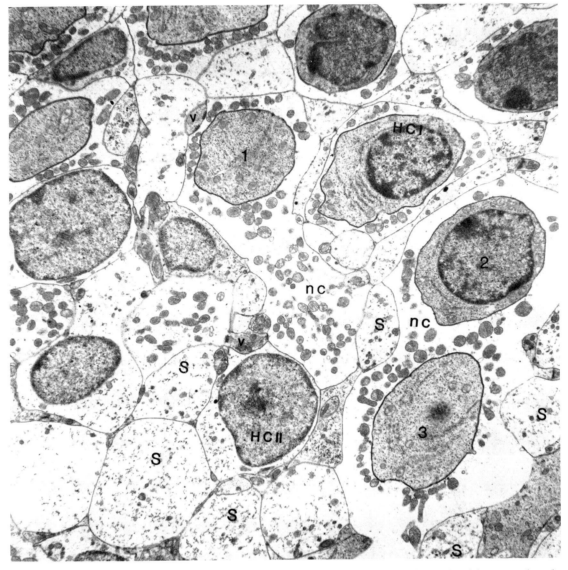

Figure 9.7 *Horizontal section at the level of the nucleus demonstrating a relationship of three type I cells (1, 2, 3) similar to that shown in Figure 9.5. Two of the type I sensory cells are seen to be enclosed by one nerve calyx (nc) and the nerve calyx of an adjacent type I cell stretches over to join it. Several vesiculated nerve endings (v) can be seen to synapse with the calyces. Rat macula utriculi.* ×5500

Type II Sensory Cell

Type II sensory or hair cells, which are considered to be genetically older than type I cells, are of irregular but generally cylindrical shape and vary considerably in length (*Figures 9.8* and *9.9*). The arrangement of the cilia is identical with that found in the type I cell described above. The location of the circular nucleus is seen to vary through the middle one-third of the cell (*Figures 9.1* and 9.8). The cytoplasm of the type II cell contains the same organelles as that of the type I cells; however, the organelles appear more evenly distributed throughout the cytoplasm.

Both afferent and efferent nerve endings synapse with type II cells (*Figure 9.8*). Nerve contact generally occurs in the lower one-third of the cell but occasional nerve endings with synapses are seen at higher levels.

Saccule

The saccule is located in the spherical recess of the vestibule. In man it is in a vertical position when the head is upright. The hook-shaped sensory epithelium of the saccule (macula sacculi) is covered by an intermediate mesh which supports the statoconial layer (*Figure 9.15*). The sensory epithelium of the saccule is composed of type I cells, type II cells, and supporting cells which generally separate and surround the sensory cells (*Figures 9.7* and *9.25*). Sensory cells extend from the

216

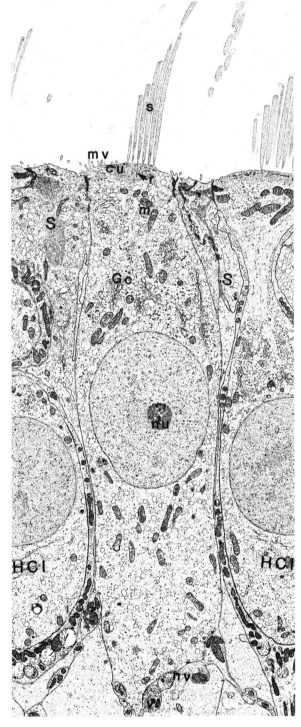

Figure 9.8 *Type II sensory cell from the utricle of the cat. The rod-shaped cell contains a finely granular cuticular plate (cu) that partially covers the upper end of the cell. The rootlets (r) of the stereocilia (s) are embedded in the cuticular plate. The nucleus of the cell is centrally placed and contains a large nucleolus (nu). The Golgi complex (Gc), the endoplasmic reticulum and a few mitochondria (m) are located at the supranuclear zone. Vesiculated (v) and non-vesiculated (nv) nerve endings make synaptic contact at the infranuclear zone of the cell. HCI = Type I sensory cell; S = supporting cell; mv = microvilli. ×4510*

Figure 9.9 *A fracture through the macula sacculi of the chinchilla demonstrates the shape and structure of a type II sensory cell (HCII) and its relationship to a supporting cell (S). A long kinocilium (k) can be seen on the sensory cell and a short rudimentary kinocilium is seen on the supporting cell. mv = Microvilli*

upper surface to varying depths but never reach the basal lamina (basement membrane). Individual supporting cells rest on the basal lamina and extend from it to the upper surface of the epithelium (*Figures 9.19* and *9.23*). Beneath the basal lamina lies a layer of nerve fibres and blood vessels surrounded by connective tissue (*Figure 9.19*). It is not unusual to find blood vessels running above the basal lamina in some species (*Figure 9.19*). Nerve fibres lose their myelin sheaths before penetrating the basal lamina (*Figures 9.1* and *9.19*) and both myelinated and unmyelinated fibres are seen beneath it (*Figure 9.19*). The cilia of the sensory cells change their orientation or polarization along a narrow area or line known as the *striola* (*Figures 9.21* and *9.22*). In the macula sacculi the tallest stereocilia and the

Figure 9.10 *The step-like arrangement of the stereocilia (s) and the difference in overall height of stereocilia on different cells is clearly shown in this scanning micrograph of the macula utriculi of a* *chinchilla. The much longer kinocilia (k) bear the remnants of their attachment to the intermediate mesh separating sensory epithelia from the statoconial layer*

kinocilium face away from the striola towards the periphery (*Figure 9.22*). Type I cells are more highly concentrated in the striolar area than in the periphery and the cilia are shorter on cells in the striola. This and the fact that the sensory cells are less dense, with typically larger surface areas in the striola, cause it to stand out as a distinct area from the remainder of the sensory epithelium.

Utricle

The utricle is located in the elliptical recess in the medial wall of the vestibule. In man it lies superior to the saccule and is in the horizontal plane with the head in an upright position. The sensory epithelium of the utricle (macula utriculi) is generally shell shaped or kidney shaped (*Figures 9.18* and *9.33*). The general structure of the macula utriculi is the same both above and below the basement membrane as that described for the macula sacculi. Orientation or polarization of the cilia is essentially opposite to that of the saccule. In the utricle the longest stereocilia and the kinocilium face each other along the striola (*Figures 9.19* and *9.20*).

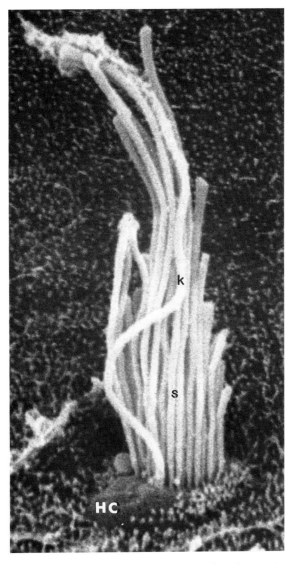

Figure 9.11 *An isolated sensory cell (HC) from the macula utriculi of the hamster demonstrates the configuration and relationship of kinocilium (k) to stereocilia (s)*

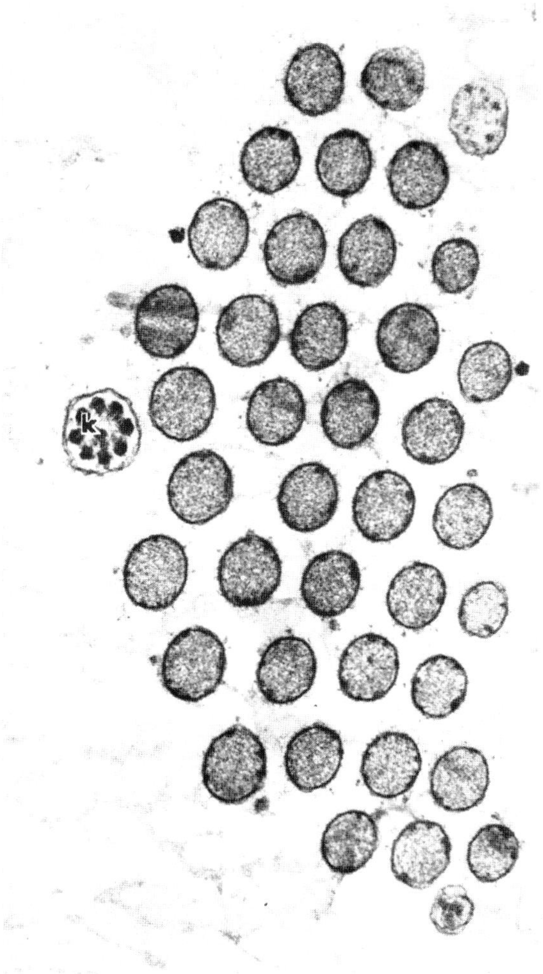

Figure 9.12 *A horizontal section through the cilia of a single sensory cell further demonstrates the kinocilium/stereocilia relationship shown in Figure 9.11. The longer stereocilia have slightly greater diameters than the shorter ones and the kinocilium has a greater diameter than any of the stereocilia. Rat. ×41 000*

Supporting Cells

The supporting cells in the sensory epithelia run from the epithelial surface down to the basal lamina (*Figure 9.23*). They completely surround and generally separate the sensory cells and it is not unusual for five or more supporting cells to be in contact with a single sensory cell (*Figures 9.6* and *9.7*). At the upper surface, the supporting cells cleanly abut the sensory cell and each other (*Figure 9.25*); however, within the macula, extensive interdigitation of the supporting cell bodies may occur.

The epithelial surface of the supporting cell has sparse, scattered microvilli and in some species a rudimentary kinocilium is present on the surface (*Figure 9.9*). At the epithelial surface, the supporting cells are joined to each other and to the sensory cells with tight junctions, which usually back on an area of dark staining, granular substance similar to that found in the cuticular plate of the sensory cells (*Figures 9.23* and *9.24*).

The upper half to two-thirds of the supporting cell is filled with large, densely packed vesicles among which are scattered mitochondria, Golgi complexes, lysosomes, and free ribosomes (*Figures 9.23* and *9.26*). The nucleus is located in the lower part of the cell near the basal lamina and generally below the nuclei of the sensory cells (*Figure 9.19*).

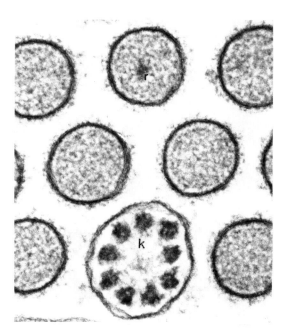

Figure 9.14 *Lower portion of several stereocilia and the kinocilium (k) sectioned horizontally, at high magnification. The membrane of the stereocilia is a continuation of the surface membrane of the sensory cell. The interior structure is composed of actin filaments. A rootlet (r) can be seen in one of the stereocilia. Original magnification ×120 000 reduced to 82%*

Crista Ampullaris

A crista with sensory epithelia is located in the ampulla of each of the lateral (horizontal), anterior and posterior canals. The sensory epithelium of the crista is not flat but curved in two directions and is best described as saddle shaped (*Figure 9.27*). Like that of the saccule and utricle, the sensory epithelium is composed of type I, type II and supporting cells resting on a basal lamina (*Figures 9.28* and *9.29*). Beneath the basal lamina lies the nerve and blood supply surrounded by connective tissue (*Figures 9.29* and *9.31*).

Considerable differences occur in the density of sensory cells over the surface of the crista. The density is greatest at the periphery and decreases towards the central area (*Figure 9.30*). This occurs in all three cristae and is accompanied by a corresponding change in the length of the cilia on the sensory cells. Stereocilia and kinocilia are considerably longer on the peripheral sensory cells than on the central sensory cells (*Figures 9.30* and *9.32*).

In each crista, the sensory cells are oriented in the same direction over the entire surface (*Figures 9.28* and *9.30*). In the lateral crista, polarization is towards the utricle; in the anterior and posterior cristae, polarization is away from the utricle.

In some species (i.e., cat, dog, rat) the anterior and posterior cristae are divided across the centre by a transverse ridge of irregular cells called the *septum cruciatum* or *eminentia cruciata* (*Figures 9.34* and *9.35*). No sensory hairs are present on these cells and the sensory epithelium is completely divided (*Figures 9.35* and *9.36*).

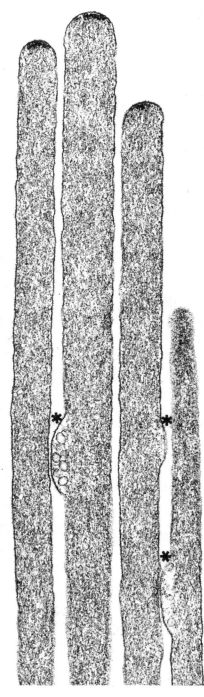

Figure 9.13 *Upper portion of four stereocilia sectioned lengthwise, at high magnification. The diameter of the longer stereocilia is greater than that of the shorter ones. In some areas along the length of the stereocilia, vesicles have accumulated at the periphery (★). At this level, the membrane bulges out and contacts the membrane of the opposite cilium. Thin filaments are seen connecting the membranes at this level. ×30 780*

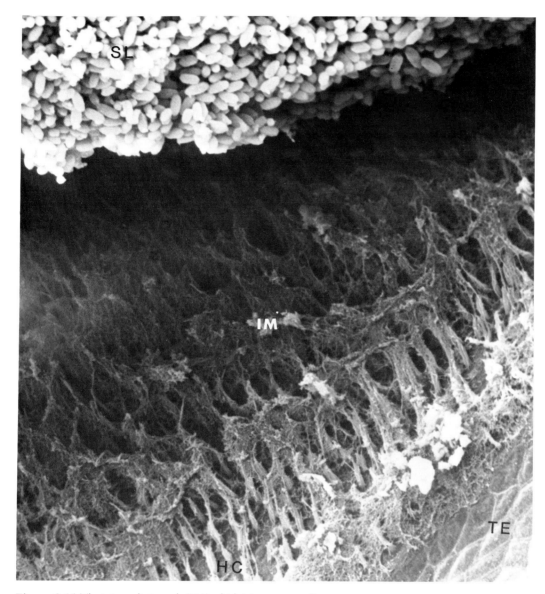

Figure 9.15 *The intermediate mesh (IM) which joins sensory cells (HC) to the statoconial layer (SL) is clearly shown in this scanning micrograph of a chinchilla utricle. TE = Transitional cells*

(continued on page 234)

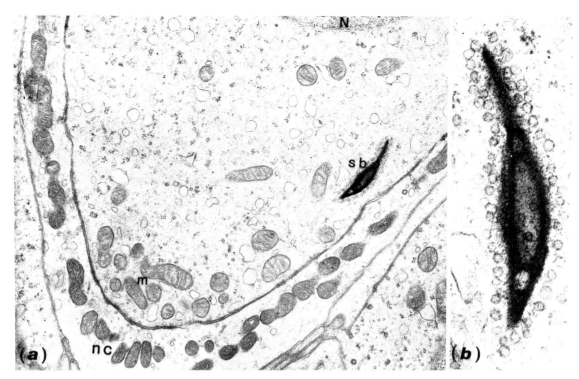

Figure 9.16 *(a) Basal portion of type I sensory cell showing the nucleus (N), numerous mitochondria (m), which are accumulated at the lower end of the cell, and a large synaptic bar (sb). nc = Nerve chalice. ×6000. (b) Synaptic bar of (a) at high magnification. Sectioning the synaptic bar transversely can produce different shapes,* *ranging from a solid bar to doughnut-shaped figures of different diameters to a solid dot. This explains the different shapes often seen in sections of synaptic bars. (An example of this is found in Figure 9.17.) ×14 000*

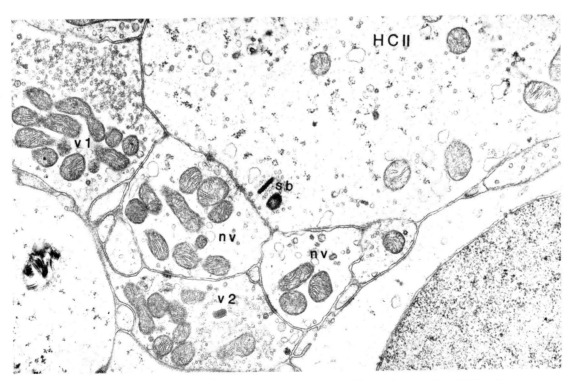

Figure 9.17 *Portion of the basal end of a type II (HCII) sensory cell of the macula of the utricle of the cat. A vesiculated (v1) and two non-vesiculated (nv) nerve endings make synaptic contact with the* *cell. Another vesiculated (v2) ending makes synaptic contact with the non-vesiculated endings. sb = Synaptic bar. ×10 000*

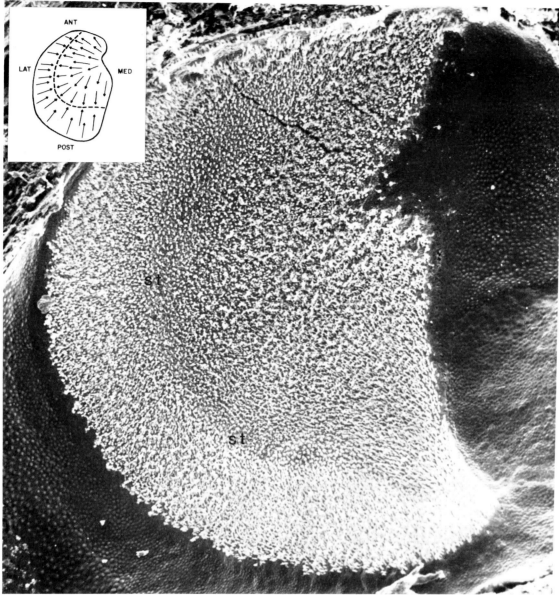

Figure 9.18 *Utricle of a chinchilla demonstrating the typical shell shape formed by the sensory cells of the macula utriculi. The striola (st) can be seen to take the general pattern shown in the inset. Supporting border cells are seen but at the low magnification transitional cells cannot be differentiated from dark cells. Inset shows the polarization or orientation of the sensory cells. In the macula utriculi polarization is toward the striola*

223

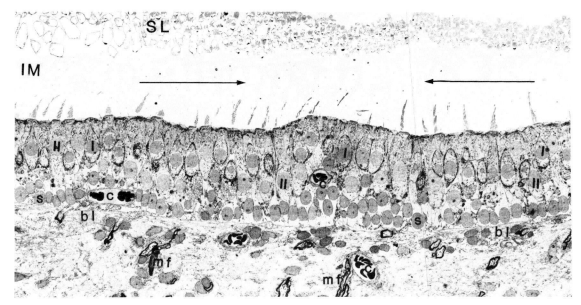

Figure 9.19 *Low-power transmission electron micrograph of the sensory epithelium and the otolithic membrane of the macula of the utricle of the cat showing the relationship between the statoconial layer (SL) and the sensory hairs of the sensory epithelium, the morphological polarization and the striola. The statoconial membrane is composed of the otoconia and the gelatinous membrane or intermediate mesh (IM). The sensory epithelium shows the distribu-* *tion of type I (I) and type II (II) cells and the supporting cells (S). It also shows two intra-epithelial capillaries (c). The morphological polarization is demonstrated by the direction of the arrows. The striola is located in the area between the arrows. Numerous myelinated nerve fibres (mf) are seen in the subepithelial connective tissue. bl = Basal lamina. ×488*

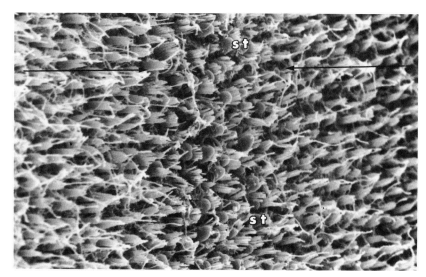

Figure 9.20 *Low-power scanning electron micrograph of the macula of the utricle of the chinchilla demonstrating the orientation or polarization (arrows) of the sensory cells toward the striola (st).* *Cilia on sensory cells in the striola area are seen to be considerably shorter than those on sensory cells peripheral to it*

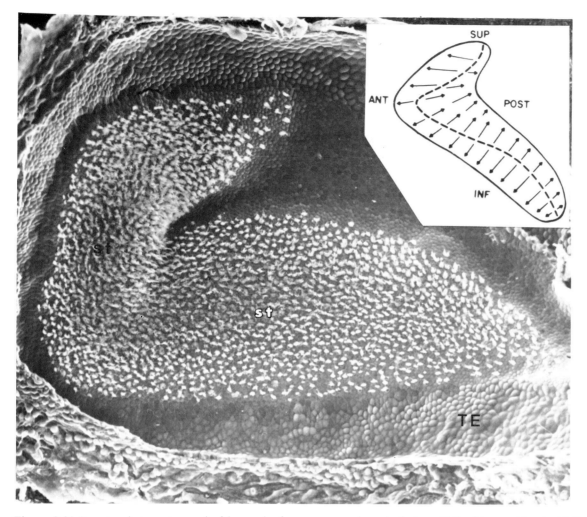

Figure 9.21 *Scanning electron micrograph of the saccule of a hamster demonstrating the typical hook shape seen in mammals. A rather wide striolar area (st) is seen corresponding in direction to that shown in the inset. Cells of the transitional epithelium (TE) stand out clearly. The inset shows the polarization or orientation of sensory cells in the macula sacculi*

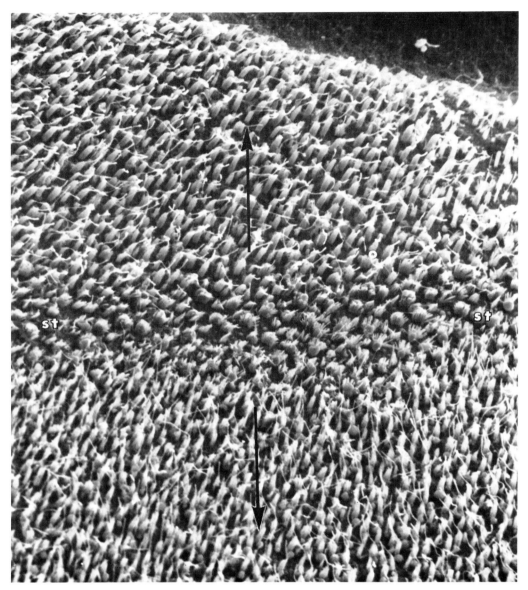

Figure 9.22 *Macula sacculi of the chinchilla showing the striola (st) and polarization (arrows) of sensory cells away from the striola toward the periphery. Cilia on cells in the striola are seen to be considerably shorter than those on sensory cells peripheral to the striola*

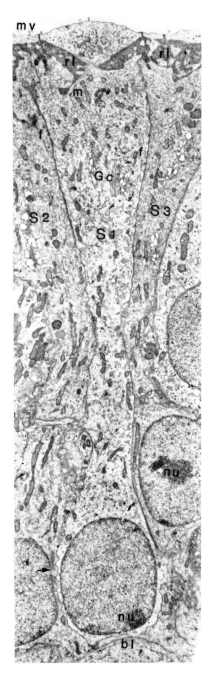

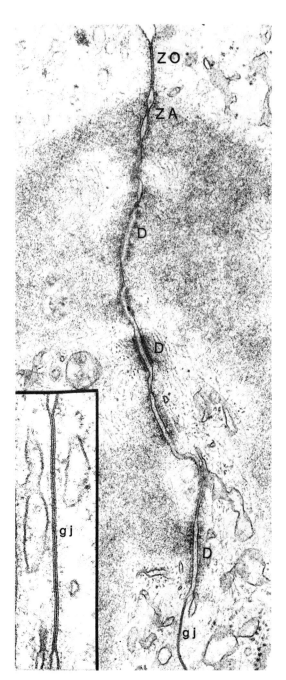

Figure 9.23 *A section through the whole length of a supporting cell (S1) of the macula of the utricle of the cat. The cell reaches from the basal lamina (bl) to the surface of the epithelium. A few microvilli (mv) protrude from the free surface of the cell. Below the surface and at some distance from the apical cell membrane is the reticular lamina (rl). A large number of lightly stained granules, the Golgi complex (Gc), numerous mitochondria (m) and a few longitudinal fibrils (f) are accumulated in the upper third of the cell. The lower two-thirds of the cell contains few cytoplasmic organelles. The nucleus is located at the basal portion and contains a large nucleolus (nu). The supporting cell (S1) is in contact with neighbouring supporting cells (S2, S3). The intercellular space between supporting cells ranges from large, irregular dilatations to gap junctions. The latter are found at different levels (arrowhead) along the length of the lateral cell membrane.* ×5090

Figure 9.24 *A portion of the apical cytoplasm of two adjacent supporting cells showing a junctional complex connecting the lateral plasmalemmas of the cells at the level of the reticular lamina. The junctional complex consists of the zonula occludens (Z0), the zonula adherens (ZA) and a series of desmosomes (D). A gap junction (gj) is shown at the inset. ×49 955. Inset: gap junction between two supporting cells at high magnification. The intercellular space measures 20 Å (2 nm). ×77 805*

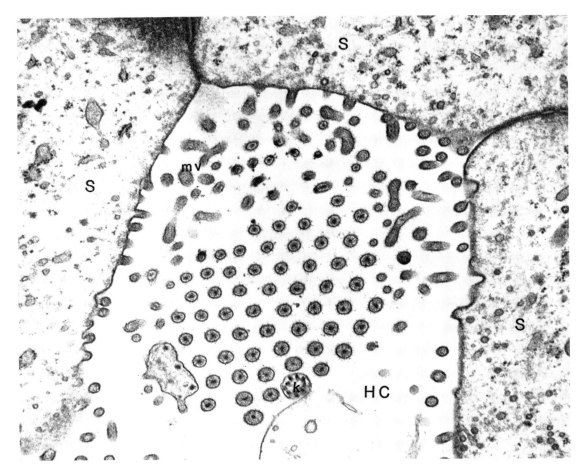

Figure 9.25 *Horizontal section at the apical surface of supporting cells and slightly above the surface of a sensory cell. At this level supporting cells (S) join cleanly to surround sensory cells (HC) with no interdigitation. Each of the stereocilia on the sensory cell has a rootlet in the centre. k = Kinocilium; mv = microvilli. ×32 000*

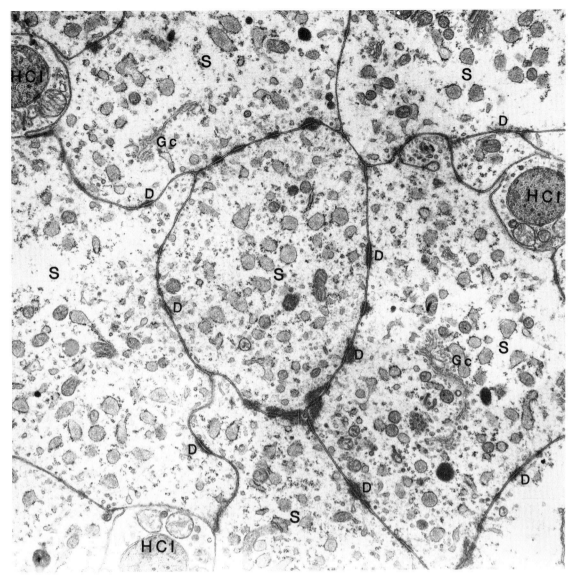

Figure 9.26 *Horizontal section at the neck level of the flask-shaped type I cells (HCI). Supporting cells (S) show a dense population of organelles at this level. Large numbers of desmosomes (D) are seen attaching the supporting cells in this area. Gc = Golgi complex. ×16 200*

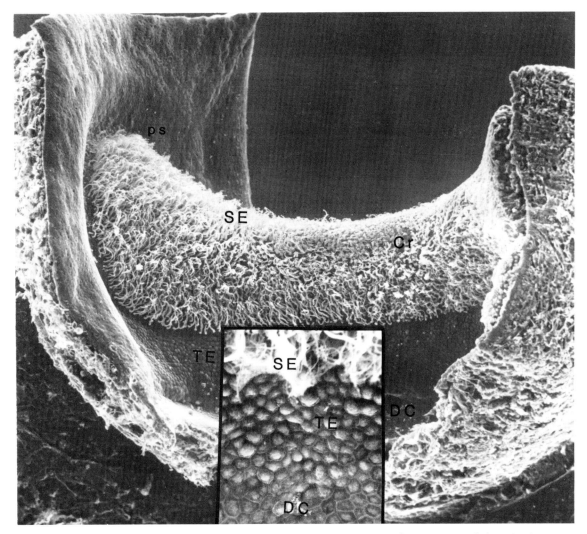

Figure 9.27 *Saddle-shaped crista (Cr) from the ampulla of a chinchilla. The cupula is taken away and cilia of sensory cells are seen to cover the crista. The higher magnification inset shows the transition from sensory epithelium (SE), to transitional epithelium (TE), to dark cells (DC), in the area that it covers. ps = Planum semilunatum*

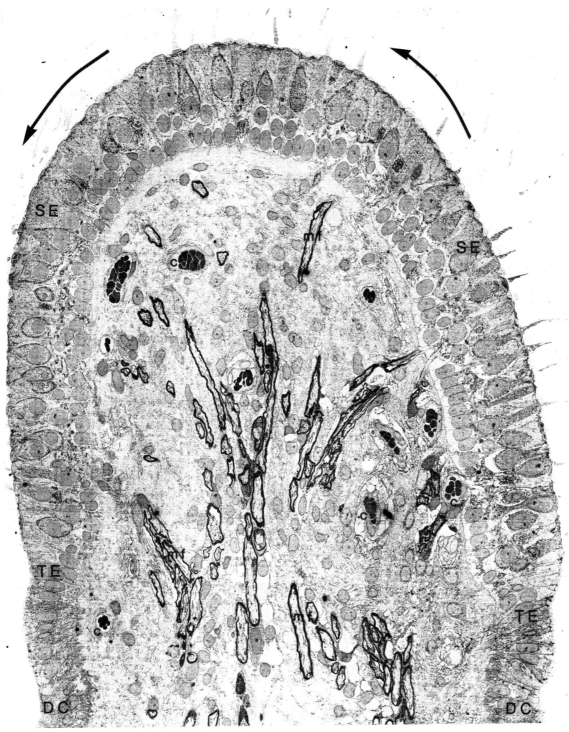

Figure 9.28 *Low-power electron micrograph of a vertical section through the crista ampullaris of a semicircular canal of the cat showing the sensory epithelium (SE), the transitional epithelium (TE) and the dark-cell zone (DC). Type I and type II sensory cells intermix with supporting cells along the sensory epithelium. The morphologic-al polarization is indicated by the arrows. Numerous thick and medium-sized myelinated nerve fibres (mf) located in the central region of the crista are seen bending towards the epithelium. c = Capillary. ×608*

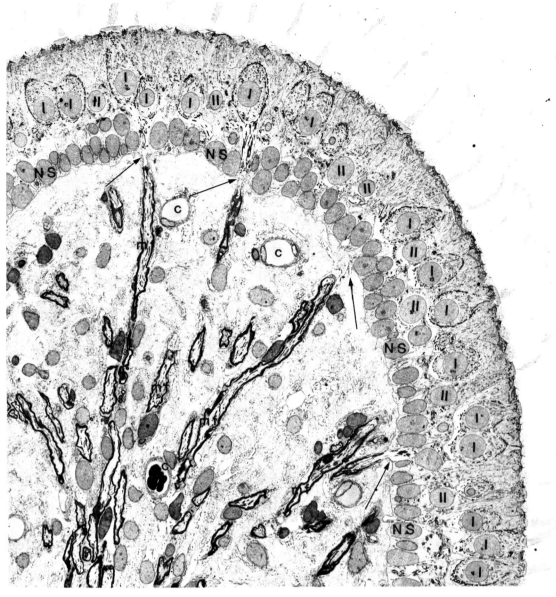

Figure 9.29 *Portion of the crista ampullaris showing the number and distribution of the type I (I) and type II (II) cells in the sensory epithelium. Thick myelinated fibres are located in the central portion of the crista. Medium-sized myelinated fibres (mf) are located at the* *periphery of the crista. Several nerve fibres (arrrows) are seen entering the epithelium after losing their myelin sheath. c = Capillary; NS = nucleus of supporting cell. ×846*

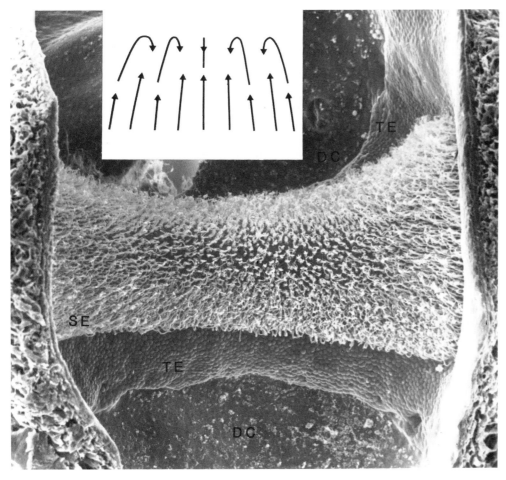

Figure 9.30 *Scanning electron micrograph of the crista showing the relative density of sensory cells on the sensory epithelia (SE). In the central area the cells are much less dense than at the periphery. Stereocilia are considerably shorter on the sensory cells in the central region. TE = Transitional epithelia; DC = dark cells. Inset shows the polarization of the sensory cells all of which are oriented in the same direction*

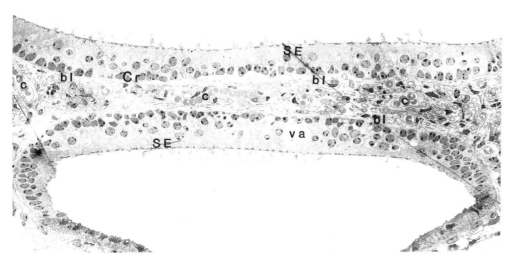

Figure 9.31 *A low-magnification transmission micrograph of a horizontal section through a specimen of a crista like that shown in Figure 9.30. The sensory epithelium (SE) is shown and the nuclei of supporting cells are seen to delineate the basal lamina (bl). Capillaries (c) generally run parallel to the cut. Note the ample blood supply to the crista. A large vacuole (va) is shown which is not unusual in cristae. Original magnification ×550 reduced to 88%*

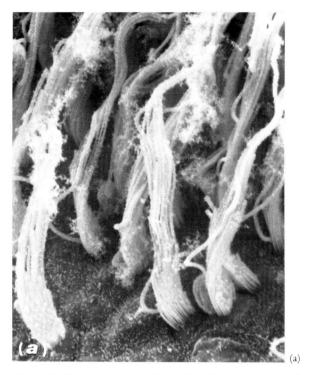
(a)

(b)

Figure 9.32 *(a) Stereocilia and kinocilia of peripheral cells of the crista ampullaris are seen to be much longer than those of cells in the* *central area shown in (b). Cilia on central-area cells, although shorter, have a larger diameter than those on peripheral cells*

Cupula

A gelatinous mass with a large number of openings to accommodate the long cilia overlies each crista (*Figures 9.33* and *9.37*). The gelatin-like structure makes determination of boundaries and features of the cupula difficult. Current findings indicate that the cupula rests firmly on the crista at either end and extends fully to the inner epithelial wall of the ampulla. In species where the crista is divided by the septum cruciatum, the cupula is also divided and is in two sections, one over each half of the sensory epithelium (*Figure 9.34*).

Transitional Epithelium

The sensory epithelia of the saccule, utricle and crista are surrounded by a group of transitional cells that are similar in appearance to the supporting cells seen in the sensory epithelia (*Figures 9.38* and *9.41*). The cells differ from the supporting cells surrounding the sensory cells in that the transitional cells are smaller and their nucleus is more central. There are fewer vesicles in the transitional cells and the organelles present are more evenly distributed. These border or transitional cells have a slightly convex top but have no dark-staining ground substance resembling a cuticular plate in the apical area.

Dark Cells

The utricle and cristae contain a further distinct group of cells peripheral to the transitional cells called dark cells (*Figures 9.38* and *9.40*). At the epithelial surface these cells are flat and multi-sided, with generally straight sides. Microvilli are few and are concentrated at the cell borders. Otoconia in various stages of degeneration are found on the dark cells both of the utricle and of the cristae (*Figures 9.40* and *9.43*). The cytoplasm of these cells stains darkly and is filled with large vacuoles and small mitochondria. A radically lobulated nucleus is located centrally (*Figures 9.38* and *9.43*). The most noteworthy feature of the dark cells is the extensive interdigitation of the cytoplasm in the basal portion of the cell (*Figure 9.43*). Dark cells are occasionally found interspersed with the transitional epithelial cells.

Planum Semilunatum

At each end of the crista the circular, slightly convex border cells abut an area of flat, very irregularly shaped, interdigitating cells (*Figure 9.45*). The shape of these cells gradually changes to the more regular shape of the dark cells. This results in an area of cells that has a crescent or half-moon shape when viewed with the light microscope, and is called the planum semilunatum.

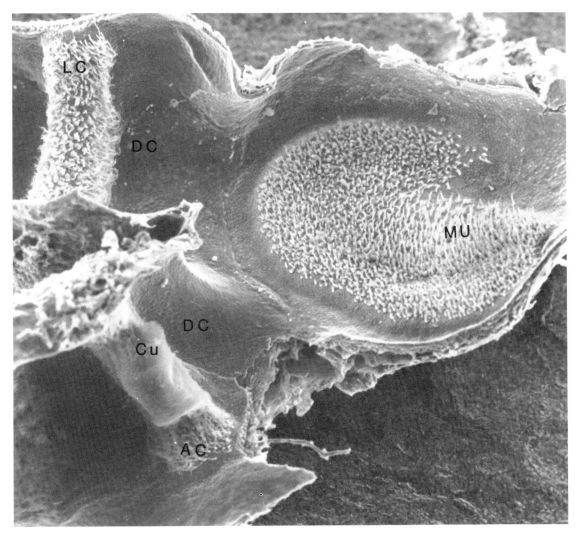

Figure 9.33 *A scanning electron micrograph of the vestibular apparatus of the hamster. Covering epithelium has been removed to demonstrate the relative position of the macula utriculi (MU), lateral crista (LC) and anterior crista (AC). The cupula (Cu) has been removed from the lateral crista but remains on the anterior crista. DC = Dark cells*

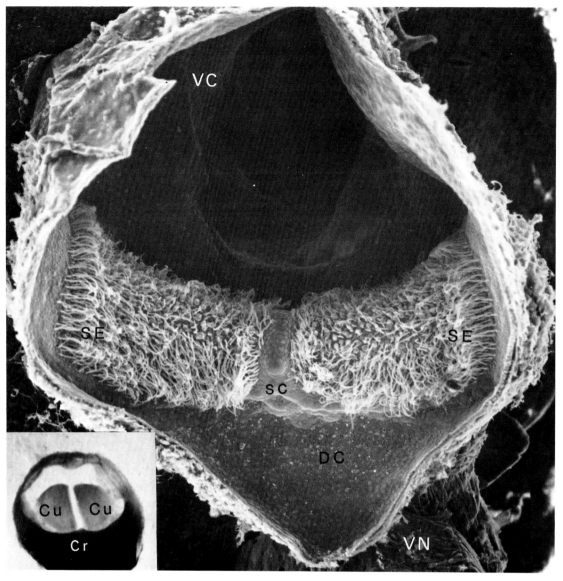

Figure 9.34 *Crista in the posterior ampulla of the rat. The septum cruciatum (SC) is shown to divide the sensory epithelium (SE) of the crista through the centre. DC = Dark cells; VN = vestibular nerve;*

VC = vestibular canal. Inset is a light micrograph of the crista (Cr) in the posterior ampulla showing how the cupula (Cu) is divided over the septum cruciatum

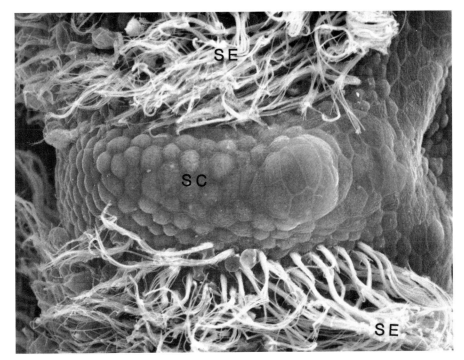

Figure 9.35 *Scanning electron micrograph showing the surface structure of cells of the septum cruciatum (SC) and the complete separation of cells of the sensory epithelium (SE)*

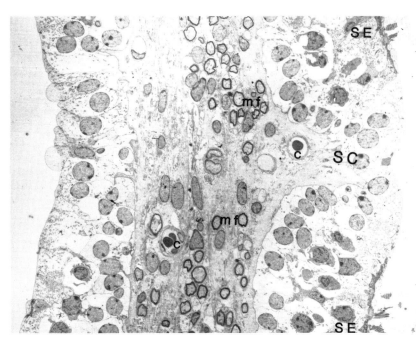

Figure 9.36 *Transmission electron micrograph through area shown in Figure 9.35 demonstrating the division of myelinated nerve fibres (mf) of the crista by the septum cruciatum (SC). c = Capillary. ×1000*

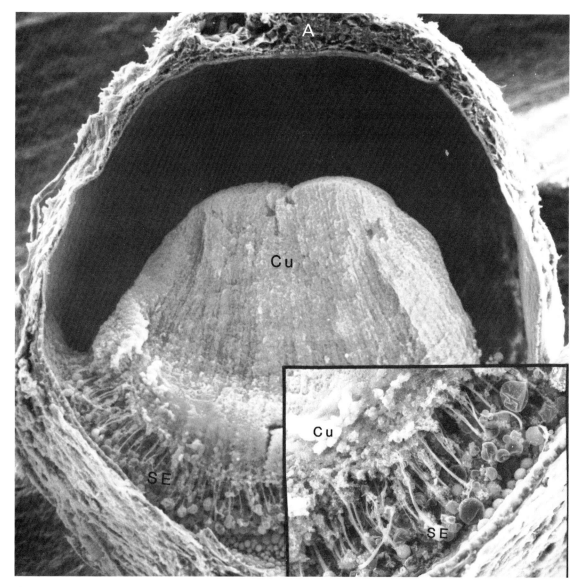

Figure 9.37 *Scanning micrograph of the ampulla (A) of a chinchilla showing the cupula (Cu). The cupula is seen to cover the sensory epithelium (SE) of the crista completely. Inset is a higher magnifica-tion and shows the contact between the cilia of the sensory cells and the cupula (Cu) in the area it covers*

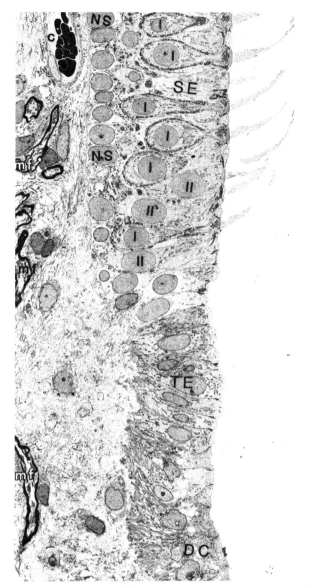

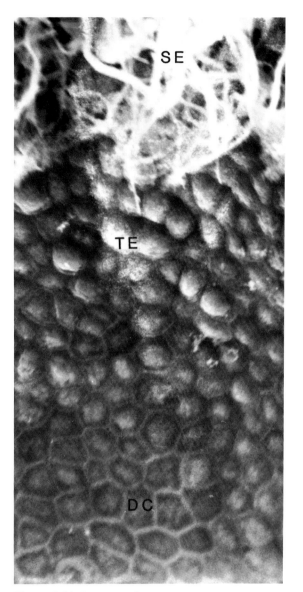

Figure 9.38 *Portion of the crista ampullaris showing the zone of connection between the sensory epithelium (SE), the transitional epithelium (TE) and the dark-cell area (DC). The transitional eipthelium consists of an area of cells located between the sensory epithelium and the dark-cell area. The sensory epithelium at this level shows numerous type I sensory cells (I) and a few type II sensory cells (II). NS = Nucleus of supporting cell; mf = myelinated fibres; c = capillary. ×980*

Figure 9.39 *A scanning electron micrograph demonstrating the surface of the area shown in Figure 9.38.*

Figure 9.40 *Macula of the utricle of a chinchilla demonstrating the change in appearance of cells from the sensory epithelium (SE) to the transitional epithelium (TE) to the dark-cell area (DC). Degenerating otoconia (o) are regularly seen on the surface of the dark cells*

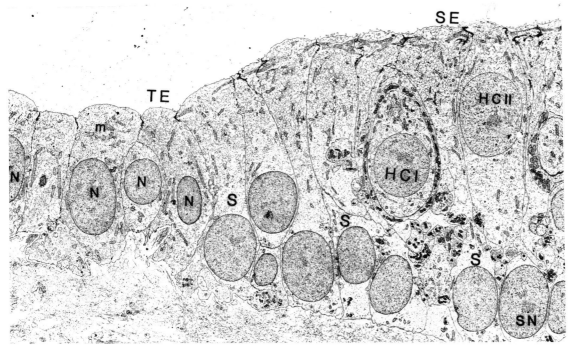

Figure 9.41 *A portion of the macula of the utricle of the cat showing the zone of connection between the sensory epithelium (SE) and the transitional epithelium (TE). The cells of the transitional epithelium are shorter than the supporting cells of the macula. The nucleus (N) is located in the central portion of the cell. The cytoplasm contains numerous clear vesicles or vacuoles distributed throughout. Mitochondria (m) and other cytoplasmic organelles are located in the apical cytoplasm. The basal portion of the cells has an irregular contour. S = Supporting cells; HCI = type I sensory cell; HCII = type II sensory cell. ×2215*

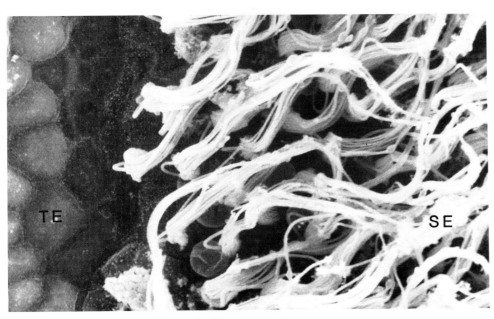

Figure 9.42 *Scanning electron micrograph of an area equivalent to that shown in Figure 9.41 from the macula of the utricle of the chinchilla showing the surface of the cells*

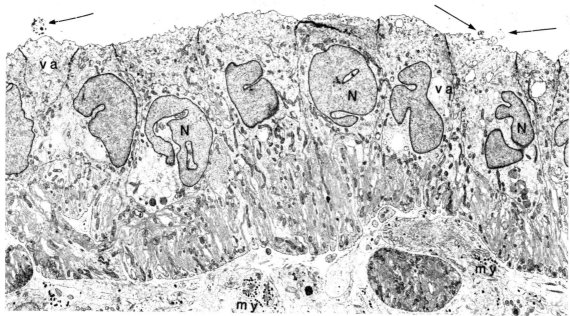

Figure 9.43 *Dark cells of the crista ampullaris of a semicircular canal of the cat. The cells have an irregular luminal surface and a large, lobulated nucleus (N) placed in the upper portion of the cell. Most of the cytoplasmic organelles are located in the upper two-thirds of the cell, including numerous irregular vacuoles (va). Deep infoldings of the lateral and basal plasmalemma divide the lower third of these cells into numerous foliate cytoplasmic compartments containing mitochondria. Numerous melanocytes (my) are found in the subepithelial connective tissue. Objects equivalent to the degenerating otoconia seen in Figure 9.40 are regularly seen on the surface of dark cells in this area (arrows). ×2873*

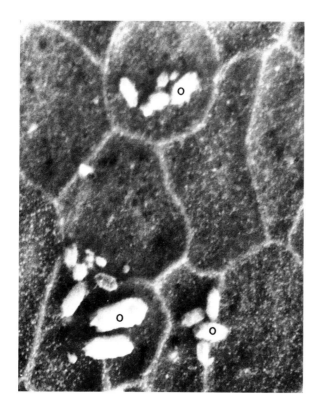

Figure 9.44 *Surface area of dark cells of the utricle showing otoconia (o) in a degenerated state. These are also found regularly in micrographs of dark cells of the cristae (see Figures 9.30 and 9.40)*

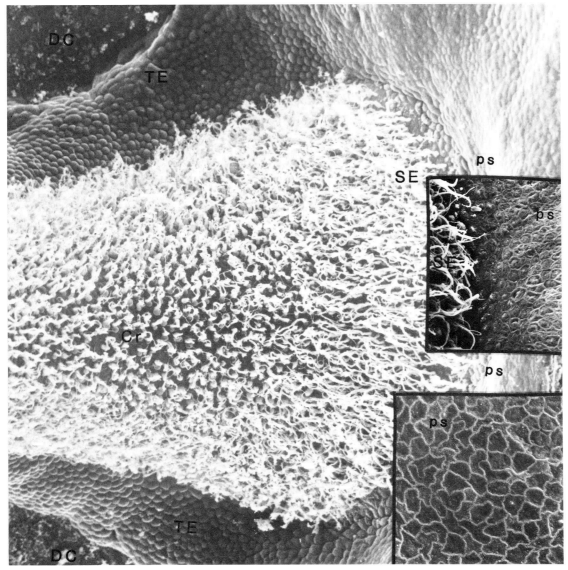

Figure 9.45 *Scanning electron micrograph of one end of a crista (Cr) showing an area named the planum semilunatum (ps) because of its unusual crescent-shaped appearance when viewed with the light microscope. TE = Transitional epithelium; DC = dark cells. Upper insert shows at higher magnification the transitional zone from the sensory epithelium to the planum semilunatum in the area it covers. Lower insert shows the peculiar interdigitation of cells at the surface in the planum semilunatum accounting for its unique appearance*

243

Acknowledgements

The technical assistance of R. Tai and R. Mount and the secretarial work of B. Poole are gratefully acknowledged. This work was partially supported by the Medical Research Council of Canada and the National Institutes of Health.

References

Dohlman, G. F. (1971) The attachment of the cupular, otolith and tectorial membranes to the sensory cell areas. *Acta Oto-Laryngologica,* **71,** 89.

Engström, H., Bergström, B. and Ades, H. W. (1972) Macula utriculi and macula sacculi in the squirrel monkey. *Acta Oto-Laryngologica,* Supplement **301,** 75.

Flock, Å. (1964) Structure and function of the macula utriculi with special reference to directional interplay of sensory responses as revealed by morphological polarization. *Journal of Cell Biology,* **22,** 413.

Harada, Y. (1980) *Middle Ear and Inner Ear Scanning EM Atlas.* Tokyo. Kimbara: (Japanese text.)

Iurato, S. (1967) *Submicroscopic Structure of the Inner Ear.* London: Pergamon Press.

Kimura, R., Lundqvist, P. -G. and Wersäll, J. (1963) Secretory epithelial linings in the ampullae of the guinea pig labyrinth. *Acta Oto-Laryngologica,* **57,** 517.

Lim, D. J. (1969) Three dimensional observations of the inner ear with the scanning electron microscope. *Acta Oto-Laryngologica,* Supplement **255.**

Lim, D. J. (1975) Fine morphology of the cupula – A scanning electron microscopic observation. In *Proceedings of the Bárány Society, Fifth Extraordinary Meeting of the Bárány Society,* Kyoto, Japan, p.390.

Lindeman, H.H. (1969) Studies on the morphology of the sensory regions of the vestibular apparatus. In *Advances in Anatomy, Embryology and Cell Biology,* vol. 42, p. 1. Berlin: Springer-Verlag.

Lindeman, H. H. (1967) Cellular pattern and nerve supply of the vestibular sensory epithelia. *Acta Oto-Laryngologica,* Supplement **224,** 86.

Smith, C. A. (1956) Microscopic structure of the utricle. *Annals of Otology, Rhinology and Laryngology,* **65,** 450.

Spoendlin, H. (1970) Vestibular labyrinth. In *Ultrastructure of the Peripheral Nervous System and Sense Organs,* ed. A. Bischoff, p. 264. Stuttgart: Thieme Verlag.

Wersäll, J. (1956) Studies on the structure and innervation of the sensory epithelium of the cristae ampullaris in the guinea pig: A light and electron microscopic investigation. *Acta Oto-Laryngologica,* Supplement **126.**

10
The Development and Structure of the Otoconia

David J. Lim

Otoconia (statoconia) or otoliths (statoliths) are found in the gravity receptor organs of many invertebrates and all vertebrates, but there are considerable species differences in the shape, size, chemical make-up and mode of formation of these ear stones (*Figure 10.1*). For example, rays take up grains of sand (foreign particles) through the open endolymphatic duct and deposit them on the gelatinous layer of the otoconial membrane. This type of otoconium is called the exogenous type, in contrast to the endogenous type, which is formed *in situ* by the organism. Reptiles, birds and mammals have numerous small otoconia (calcium carbonate in the form of aragonite or calcite) covering the gravity receptors (*Figure 10.2*); bony fish, on the other hand, form a single large otolith (*see Figure 10.1(a)*) (Breschet, 1936). In the fish otolith there are alternating bands known as *annular rings* which coincide with seasonal patterns of growth (Kreiger, 1840; Reibisch, 1899; Pannella, 1971) and are used in determining the age of a fish.

Crystalline calcium salts occur naturally in several forms: the phosphate salts in apatite and hydroxyapatite, found in bone and teeth, and the carbonates in calcite, aragonite and vaterite, three different forms of $CaCo_3$ but with different crystal lattices. In some primitive fishes, such as the gar (ganoid fish), discoid vaterite crystals have been reported in addition to the aragonite crystals (Carlström, 1963). Vaterites may also occur as aberrant otoliths in teleost fish (Palmork, Taylor and Coates, 1963). According to the X-ray diffraction study of Carlström and Engström (1955), the otoconia consist of calcium carbonate crystallized in the form of calcite in mammals, birds, and sharks; of calcium carbonate in the form of aragonite in amphibia; and of calcium phosphate in the lamprey. The otoconia of the salamander contain carbonate hydroxyapatite as well as aragonite (Hastings, 1935). The otoliths and otoconia of invertebrates also vary, not only in their morphology but also in their chemical make-up. The otoliths of some jellyfish (Scyphomedusa) are formed by hydrated calcium sulphate, whereas other medusan otoliths (Ctenophora) contain mainly $BeMg(CO_3)_2$, and those of the Arthropoda contain calcium fluoride (Vinnikov *et al.*, 1981).

According to Iurato and de Petris (1967), the density of otoconia composed of calcite is 2.71, and the density of those composed of aragonite is 2.93. Therefore, the otoconia increase the specific gravity of the statoconial membrane, which has a density of only 1.9–2.2. This value is higher than that of the cupula and twice as high as that of endolymph (1.02–1.04), as measured by Trincker (1967). Thus, the otoconia increase the specific weight of the statoconial membrane and thereby make the gravity receptors sensitive to the force of gravity.

Calcium salts of the otoconia may take part in the process of bone calcification and the storage of calcium in certain species (Guardabassi, 1952). In amphibia a large number of crystals resembling otoconia are contained in the paravertebral calcareous sac at the level of the intervertebral foramen (Iurato and de Petris, 1967), and the calcium content of amphibian statoconia varies with the experimental conditions (Guardabassi, 1953). Whether the same phenomenon occurs in mammals is not fully known, but, based on data obtained from isotope-labelled calcium uptake in the otoconia, Ross and her co-workers have suggested that otoconia serve as a calcium reservoir of the inner ear (Ross, 1979; Ross and Williams, 1979; Ross *et al.*, 1980).

(*continued on page 248*)

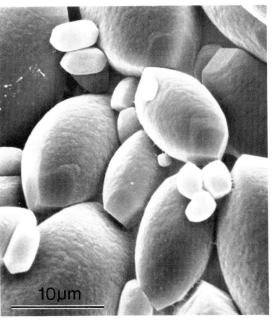

Figure 10.1 *(a) Scanning electron micrograph (SEM) of a goldfish (Carassium auratus) otolith. An arrow points to the depression to which the sensory epithelium was attached. (b) SEM of frog otoconia in an aragonite form. (c) SEM view of pigeon otoconia in a calcite form.*

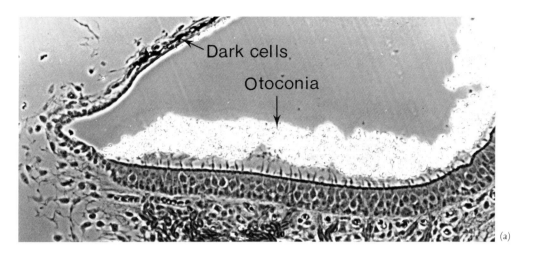

Dark cells

Otoconia

(a)

100μm

(b)

Figure 10.2*(a) A phase-contrast light-microscope view of the utricular otoconia of a guinea-pig. Some otoconia are adherent to the dark cells of the utricular wall. ×390. (b) Low-power scanning electron micrograph of the guinea-pig utricular otoconia of varying sizes*

Morphology of the Mammalian Statoconial Membrane

Since the detailed morphology of mammalian gravity receptors was first described (Igarashi and Kanda, 1969; Lindeman, 1969; Marco, Sanchez-Fernandez and Rivera-Pomar, 1971; Sanchez-Fernandez *et al.*, 1972; Lim, 1973; Lindeman, 1973; Lim, 1977), various terminologies have been used to denote specific structures. To minimize this confusion, the following terminology is used in this chapter. The whole membrane is called the *statoconial membrane* (synonymous with *statolithic mem-*brane in bony fish). It is composed of the *otoconial layer, gelatinous layer* and *subcupular meshwork* (*Figures 10.3 –10.6*) (Lim, 1980). Some otoconia are partly embedded in the gelatinous layer, and the rest are held together by the gluey, gelatinous substance (*Figure 10.7*). In fixed specimens, the gelatinous layer is formed of amorphous and fibrillar material (Johnsson and Hawkins, 1967; Lim, 1973) and is perforated, particularly near the striolar zone (Lindeman, 1969; Lim, 1979). It is formed of two distinct parts: the upper layer, to which the otoconia are attached; and the *honeycomb layer*, which houses the tall ciliary bundles of sensory cells (*see Figures*

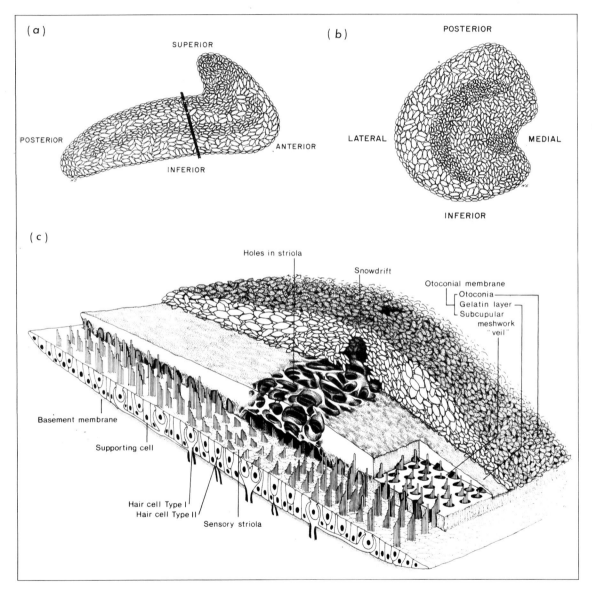

Figure 10.3(a) A surface-projected view of the saccule illustrating otoconial size distribution. The large crystals are found mainly in parastriolar areas, while small crystals are found in the superficial layer, in the striola and along the periphery. A dark line indicates the cross-section shown in (c). (b) In the utricle, the largest crystals are found in the lateral and posterior parts, whereas the small crystals are found mainly in the striola, medial part, along the margins and on the surface. (c) Artist's view of a cross-section of the saccule showing the substructures of the otoconial (statoconial) membrane with its striolar holes and snowdrift zone (From M. M. Paparella and D. A. Shumrick (eds) (1980) Textbook of Otolaryngology, *Vol. 1, Basic Sciences and Related Disciplines, p. 459. Philadelphia: W. B. Saunders. Reproduced by kind permission of publishers.*)

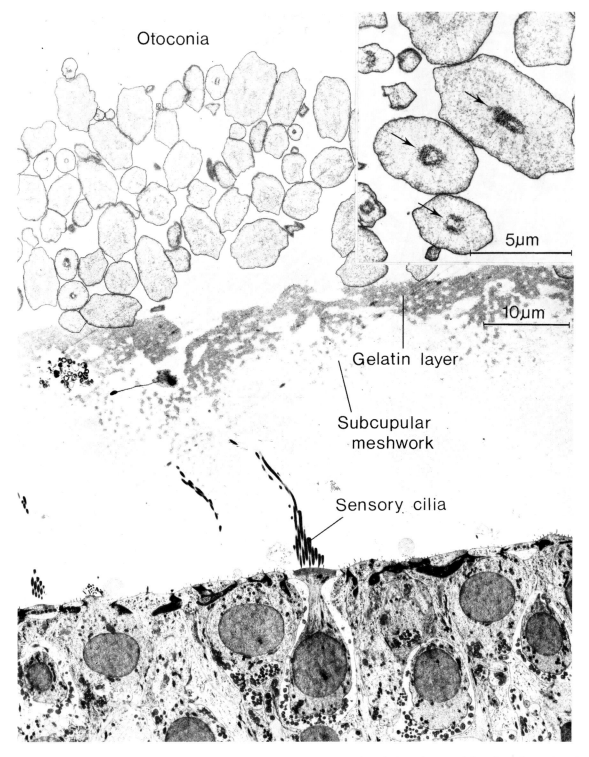

Otoconia

5μm

10μm

Gelatin layer

Subcupular meshwork

Sensory cilia

Figure 10.4 *Transmission electron micrograph of an EDTA-decalcified, guinea-pig statoconial membrane which shows otoconia, the gelatinous layer, and the subcupular meshwork. An inset shows the nucleus (central core) of the otoconia (arrows)*

10.5, 10.6 and 10.8) (Lim, 1979). The underside of the honeycomb layer is attached to the supporting cells of the sensory epithelium by a gelatinous meshwork (*Figure 10.6*). This meshwork in mammals is similar to the *subcupular meshwork* described in pigeons by Dohlman (1971). There appear to be some species differences: the honeycomb layer is well developed in rats, mice and chinchillas but poorly developed in guinea-pigs. The honeycomb layer appears to be absent in the striolar zone, leaving a wide open space. The open space between the gelatinous layer and the sensory epithelium is known as the *subcupular space* (Igarashi and Kanda,

1969). The meshwork is often missing in decalcified specimens, as shown in *Figure 10.9*. Histochemical investigations (Wislocki and Ladman, 1955; Veenhof, 1969) have shown that the gelatinous layer and subcupular meshwork are formed of acidic mucopolysaccharides and glycoproteins. These substances appear to be secreted by the supporting cells, based on an ultrastructural study and a light-microscope autoradiographic study using tritiated leucine and glucose (Lim, unpublished observations).

(*continued on page 253*)

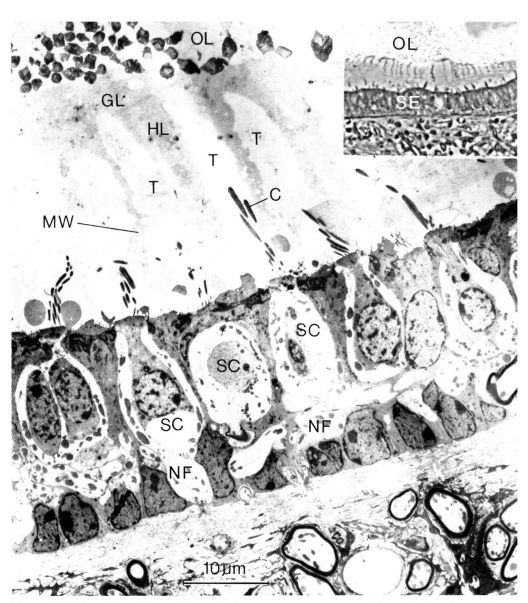

Figure 10.5 *Transmission electron micrograph of a mouse statoconial membrane, showing the otoconial layer (OL), gelatinous layer (GL), tubules (T), and meshwork (MW). The sensory ciliary bundles (C) penetrate part of the tubules. SC = Sensory cells; NF = nerve fibres. An inset shows a phase-contrast light-microscopic view of the tentacle-like substructures of the honeycomb layer (HL) and meshwork (MW). SE = Sensory epithelium. ×250*

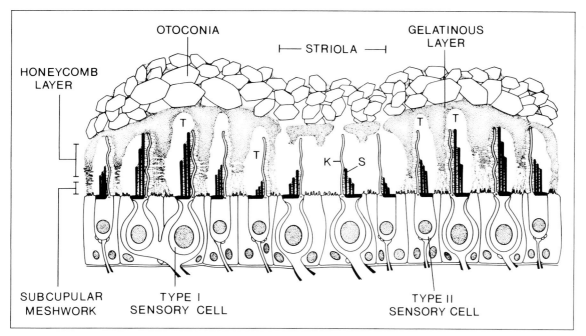

Figure 10.6 *Artist's conception of the gravity receptor, modelled after the mouse, depicting the substructures of the statoconial membrane: otoconial layer, gelatinous layer, honeycomb layer and sub-* *cupular meshwork. The honeycomb layer contains numerous tubules (T) in which sensory ciliary bundles (S) are housed. K = kinocilium*

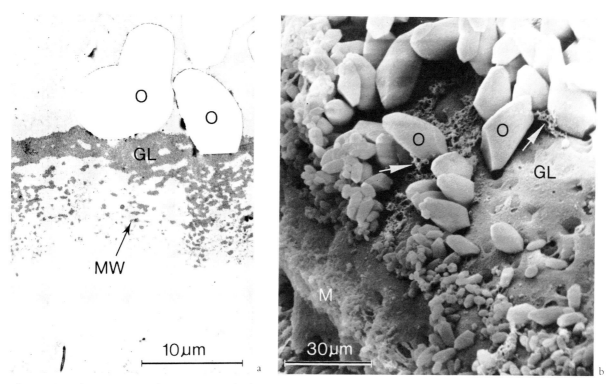

Figure 10.7*(a) Transmission electron micrograph of the mineralized mouse otoconia (O) that are partially embedded in the gelatinous layer (GL) with its underlying meshwork (MW). (b) Scanning electron micrograph of the surface of the statoconial membrane of a* *chinchilla saccule, showing the gelatinous layer (GL) to which the otoconia (O) adhere. A lacy substance (arrows) covers some of the otoconia and may aid in their attachment. M = outer margin of the gelatinous layer*

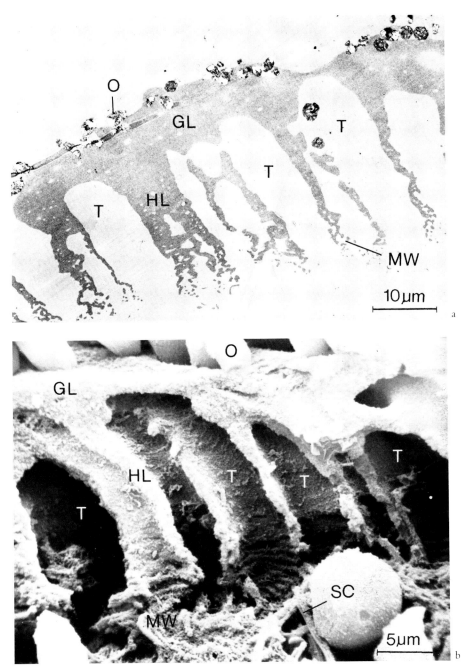

Figure 10.8(a) *Transmission electron micrograph of a mouse statoconial membrane which is formed by otoconia (O), gelatinous layer (GL), honeycomb layer (HL) and meshwork (MW). T = Tubule. (b) Scanning electron micrograph of a mouse statoconial membrane which shows otoconia (O), gelatinous layer (GL),* *honeycomb layer (HL) and meshwork (MW). The honeycomb layer contains numerous tubules (T) into which the ciliary bundles (SC) are partially inserted. In this micrograph, ciliary bundles are pulled away from the tubules due to fixation artefacts*

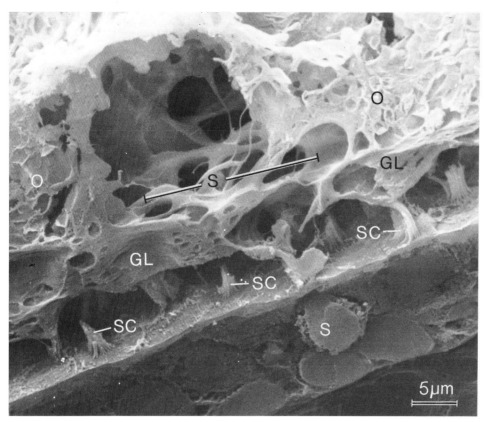

Figure 10.9 *A fractured view of EDTA decalcified, guinea-pig saccule near the striola (S), which shows striolar holes. Decalcified otoconia (O) are fused into a mass. The tips of some sensory ciliary bundles (SC) are attached to the undersurface of the gelatinous layer (GL), while some of them are free-standing in the open subcupular space*

Morphology of the Otoconia

Under the light microscope the decalcified otoconia appear as globular structures that adhere to the gelatinous layer of the otoconial membrane, whereas the undecalcified otoconia appear as a chalky white powder when viewed with the dissection microscope, and as highly refractile crystals (*see Figure 10.2(a)*) by phase-contrast or polarized-light microscopy. Although much valuable information has been gathered by phase-contrast and polarized-light microscopy (Lindeman, 1969; Istenič and Bulog, 1976) and transmission electron microscopy (Carlström and Engström, 1955; Iurato and de Petris, 1967; Marco, Sanchez-Fernandez and Rivera-Pomar, 1971; Lim, 1973), the advent of the scanning electron microscope has provided many additional ultrastructural details (Lim, 1969; Kellerhals, Marti and Villiger, 1970; Lim, 1973; Lindeman, Ades and West, 1973; Ross *et al.*, 1976; Wright and Hubbard, 1978).

The lengths of the otoconia range from 0.1 µm to 25 µm in guinea-pigs and up to 30 µm in humans. In rats, the saccular otoconia are much larger than those in the utricle. The distribution of these different otoconia follows a distinct pattern in that the small crystals are located mainly on the surface and outer margin, as well as in the striolar zone of the statoconial membrane (*see Figures 10.3(a) and (b) and 10.10*). The latter area is described as the 'snowdrift' zone because here the statoconia are piled higher than in surrounding areas and give the impression of drifted snow (*see Figures 10.3(c) and 10.10(b)*) (Ades and Engström, 1965; Lindeman, 1969; Wright and Hubbard, 1978; Wright, Hubbard and Clark, 1979). These peculiar arrangements, with minor differences, are quite consistent among all the mammals examined: guinea-pigs, rats, chinchillas, mice, squirrel monkeys, cats and humans. Neither the mechanism of these peculiar anatomical arrangements nor its functional significance is known. It has been suggested that the snowdrift zone will allow the free flow of endolymph through the perforated holes present in the striola of the gelatinous layer, thus making the sensory cells of the central (striolar) zone of the macula more susceptible to ciliary deflection by the endolymph drag than those of the peripheral zone (Lim, 1977).

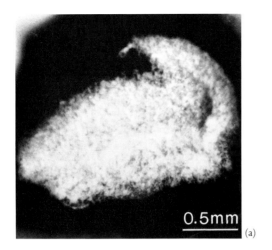

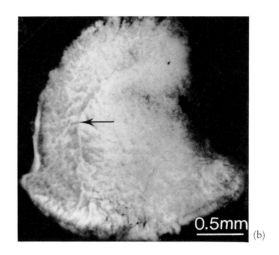

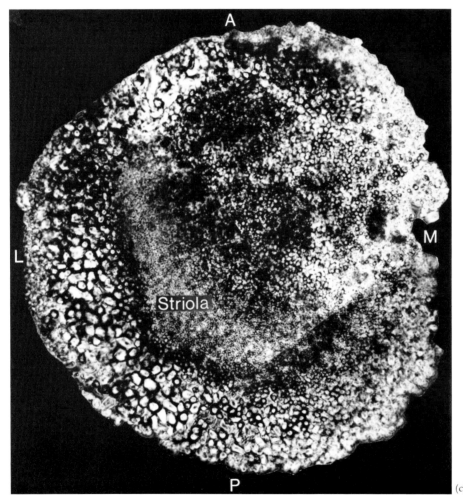

Figure 10.10*(a) A surface view of a human (2-year-old) saccular statoconial membrane. (From Wright and Hubbard, 1979, by kind permission of authors and publishers.) (b) A surface view of a human (36 weeks gestational age) utricle which shows the snowdrift line (arrow). (From Wright, Hubbard and Clark, 1979, by kind permission of authors and publishers.) (c) A phase-contrast light-microscope view of a rat utricle showing the distinct otoconial size distribution patterns. Small crystals are in the striola and medial (M) part of the utricle and large crystals are in the lateral (L) and posterior (P) parts. Anterior (A) parts contain medium-sized crystals*

Under the scanning electron microscope, the crystals also appear powdery in a low-power view (see Figure 10.2(b)). In a high-power view, the majority of mammalian calcite otoconia have rounded bodies with pointed tips that have three planar (rhombohedron) surfaces (Figures 10.11 and 10.12). However, a few odd-shaped crystals have been reported; these are barrel shaped – a rounded body with two flat terminal faces (Figure 10.13) – or multifaceted, dumb-bell shaped, or even cross shaped (Kellerhals, Marti and Villiger, 1970; Ross and Peacor, 1975; Johnsson et al., 1982).

The ultrastructural morphology of mammalian otoconia, as demonstrated by the transmission electron microscope, varies only in minor details in different reports. The differences may be due largely to the methods of specimen preparation (Igarashi and Kanda, 1969; Veenhof, 1969; Marco, Sanchez-Fernandez and Rivera-Pomera, 1971; Sanchez-Fernandez et al., 1972; Lim, 1973; Nakahara and Bevelander, 1979; Salamat, Ross and Peacor, 1980; Anniko, 1980). It has been suggested that such differences are most likely due to the loss of minerals (decalcification effect) and/or organic substances during specimen preparation (Peacor, Rouse and Ross, 1980). This contention is supported by the considerable morphological differences observed in otoconia after EDTA decalcification and in those that have not been decalcified but are presumed to be slightly demineralized as a result of the loss of labile mineral salts during tissue preparation. Therefore, caution must be exercised in interpreting the ultrastructural morphology of the otoconia.

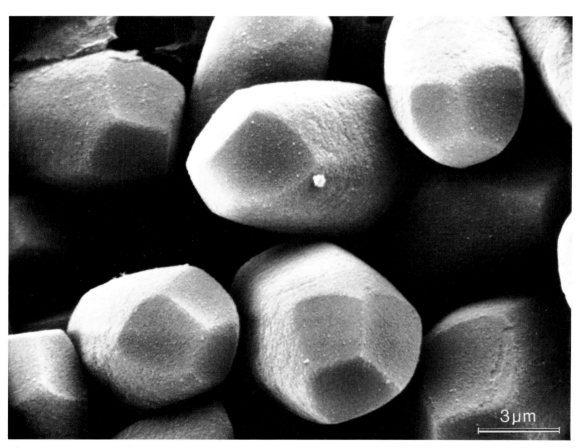

Figure 10.11 *Scanning electron micrograph of squirrel-monkey otoconia*

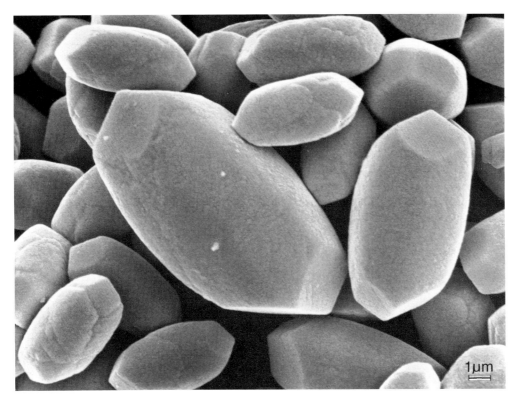

Figure 10.12 *Scanning electron micrograph of human (10-month-old infant) utricular otoconia (Reproduced by kind permission of Dr C. G. Wright.)*

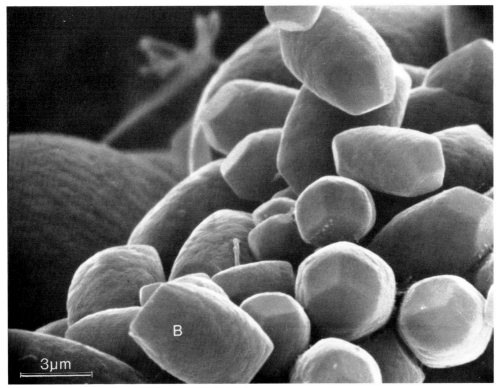

Figure 10.13 *Scanning electron micrograph of adult human otoconia, one of which is barrel shaped (B)*

Transmission electron microscopy has shown that decalcified otoconia in mammals retain the original shape of the calcite crystals formed by residual organic substances with dense nuclei (*see Figure 10.4*). The ground substance of the decalcified crystals is essentially similar to the ground substance of the gelatinous layer of the statoconial membrane (*see Figure 10.4*). The otoconial ground substance is periodic acid–Schiff (PAS) positive, thus indicating glycoprotein and muco-polysaccharide components, similar to the gelatinous layer (Wislocki and Ladman, 1955). One can speculate, however, that there must be a difference between the chemical composition of the ground substance of the otoconia and that of the gelatinous layer, since the latter is not known to be calcified under physiological conditions. Apparently, the mucoprotein in the otoconia attracts calcium and carbonate. Shrader, Erway and Hurley (1973) demonstrated that the otoconial ground substance incorporates sulphur; therefore it is considered to consist of sulphated mucopolysaccharides. The fish's otolith, unlike the mammalian otoconia, is formed by alternating calcium-rich and gelatin-rich layers representing fast- and slow-growth zones (*Figure 10.14*).

The ultrastructure of mineralized (and partly mineralized) otoconia shows an internal organization of fine spindle-shaped calcite crystallites and fibrillar organic substance. The arrangement of crystallites closely mimics the fibrillar arrangement of the organic substance. In the central core (nucleus), the spindles, when present, are arranged almost parallel to the long axis of the crystal, or they may appear to be amorphous (*Figures 10.15* and *10.16*). In developing rat otoconia, these spindles are less obvious and when found they appear to follow the outline of the crystals (Salamat, Ross and Peacor, 1980). There are also distinct differences in the organic material (and calcite) present in the different locations (such as the nucleus, terminal faces and body) within the crystal. For example, the pointed end-portions of the otoconia are often more electron lucent than the rounded sides of the crystals. The direction of the spindle arrangement appears to follow the geometric configuration of the internal structure of the crystal. In the rounded body, the spindles are arranged tangentially to the outer surface of the crystal, whereas in the terminal faces they are at right angles to the surface (*Figures 10.15(b)* and *10.16*).

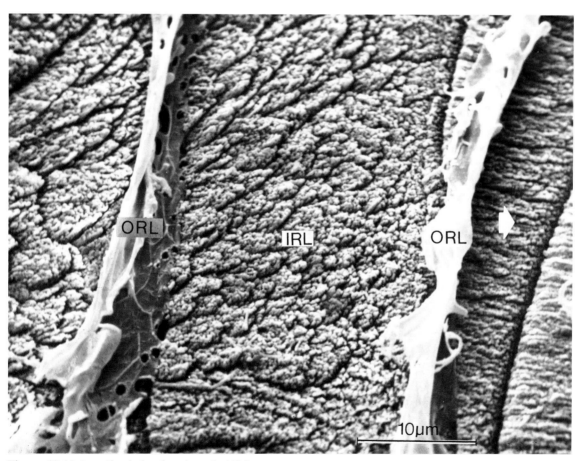

Figure 10.14 *An acid-etched surface view of goldfish otoconia showing the alternating organic substance-rich layers (ORL) and inorganic substance-rich layers (IRL). The crystallites within the IRL appear to be arranged in a radial direction (large arrow)*

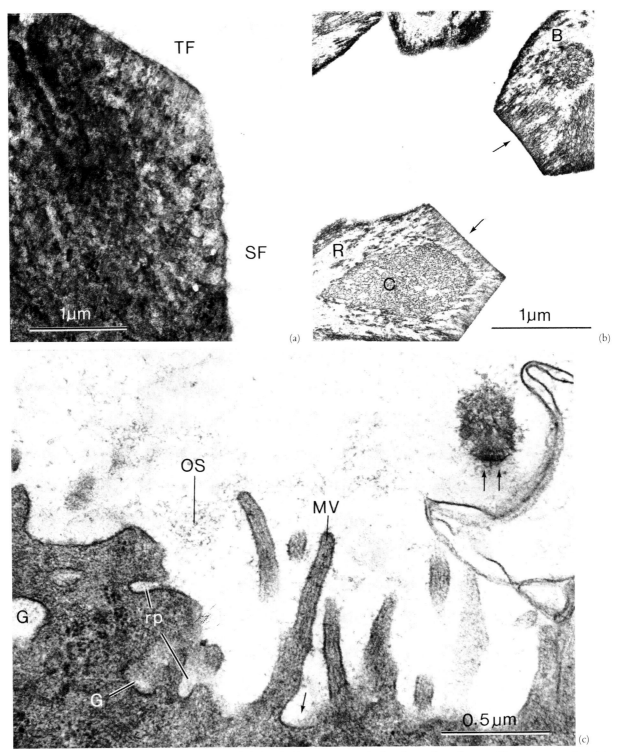

Figure 10.15*(a) Transmission electron micrograph (TEM) of a well-mineralized adult mouse otoconium, showing a smooth terminal face (TF) and irregular side face (SF). Observe the fine spindles (crystallites) which are arranged at right angles to the plane of the terminal face. Fuzzy-appearing organic substances are in close contact with the crystal surface. (b) TEM from a 1-day-old mouse showing a rhombohedral crystal (R) with its central core (C) and a barrel-shaped crystal (B). These crystals are not fully mineralized (or were demineralized during specimen preparation). Arrows point to the terminal surfaces where the crystallites are arranged at right angles to the surface. (c) A small primitive crystal (double arrows) with a recognizable terminal face and crystallites. It is being formed near the long microvilli (MV) of the supporting cells in this 16½-day-old mouse embryo. The specific granules (G) appear to be within pinocytotic vesicles that eventually fuse with the cytoplasmic membrane for release of the granules (arrow). Organic substances (OS), which appear to be secreted by reverse pinocytosis (rp), are scattered near the supporting cells*

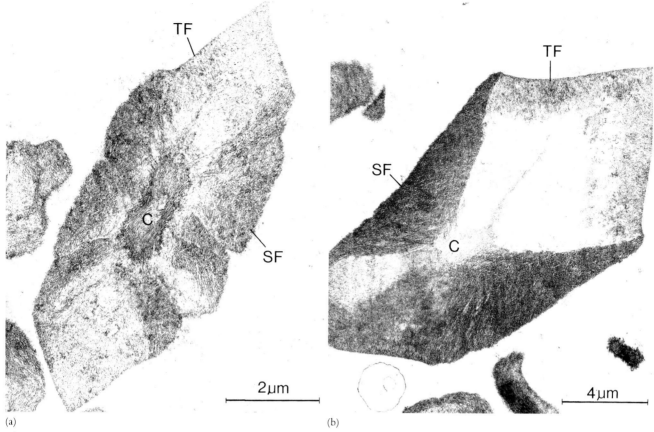

Figure 10.16(a) *Transmission electron micrograph (TEM) of a maturing otoconium from a 1-day-old mouse which shows the central core (C), irregular side face (SF) and well-defined terminal face (TF). Observe the contrasting arrangement of crystallites in the central core in relation to those of the body of the crystal. (b) TEM of a mature otoconium from a 1-day-old mouse which shows the central core (C) and smooth side face (SF) and terminal face (TF). Also observe the well-defined internal crystal substructure composed of different organic and inorganic substances*

Whether the otoconia are true single crystals or regular assemblies of many small iso-oriented crystallites has been debated. Carlström and Engström (1955), on the basis of their X-ray diffraction study of human otoconia, and Iurato and de Petris (1967), on the basis of their electron diffraction study on single rat otoconia, have proposed that the otoconia act as a single crystal. Carlström and Engström suggested that the crystallites which form a 'single crystal hexagonal prism' are formed by the 'hexagonal platelets' obtained by crushing the otoconia. Recently, Ross *et al.* (1982) investigated fish otoliths and otoconia of rats and frogs using high-resolution electron microscopy and electron-beam diffraction to resolve the question of whether the final 'crystals' (otoconia and otoliths) achieve the three-dimensional periodicity of a single crystal during maturation or remain microcrystalline at some level. They consistently obtained single crystal electron beam diffraction patterns from the fish otoliths. High-resolution TEM showed that all of the biological samples (otoliths and otoconia) consisted of 'highly ordered domains', apparently arranged in sheets. On this basis, these authors suggested that neither 'single crystal' nor 'polycrystalline' should be used to describe otoconia or otoliths, as they appear to be characterized by ordered domains.

The question has also been raised as to whether the otoconia, once formed, are stable, renewable, or can be regenerated. It has been generally believed that mature otoconia have limited calcium turnover and that this turnover occurs along the outer surfaces of the crystals (Belanger, 1960; Balsamo, de Vincentiis and Marmo, 1969; Veenhof, 1969). Preston *et al.* (1975) measured the rate of calcium incorporation into the gerbil statotoconial membrane with scintillation spectrometry 24 hours after a single intraperitoneal injection of $^{45}CaCl_2$. They found the incorporation was 0.06–0.07 mN Ca^{2+} per day, which represents a fractional rate of 0.1 per cent with a half-life of approximately 11 days. The incorporation rate in the statotoconial membrane was 5–7 times lower than in the middle ear ossicles, otic capsule and skeletal bone. Ross (1979), using rats, observed a significantly higher level of ^{45}Ca uptake by the saccular otoconial membranes than by those of the utricle. On the basis of these experimental data, these authors suggested that the otoconia are in a dynamic state. It is known that the otoconia may degenerate or dissolve with age (Ross *et al.*, 1976) or may fuse to form giant crystals after intoxication by ototoxic drugs (Lim, 1980).

Whether regeneration of otoconia can occur after their loss in adult animals has not been resolved. Some authors have reported the appearance of new otoconia, presumably by neogenesis, after removal in frogs, rabbits, guinea-pigs (Werner, 1933) and carp (von Frisch, 1938). On the other hand, James (1962) failed to observe any evidence of generation of new otoconia after their removal in rabbits.

Morphogenesis of Otoconia

Little is known about the mechanism of formation of otoconia or of the factors regulating their growth and final dimensions. As for their morphogenesis, four hypotheses have been proposed: (1) that the otoconia are formed *in situ* on the maculae in mammals (Stricker, 1928; Belonoschkin, 1931; Lyon, 1955; Wright and Hubbard, 1983); (2) that the otoconia are secreted by the supporting cells of the gravity receptor organs in the statocyst of *Aplysia* (Geuze, 1968) or formed inside the otoconioblasts in Ctenophora (Vinnikov *et al.*, 1981); (3) that the otoconia are produced by the dark cells of the vestibule (Wright and Hubbard, 1983); and (4) that the otoconia are formed in the endolymphatic sac and transported to the gravity receptor organs in the chick (Balsamo, de Vincentiis and Marmo, 1969) and the shark (Vilstrup, 1951).

Veenhof (1969) – using autoradiography, cytochemistry and electron microscopy – and Lyon (1955) – using light microscopy with polarizing light – studied the formation of otoconia in the mouse. It is reported that the otoconia start to form at about 14 days of gestation and complete their formation by 7 days after birth, and the maximum activity in the formation of crystals appears to occur between the 15th and 16th days of gestation (Veenhof, 1969). This finding has been essentially confirmed in our laboratory (Lim and Erway, unpublished observations) and by others (Anniko, 1980) using electron microscopy.

It is generally believed that the organic substance is related to the growth of otoconia, possibly serving as a nucleation site during formation, although the exact mechanism of calcite seeding and growth is not fully understood (Lyon, 1955; Salamat, Ross and Peacor, 1980). Veenhof (1969) observed a transient increase in periodic acid-Schiff (PAS) staining and a transient decrease in intensity of staining of the otoconial membrane during the critical period of otoconial formation. Since PAS stains neutral sugars and toluidine blue and Alcian blue stain acid protein polysaccharides, the above results indicate that there are transient changes in the composition of carbohydrates and related substances which may be critical for initiating the formation of crystals. These chemical changes are coupled with the presence of alkaline phosphatase activity up to the 13th day of gestation, after which activity begins to fade rapidly (Veenhof, 1969), thus indicating potential biochemical events that make the gelatinous membrane favourable for crystallization during the critical period of otoconial formation. Such biochemical changes in reverse may explain the deterrent factor to otoconial formation in similar but non-otoconia-forming structures of the inner ear, such as the tectorial membrane and cupula, as discussed by Veenhof (1969). His observations also indicate that there is no saturation of calcium in the tissue, as judged by autoradiography, although he could not determine whether a high concentration of calcium in the endolymph occurs before or during the crystallization period. In a recent study of endolymph ionic concentration in the developing inner ear, using X-ray analysis and frozen sectioned tissue, Anniko and Wroblewski (1981) found no significant concentration of calcium in the endolymph during the otoconial forming period. However, Anniko (1980) noted a concentration of calcium in the apical portion of the epithelial cells of the macula during this critical period, but he was unable to determine whether this high calcium concentration occurred in the sensory cells or in the supporting cells or in both.

The 'critical nucleation' concept of the initiation of crystal formation has been supported by several investigators (Veenhof, 1969; Lim, 1973). According to this concept, pre-existing heterogeneous nucleation initiates the growth of crystals. This nucleus in the crystal, described by various investigators who used the light microscope (Henle, 1873; Kolmer, 1927; Lindeman, 1969), has proved to be an organic substance (Lim, 1973) that may serve as a nucleation site at which the otoconium is formed. On the other hand, Nakahara and Bevelander (1979) proposed that primitive otoconia form a non-mineralized organic mass that serves as a 'template' on which calcium carbonate is deposited after birth.

A recent study on the development of otoconia in rats (Salamat, Ross and Peacor, 1980) demonstrated the multinucleation sites of crystal seeding during the early development of the calcite crystals. This observation is essentially in agreement with our ultrastructural data on fetal mice, except for some differences in detail. These authors described several different shapes of immature otoconia in fetal rats: spindle, trigonal, dumb-bell, and quadrilobed with stepped terminals (*Figure 10.17*); however, the last formation has been interpreted as a type of twinning or of otoconial intergrowth formed by the interpenetration of two crystals at 90 degrees (Ross *et al.*, 1976; Johnsson *et al.*, 1982). Salamat, Ross and Peacor (1980) suggested that in the rat the trigonal crystals become dumb-bell-shaped crystals. In mice and rats the immature otoconia assume similar shapes, although mouse otoconia assume a trigonal shape more often than any other species.

Salamat, Ross and Peacor (1980) observed that the central core in a thin-sectioned dumb-bell-shaped otoconium is made up of arrays of orderly, fine needle-shaped mineral deposits. Since X-ray analysis showed that this core contains phosphorus and calcium, indicating the presence of calcium carbonate and organic substance, they suggested that primitive otoconia are calcite crystals from the beginning. Our data on mice are in agreement with theirs (*see Figure 10.16(b)*). Calcium was distinctly detected in the otoconia of 7-week-old human fetuses, further supporting the idea that the immature otoconia are calcium carbonate crystals (Wright and Hubbard, 1983).

The majority of the otoconia in the newborn are mature forms, suggesting that the immature otoconia develop into the mature form during the later part of the gestational period when calcium incorporation into the otoconia is most active (Veenhof, 1969). Since the crystallite (spindle) arrangement and central-core arrangement are different in the primitive and mature forms, it can be suggested that an internal rearrangement of crystallites must have occurred during their development.

The morphology of the primitive otoconia is closely related to that of the nuclei (or cores) of the crystals (*Figure 10.18*). Several investigators have suggested that the central core of the otoconium contains less $CaCO_3$ and thus is more vulnerable to decalcification during tissue preparation (Salamat, Ross and Peacor, 1980; Vinnikov *et al.*, 1980). The cores are empty in most immature otoconia of mice (*Figure 10.18*) and rats (*Figure 1* in Salamat, Ross and Peacor, 1980), and it can be interpreted that this emptiness is due to decalcification. The other possibility is that the cores contain more minerals than the surrounding bodies, and these hard cores may fall off from the surrounding epoxy resin during sectioning because of poor penetration. On the other hand, these empty cores can also be interpreted as evidence that calcite crystal seeding may be accomplished without organic substances. This mechanism would require a supersaturation of endolymphatic calcium, but there is no evidence of a significant increase of calcium in the endolymph during the critical period of otoconia formation in mice (Anniko and Wroblewski, 1981).

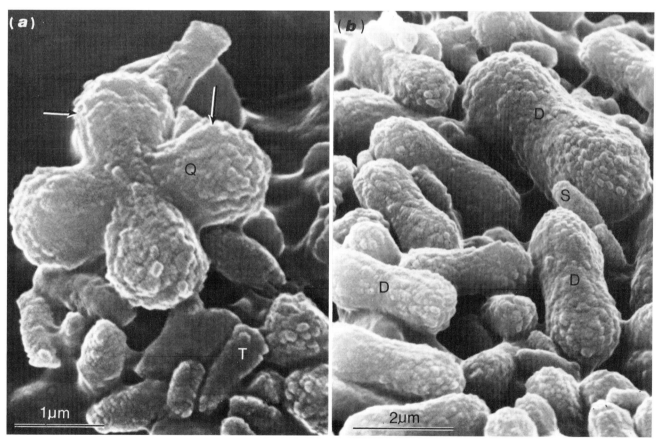

Figure 10.17(a) Scanning electron micrograph (SEM) of the utricle of a rat fetus (18 days old) showing immature types of otoconia: trigonal (T) and twinned or quadrilobed (Q). Arrows point to the stepped arrangement. (From Salamat, Ross and Peacor, 1980, by kind permission of authors and publishers.) (b) SEM of the utricle of a rat fetus (18 days old) showing dumb-bell-shaped (D) and spindle-shaped (S) primitive otoconia. (From Salamat, Ross and Peacor, 1980, by kind permission of authors and publishers.)

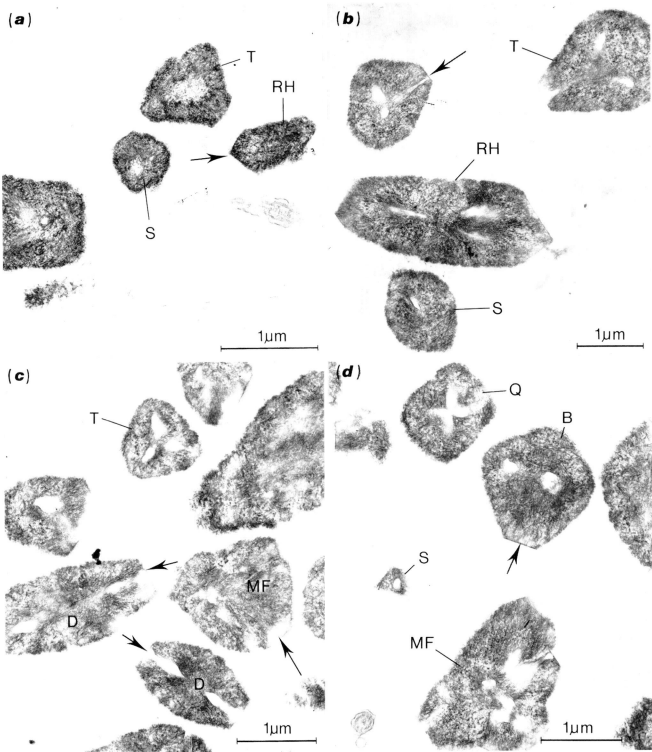

Figure 10.18*(a) Immature forms of mouse utricular otoconia in singlet (S), trigonal (T), and rhombohedral (RH) shapes with pointed ends (arrow) (undecalcified, 16½ days of gestation). Observe the less dense nuclei in the centre of the otoconia. (b) Same specimen as (a). Immature otoconia show trigonal (T), rhombohedral (RH) and singlet (S) types with clear (empty) nuclei. The nuclei in the trigonal-shaped crystal extend to the terminal face (arrow). Observe the well-defined terminal faces in which the crystallites (spindles) are* arranged at right angles to the surface. (c) Same specimen as in (a). Immature forms of otoconia – trigonal (T), doublet (D) and multifaceted (MF) – are shown. The doublet type and multifaceted type have nuclei (arrows) which extend to the terminal face of the crystal. (d) Same specimen as in (a). Mononuclear (S), binuclear (B), quadrinuclear (Q) and polynuclear (MF) crystals are shown. Observe the well-defined terminal face of one of the crystals (arrow)*

It has been suggested that the primitive otoconia are formed from cytoplasmic blebs of supporting cells (Harada, 1979; Anniko, 1980). However, these blebs are more numerous when fixation of the tissue is poor. Instead, organic substance appears to be expelled by exocytosis (secretory process) of the supporting cell (see Figure 10.15), and specific granules appear to be intimately involved in this process (Lim and Erway, unpublished observations). Recently, on the basis of a study of chick and quail embryos, Il'Inskaya (1981) has suggested that the initial stage of otoconia lies in the formation of acid mucopolysaccharides that are derived from the glycocalyx.

We have considered whether otoconia may be formed outside the sensory epithelium and then transported to the gravity receptors. The vestibular dark cells sometimes have otoconia adhering to their surfaces in adult animals (Figure 10.19) (Lim, 1973; Harada and Sugimoto, 1977), and in the human newborn (Figure 10.20) (Wright, personal communication). Preston et al. (1975) demonstrated that the vestibular dark cells contain higher concentrations of radioisotope-labelled calcium than the surrounding tissue, thus suggesting that the dark cells are involved in active calcium transport. It has been suggested that these otoconia have been dislodged from the gravity receptors and have been demineralized (Lim, 1973). On the other hand, since immature otoconia are also found on the surface of the dark cells in fetal mice and human infants (Figure 10.20), Wright and Hubbard (1983) suggested that otoconia may be formed by the dark cells in the early developmental stages but would be later reabsorbed by these cells. However, in our material (over 100 fetal mice examined by light and electron microscopy), the otoconia were restricted with rare exceptions to the sensory epithelial area of both the utricle and the saccule (Figure 10.21). Furthermore, the gelatinous layer in fetal mice has not been fully developed; consequently, the concept that otoconia formed elsewhere (as in the endolymphatic sac) may be transported and attached to the gelatinous layer is rather unlikely. In fact, the gelatinous layer is absent in the gravity receptors in 18½-day-old mice, although the cupula has developed (Lim and Erway, unpublished observations). The gelatinous layer is also absent during the early stages of otoconial formation in man (Wright and Hubbard, 1983).

There is some compelling evidence that trace elements such as Mn and Zn are essential for the formation of otoconia during the critical stages of their development (Purichia and Erway, 1972). Mutant pallid mice are known to produce offspring with congenital otoconia deficiencies, but Mn supplementation during pregnancy prevents this deficiency (Erway, Fraser and Hurley, 1971). It has been suggested that Mn is essential for the normal biosynthesis of mucopolysaccharides, which are required for otoconial seeding and growth, and that Zn is an essential component of the enzyme carbonic anhydrase, which contains a Zn molecule. It has been further shown that the administration of carbonic anhydrase inhibitors prevents the formation of otoconia in chicks (de Vincentiis and Marmo, 1968) and in mice (Purichia and Erway, 1972). Therefore, it can be suggested that mucopolysaccharides and carbonic anhydrase are essential for the formation of otoconia (de Vincentiis and Marmo, 1968; Erway, Fraser and Hurley, 1971; Purichia and Erway, 1972). Furthermore, the sulphation of mucopolysaccharides is critically related to the calcification process of the otoconia (Shrader, Erway and Hurley, 1973).

Degenerated and Malformed Otoconia

Although the purpose of this chapter is to discuss normal otoconial morphology, some recently recognized otoconial abnormalities have provided considerable information regarding basic otoconial biology as well as the profound clinical implications of otoconial pathology. Therefore, a short discussion of these otoconial pathologies and abnormalities is included here.

Contrary to the popular belief that otoconia are very stable, recent evidence suggests that they may degenerate after displacement (Lim, 1973), with ageing (Ross et al., 1976), and under various pathological conditions in man (Johnsson et al., 1981, 1982) and laboratory animals (Lim, Erway and Clark, 1978; Lim, 1980). The degeneration of otoconia may occur when they are displaced from the macula and attached to the surface of the dark cells, in which case the degenerating otoconia appear collapsed, probably due to demineralization (see Figure 10.19). Cupulolithiasis is characterized clinically by a positional vertigo, and its pathology is characterized by a deposition of inorganic substance on the cupula of the posterior crista. It is suggested that this deposit is composed of displaced, degenerated utricular otoconia (Schuknecht, 1964). Interestingly, there is a high incidence of temporal bone fracture (47 per cent), head injury (21 per cent), and otitis media (26 per cent) among those who suffer from cupulolithiasis (Schuknecht, 1974). Wright and Hubbard (1978) also reported a significantly higher incidence of separated statoconial membranes among human infants with middle ear effusions than in those without effusion. Another type of otoconial degeneration is seen in aged individuals where the otoconial body is pitted or hollowed out, probably due to the loss of mineral salts as a result of ageing (Ross et al., 1976). An increased incidence of abnormal otoconia has also been reported under certain pathological conditions, such as ethacrynic acid, neomycin, and streptomycin ototoxicity (Lim, 1980; Harada and Sugimoto, 1977). Lim (1980) suggested that the abnormal otoconia, particularly the giant ones, are formed by the fusion of smaller crystals. Biochemical changes may occur as a result of ototoxicity and may be responsible for the formation of abnormal otoconia. Wright and Hubbard (1978) also observed an increased incidence of giant crystals in their infant patients with otitis media with effusion.

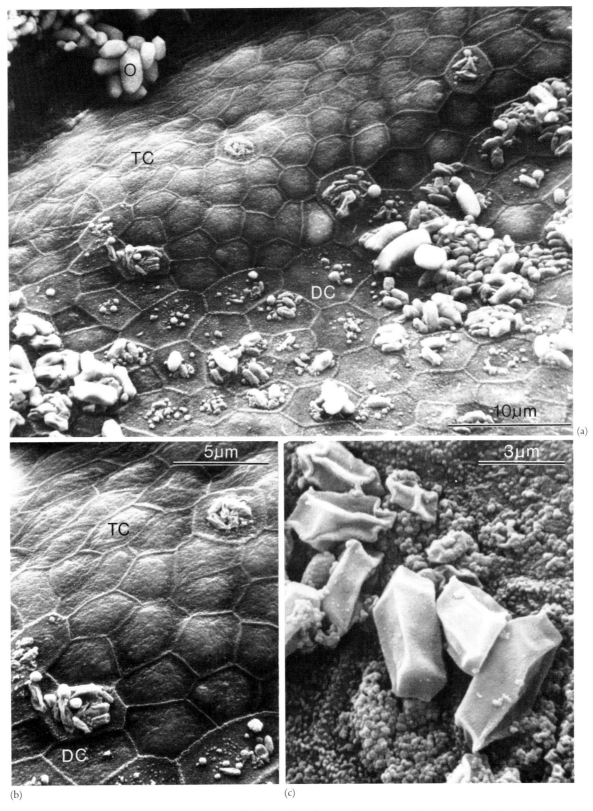

Figure 10.19(a) Degenerating otoconia that are adhering to the dark cells (DC) of a guinea-pig utricle. Otoconia (O) that are in place on the sensory epithelium appear normal. Observe the transitional cells (TC) that are free of otoconia. (b) Close-up view of apparently collapsed otoconia adhering to the dark cells. TC = Transitional cells. (c) Collapsed otoconia that are adhering to the surface of the dark cells. The collapse of the crystals is thought to be due to demineralization

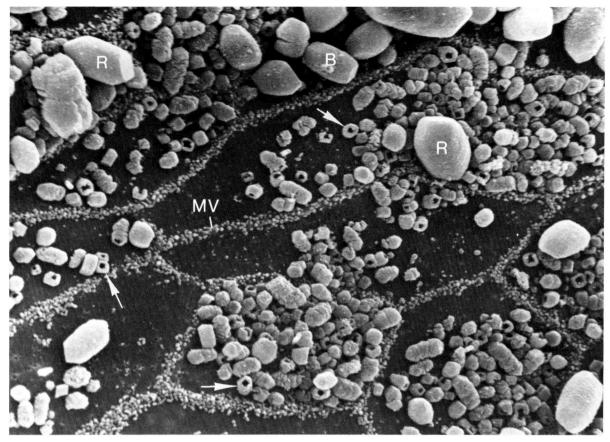

Figure 10.20 *Numerous small immature crystals, some of which have holes (arrows) and displaced mature crystals adhering to the dark cells of the inferior utricular wall of a human infant, are shown. Some crystals have rhombohedral terminal faces (R) and some have flat terminal faces (B). MV = Microvilli. (Reproduced with kind permission of Dr C. G. Wright.)*

Occurrence of aggregated apatite spherulites has been reported among gravity receptors in association with the collapsed wall of the membranous labyrinth (Johnsson *et al.*, 1982). These authors also found an admixture of calcite and apatite crystals in the same specimen. Interestingly, the spherulitic apatite described by Johnsson *et al.* (1981) was found in rat statoconial membranes that had been stored in glutaraldehyde in phosphate buffer for a prolonged period. (It is important to note that the human specimens that Johnsson *et al.* (1982) used had never been exposed to phosphate buffer.) Johnsson and his co-workers suggested that some of these abnormal deposits are secondary and considered possible explanations for their occurrence: (1) a layer of calcium phosphate was deposited over an aggregate of normal calcite otoconia; (2) pre-existing calcite otoconia were removed by solution, followed by calcium phosphate deposition; (3) calcite otoconia were transformed *in situ* into apatite by reaction with phosphate solution.

Certain mutant mice may have either otoconial deficiency (pallid mice) or abnormality (tilted-head mice) (Lim and Erway, 1974; Lim, Erway and Clark, 1978). The abnormal otoconia of tilted-head mice are bow-tie shaped and appear to be composed of leaflets formed by uneven condensation (or incorporation) of organic substances within the crystal (Lim, 1978).

Recently, Wright, Hubbard and Graham (1979) and Wright *et al.* (1981) have described the absence of otoconia in human infants; this resembles the condition seen in pallid mice. These investigators have also observed discoid vaterite otoconia (*Figure 10.22*) in a second-trimester fetus and in an infant with Potter's syndrome (Wright *et al.*, 1982). This vaterite crystal appears identical with the vaterite described in gar fish (Wright, personal communication).

Environmental influences on otoconial development and maintenance have attracted considerable research interest in recent years. As a result of the successful space programme, the influence of gravity on the inner ear is of particular interest. Apparently, rats that were born and raised in hypergravity (Lim *et al.*, 1974) and chicks that were hatched in hypergravity (Howland and Ballarino, 1978) did not show any significant alteration in the otoconia, although the gravity receptors were reported to be slightly smaller under such conditions in the chick. Rats that were flown in space failed to show any significant change in their otoconial morphology,

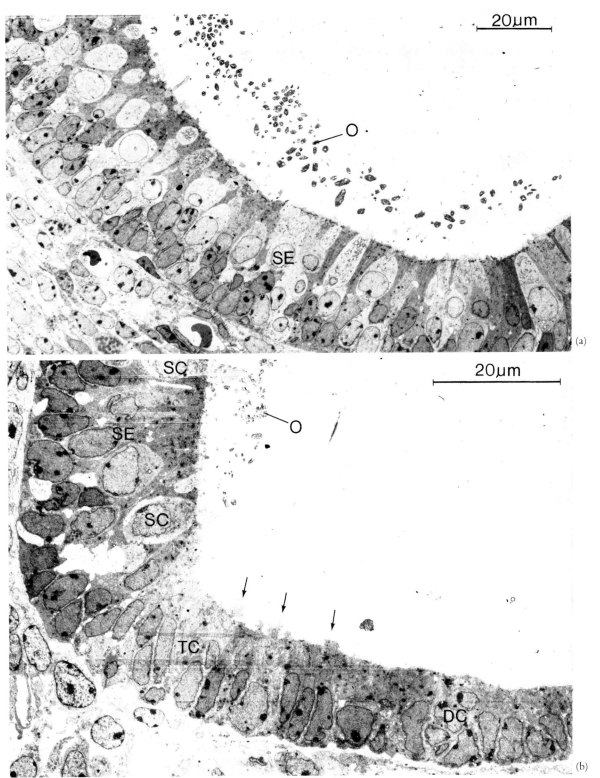

Figure 10.21(a) Transmission electron micrograph of the developing saccular sensory epithelium (SE) of the mouse (16½ days' gestation) shows numerous sensory cells and the otoconia (O). In the absence of the gelatinous layer, however, the fact that the newly formed otoconia are remaining together over the sensory epithelium suggests some substances are holding them together but were lost in tissue preparation. The sensory cells are not yet fully mature and the nerve calyx which characterizes the type I sensory cell is also not yet identified. (b) The utricle of a mouse (16½ days' gestation) shows otoconia (O) that are restricted to the sensory epithelium (SE). No appreciable primitive otoconia are found on the transitional cells (TC) or the dark cells (DC). Arrows point to protrusions of the transitional cells. SC = Sensory cells

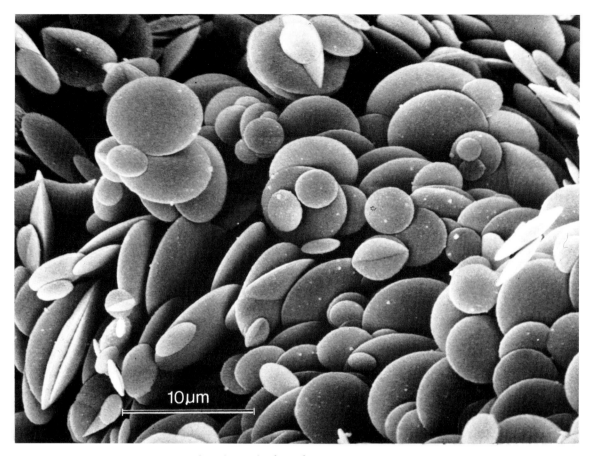

Figure 10.22 *Discoid vaterite otoconia from the utricle of an infant suffering from Potter's syndrome (From Wright et al. (1981) by kind permission of the authors and publishers.)*

and the otoconia of frogs that were hatched in a spaceship also developed normally (Vinnikov *et al.*, 1980). However, their earlier results indicated some noticeable otoconial changes in rats exposed to zero gravity for 20 days (Vinnikov *et al.*, 1979). In toads (*Xenopus laevis*) and fishes (*Branchyodanio rerio*) that were hatched in a spaceship, some disturbances in the formation of the otolith membrane and otoconia have also occurred (Vinnikov *et al.*, 1976).

The functional consequence of these otoconial pathologies is not yet fully understood; however, it is apparent that otoconial pathologies do exist both in animals and in man. Mutant mice with otoconial deficiency show head tilting, circling behaviour, absent or reduced air righting reflex, and an inability to swim (Lim and Erway, 1974; Lim, Erway and Clark, 1978). It can be speculated, therefore, that otoconial deficiency and severe degenerative changes of otoconia in man would also influence balance and co-ordination. Therefore, it is vitally important to understand the mechanisms of otoconia formation and maintenance in order to understand the processes of otoconial malformation and degeneration in pathological conditions.

Acknowledgements

I thank Drs Dennis R. Trune, Muriel D. Ross and C. Gary Wright for their critical reviews of the manuscript; Katherine Adamson and Carolyn Steely for manuscript preparation; Steven Lee McBride for photography; Nancy Sally for illustration; and Ilija Karanfilov for electron microscope photography.

Support from NASA (NSG-2220) and The Deafness Research Foundation for a large portion of the work is gratefully acknowledged.

References

Ades, H. W. and Engström, H. (1965) Form and innervation of the vestibular epithelia. In *The Role of the Vestibular Organs in the Exploration of Space.* NASA SP-77, pp. 23–41. NASA.

Anniko, M. (1980) Development of otoconia. *American Journal of Otolaryngology,* 1, 400–410.

Anniko, M. and Wroblewski, R. (1981) Elemental composition of the developing inner ear. *Annals of Otology, Rhinology and Laryngology,* 90, 25–32.

Balsamo, G., de Vincentiis, M. and Marmo, F. (1969) The effect of tetracyclin on the processes of calcification of the otoliths in the developing chick embryo. *Journal of Embryology and Experimental Morphology,* 22, 327–332.

Belanger, L. F. (1960) Development, structure and composition of the otolithic organs of the rat. In *Calcification in Biological Systems,* ed. R. F. Sognnaes, pp. 151–162. American Association for the Advancement of Science: Washington, D.C.

Belonoschkin, B. (1931) Beitrag zur Frage der Natur und der Entstehung der Otolithen. *Archiv für Ohrenheilkunde,* 128, 208–224.

Breschet, G. (1936) Recherches anatomiques et physiologiques sur l'organe de l'audition des oiseaux (as cited by Veenhof, 1969).

Carlström, D. and Engström, H. (1955) The ultrastructure of statoconia. *Acta Otolaryngologica (Stockholm),* 45, 14–18.

Carlström, D. (1963) A crystallographic study of vertebrate otoliths. *Biological Bulletin,* 125, 441–463.

de Vincentiis, M. and Marmo, F. (1968) Inhibition of the morphogenesis of the otoliths in the chick embryo in the presence of carbonic anhydrase inhibitors. *Experientia,* 24, 218–280.

Dohlman, G. F. (1971) The attachment of the cupulae, otolith and tectorial membranes to the sensory cell areas. *Acta Otolaryngologica (Stockholm),* 71, 89–105.

Erway, L. C., Fraser, A. S. and Hurley, L. S. (1971) Prevention of congenital otolith defects in pallid mice by manganese supplementation. *Genetics,* 67, 90–108.

Guardabassi, A. (1952) L'organo endolinfatico degli Anfibi Anuri. *Archivio Italiano di anatomia e di Embriologia,* 57, 242–294.

Guardabassi, A. (1953) Les sels de Ca du sac endolymphatique et les processus de calcification des os pendant le metamorphose normale et experimentale chez les tetards de *Bufo vulgaris, Rana dalmatina, Rana esculenta. Archives d'Anatomie Microscopique et de Morphologie Expérimentale,* 42, 143–167.

Geuze, J. J. (1968) Observations on the function and the structure of the statocysts of *Lymnaea stagnalis* (L). *Netherlands Journal of Zoology,* 18, 155–204.

Harada, Y. (1979) Formation area of the statoconia. SEM/1979/III, pp. 963–966. SEM Inc.: AMF O'Hare, Ill.

Harada, Y. and Sugimoto, Y. (1977) Metabolic disorder of otoconia after streptomycin intoxication. *Acta Otolaryngologica (Stockholm),* 84, 65–71.

Hastings, A. B. (1935) Chemical analysis of otoliths and endolymphatic sac deposits in *Amblystoma tigrinum. Journal of Comparative Neurology,* 61, 295–296.

Henle (1873) Quoted by Rudinger in Strickler's *Manual of Human and Comparative Histology,* vol. 3, p. 122. London: New Sydenham Society.

Howland, H. C. and Ballarino, J. (1978) Is the growth of the otolith controlled by its weight? Presented at the *Satellite Symposium of the Neuroscience Meeting on Vestibular Function and Morphology,* October 30–November 1. Pittsburgh, Penn.

Igarashi, M. and Kanda, T. (1969) Fine structure of the otolithic membrane in the squirrel monkey. *Acta Otolaryngologica (Stockholm),* 68, 43–52.

Il'Inskaya, E. V. (1981) The development of otolithic apparatus in embryonic chicks and quail *Coturnix japonicus. Journal of Evolution, Biochemistry and Physiology (Leningrad),* 17, 80–83.

Istenič, L. and Bulog, B. (1976) *Anatomical Investigations of Membranous Labyrinth in Proteus* (Proteus anguinus Laurenti, Urodela, Amphibia). Dissertation, Slovenska Akademija Znanosti in Umetnosti, Ljubljana.

Iurato, S. and de Petris, S. (1967) Otolithic membranes and cupulae. In *Submicroscopic Structure of the Inner Ear,* ed. S. Iurato, pp. 210–218. Oxford: Pergamon.

James, J. (1962) Some experiments on the function of the labyrinth. II. *Practica Oto-rhino-laryngologica,* 24, 348–350.

Johnsson, L. -G. and Hawkins, J. E., Jr. (1967) Otolithic membranes of the saccule and utricle in man. *Science,* 157, 1454–1456.

Johnsson, L. -G., Rouse, R. C., Hawkins, J. E., Jr., Kingsley, T. C. and Wright, C. G. (1981) Hereditary deafness with hydrops and anomalous calcium phosphate deposits. *American Journal of Otolaryngology,* 2, 284–298.

Johnsson, L. -G., Rouse, R. C., Wright, C. G., Henry, P. J. and Hawkins, J. E., Jr. (1982) Pathology of neuroepithelial suprastructures of the human inner ear. *American Journal of Otolaryngology,* 3, 77–90.

Kellerhals, B., Marti, E. and Villiger, W. (1970) Surface view of the guinea pig otolithic membrane. *Practica Oto-rhino-laryngologica,* 32, 65–73.

Kolmer, W. (1927) Gehörorgan. In *Handbuch der Mikroskopischen Anatomie des Menschen,* ed. W. Möllendorff. pp. 250–478. Berlin: Springer-Verlag.

Krieger, E. (1840) *De Otolithis.* Thesis. Berlin.

Lim, D. J. (1969) Three dimensional observation of the inner ear with the scanning electron microscope. *Acta Otolaryngologica (Stockholm),* Supplement 255.

Lim, D. J. (1973) Formation and fate of the otoconia: Scanning and transmission electron microscopy. *Annals of Otology, Rhinology and Laryngology,* 82, 23–36.

Lim, D. J. (1974) The statoconia of the non-mammalian species. *Brain Behaviour and Evolution,* 10, 37–41.

Lim, D. J. (1977) Ultra anatomy of sensory end-organs in the labyrinth and their functional implications. In *Proceedings of the Shambaugh Fifth International Workshop on Middle Ear Microsurgery and Fluctuant Hearing Loss,* eds. G. E. Shambaugh, Jr. and J. J. Shea, pp. 16–27. Huntsville, Ala.: Strode Publishers.

Lim, D. J. (1979) Fine morphology of the otoconial membrane and its relationship to the sensory epithelium. *SEM Inc./1979/III,* pp. 929–938. SEM: AMF O'Hare, Ill.

Lim, D. J. (1980) Morphogenesis and malformation of otoconia: A review. In *Birth Defects: Original Article Series,* Vol. XVI, No. 4, pp. 111–146. March of Dimes Birth Defects Foundation.

Lim, D. J. and Erway, L. C. (1974) Influence of manganese on genetically defective otolith: A behavioral and morphological study. *Annals of Otology, Rhinology and Laryngology,* 83, 565–581.

Lim, D. J., Erway, L. C. and Clark, D. L. (1978) Tilted-head mice with genetic otoconial anomaly. Behavioral and morphological correlates. In *Vestibular Mechanisms in Health and Disease,* ed. J. D. Hood, pp. 195–206. London: Academic Press.

Lim, D. J., Stith, J. A., Stockwell, C. W. and Oyama, J. (1974) Observations on saccules of rats exposed to long-term hypergravity. *Aerospace Medicine,* 45, 705–710.

Lindeman, H. H. (1969) Studies on the morphology of the sensory regions of the vestibular apparatus. *Advances in Anatomy, Embryology and Cell Biology,* 42, 1.

Lindeman, H. H. (1973) Anatomy of the otolith organs. *Advances of Otorhinolaryngology,* 20, 405–433.

Lindeman, H. H., Ades, H. W. and West, R. W. (1973) Scanning electron microscopy of the vestibular end organs. In *Fifth Symposium on the Role of the Vestibular Organs in Space Exploration,* NASA SP-314, pp. 145–156. NASA.

Lyon, M. F. (1955) The development of the otoliths of the mouse. *Journal of Embryology and Experimental Morphology,* 3, 213–229.

Marco, J., Sanchez-Fernandez, J. Ma. and Rivera-Pomar, J. Ma. (1971) Ultrastructure of the otoliths and otolithic membrane of the macula utriculi in the guinea pig. *Acta Otolaryngologica (Stockholm),* 71, 1–8.

Nakahara, H. and Bevelander, G. (1979) An electron microscope study of crystal calcium carbonate formation in the mouse otolith. *Anatomical Record,* 193, 233–242.

Palmork, K. H., Taylor, M. E. U. and Coates, R. (1963) The crystal structure of aberrant otoliths. *Acta Chemica Scandinavica*, **17,** 1457–1458.

Pannella, G. (1971) Fish otoliths: Daily growth layers and periodical patterns. *Science*, **173,** 1124–1127.

Peacor, D. R., Rouse, R. C. and Ross, M. D. (1980) Critique of 'An electron microscope study of crystal calcium carbonate formation in the mouse otolith'. *Anatomical Record,* **197,** 375–376.

Preston, R. E., Johnsson, L. -G., Hill, H. J. and Schacht, J. (1975) Incorporation of radioactive calcium into otolithic membranes and middle ear ossicles of the gerbil. *Acta Otolaryngologica (Stockholm),* **80,** 269–275.

Purichia, N. and Erway, L. C. (1972) Effects of dichlorophenamide, zinc, and manganese on otolith development in mice. *Developmental Biology,* **27,** 395–405.

Reibisch, A. (1899). Über die Eizahl bei Pleurobectes platessa und die Altersbestimmung dieser Form aus den Otolithen. Abt. Keil N. F. Bnd. 4, der Wissensch. Meeres Untersuchungen (as cited by Veenhof (1969)).

Ross, M. D. (1979) Calcium ion uptake and exchange in otoconia. *Advances in Oto-rhino-laryngology,* **25,** 26–33.

Ross, M. D. and Peacor, D. R. (1975) The nature and crystal growth of otoconia in the rat. *Annals of Otology, Rhinology and Laryngology,* **84,** 22–36.

Ross, M. D. and Williams, T. J. (1979) Otoconial complexes as ion reservoirs in endolymph. *Physiologist,* **22,** S63–S64.

Ross, M. D., Johnsson, L. -G., Peacor, D. and Allard, L. F. (1976) Observations on normal and degenerating human otoconia. *Annals of Otology, Rhinology and Laryngology,* **85,** 310–326.

Ross, M. D., Pote, K. D., Cloke, P. L. and Corson, C. (1980) *In vitro* $^{45}Ca^{++}$ uptake and exchange by otoconial complexes in high and low K^+/Na^+ fluids. *Physiologist,* **23,** S129–S130.

Ross, M. D., Mann, S., Parker, S., Skarnulis, A. J. and Williams, R. J. P. (1982) High resolution electron microscopy and electron beam diffraction studies of otoconia and otoliths: Are otoconia single crystals? In *Abstracts of the Fifth Midwinter Research Meeting*, pp. 47–48. Association for Research in Otolaryngology.

Salamat, M. D., Ross, M. D. and Peacor, D. R. (1980) Otoconial formation in the fetal rat. *Annals of Otology, Rhinology and Laryngology,* **89,** 229–238.

Sanchez-Fernandez, J. Ma., Marco, J., Rivera-Pomar, J. Ma. and Delgado, R. M. (1972) Electron diffraction studies on otolith organization in the macula utriculi of the guinea pig. *Acta Otolaryngologica (Stockholm),* **73,** 267–269.

Schuknecht, H. F. (1964) The pathology of several disorders of the inner ear which might cause vertigo. *Southern Medical Journal,* **57,** 1161–1167.

Schuknecht, H. F. (1974) *Pathology of the Ear*, pp. 465–573. Cambridge, Mass: Harvard University Press.

Shrader, R. E., Erway, L. C. and Hurley, L. S. (1973) Mucopolysaccharide synthesis in the developing inner ear of manganese-deficient and pallid mutant mice. *Teratology,* **8,** 257–266.

Stricker, W. (1928) Die Otolithen und Cupulae terminalis im Gehörorgan. *Archiv für Mikroskopische Anatomie (und Entwicklungsmechanik),* **103,** 259.

Trincker, D. (1972) The transformation of mechanical stimulus into nervous excitation by the labyrinthine receptors. In *Biological Receptor Mechanisms*, ed. J. W. L. Beament, pp. 289–316. Cambridge: The University Press.

Veenhoff, V. B. (1969) *The Development of Statoconia in Mice*. Amsterdam: North-Holland.

Vilstrup, T. (1951) On the formation of the otoliths. *Annals of Otology, Rhinology and Laryngology,* **60,** 974–981.

Vinnikov, Ya. A., Gazenko, O. G., Titova, L. K., Bronstein, A. A., Govardovskii, V. I., Pal'mbach, L. P., Pevzner, R. A., Gribakin, F. G., Aronova, M. Z., Kharkeevich, T. A., Tsirulis, T. P., Pyatkina, G. A. and Semak, T. V. (1976) Formation of vestibular apparatus in the weightless condition. *Minerva Otorinolaringologica,* **25,** 69–75.

Vinnikov, Ya. A., Gazenko, O. G., Titova, L. K., Bronstein, A. A., Govardovskii, V. I., Gribakin, F. G., Pevzner, R. A., Aronova, M. Z., Kharkeevich, T. A., Tsirulis, T. P., Pyatkina, G. A., Lichakov, D. V., Pal'mbach, L. P. and Anichin, V. F. (1979) The structural and functional organization of the vestibular apparatus of rats exposed to weightlessness for 20 days on board the Sputnik 'Kosmos-782'. *Acta Otolaryngologica (Stockholm),* **87,** 90–96.

Vinnikov, Ya. A., Lychakov, D. V., Pal'mbach, L. R., Bovardovskiy, V. I., Andanina, V. O., Allakhverdov, B. L. and Pogorelov, A. G. (1980) Studies on the vestibular apparatus of the clawed toad *Xenopus laevis* and rat under the conditions of prolonged weightlessness. *Journal of Evolution, Biochemistry and Physiology (Leningrad),* **16,** 574–579.

Vinnikov, Ya. A., Aronova, M. Z., Kharkeevich, T. A., Tsirulis, T. P., Lavrova, E. A. and Natochin, Yu. V. (1981) Structural and chemical features of the invertebrate otoliths. *Zeitschrift für Mikroskopische – Anatomische Forschung,* **1,** S.127–140.

von Frisch, K. (1938) Über die Bedeutung des Sacculus und der Lagena für den Gehörsinn der Fische. *Zeitschrift für Vergleichende Physiologie,* **25,** 703.

Werner, C. F. (1933) Die Differenzierung der Maculae im Labyrinth, insbesondere bei Saugetieren. *Acta Anatomica,* **99,** 696–709.

Wislocki, G. B. and Ladman, A. J. (1955) Selective and histochemical staining of the otolithic membranes, cupulae and tectorial membrane of the inner ear. *Journal of Anatomy,* **78,** 3–12.

Wright, C. G. and Hubbard, D. G. (1978) Observations of otoconial membranes from human infants. *Acta Otolaryngologica (Stockholm),* **86,** 185–194.

Wright, C. G. and Hubbard, D. G. (1983) SEM observations on development of human otoconia during the first trimester of gestation. *Acta Otolaryngologica (Stockholm).* (In press).

Wright, C. G., Hubbard, D. G. and Clark, G. M. (1979a) Observations of human fetal otoconial membranes. *Annals of Otology, Rhinology and Laryngology,* **88,** 267–274.

Wright, C. G., Hubbard, D. G. and Clark, G. M. (1979) Observations of human fetal otoconial membranes. *Annals of Otology, Rhinology and Laryngology,* **88,** 267–274.

Wright, C. G., Hubbard, D. G. and Graham, J. W. (1979) Absence of otoconia in a human infant. *Annals of Otology, Rhinology and Laryngology,* **88,** 779–783.

Wright, C. G., Rouse, R. C., Johnsson, L. -G., Weinberg, A. G. and Hubbard, D. G. (1982) Vaterite otoconia in two cases of otoconial membrane dysplasia. *Annals of Otology, Rhinology and Laryngology,* **91,** 193–199.

269

11
The Primary Vestibular Neurons

Björn Bergström

Gross and Topographic Anatomy

The primary vestibular neurons originate from bipolar cells in the vestibular ganglion; the dendrites enter the vestibular sensory regions where they make synaptic contacts with the hair cells, while the neurites project themselves centrally to the vestibular nuclei and the cerebellum. They are the afferent fibres of the vestibular nerve. There are also efferent fibres to the vestibular end-organs and to the cochlea as well as adrenergic fibres.

The vestibular nerve divides into one superior and one inferior division: the superior division passes with the facial nerve over the transverse crest, a bony ridge that divides the fundus region in two halves; the inferior division runs with the cochlear nerve below the transverse crest, with the latter in an anterior–inferior position.

The vestibular ganglion is situated in the lateral part of the internal auditory meatus, partly overriding the transverse crest, and a superior and an inferior portion can be distinguished. These two portions are connected by the isthmus ganglionaris (Alexander, 1899). Distal to the ganglion, the superior vestibular division gives off the lateral and anterior ampullary nerves and the utricular nerve. The two ampullary nerves are in fact virtually inseparable up to their entrance into the otic capsule and may be regarded as one nerve which divides peripherally into two branches (Gacek, 1969).

The inferior division emits the saccular nerve and the posterior ampullary nerve, which nearly always has a thin accessory branch (Gacek, 1961; Montandon, Gacek and Kimura, 1970; Bergström, 1973a; Okano, Sando and Meyers, 1980).

The nerve fibres from the superior division innervate the lateral and anterior ampullary crests and the macula utriculi, while the fibres from the inferior division innervate the macula sacculi and the posterior ampullary crest. However, a small part of the macula sacculi is innervated by a branch from the superior division (Retzius, 1881; Voit, 1907), usually called Voit's nerve. There is also a connection between the lateral ampullary and utricular nerves which is believed to contribute to the innervation of the macula utriculi (Shute, 1951). In mammals a small nerve branch from the inferior division of the vestibular nerve to the cochlear nerve, the vestibulo-cochlear anastomosis, has been described by Oort (1918) and it has been shown to contain efferent and afferent fibres to the cochlea (Rasmussen, 1946, 1953). Fibres belonging to the nervus intermedius are carried by a connection between the facial and vestibular nerves (Gacek and Rasmussen, 1961).

During their course in the meatus the vestibular nerve fibres rotate, so that in the vestibular ganglion the cells of the two superior ampullary nerves will be found in the superior portion and those of the utricular nerve directly beneath, while the ganglion cells of the saccular and posterior ampullary branches are located in the inferior portion of the ganglion below the transverse crest. These relations between the different nerve branches are then maintained distally to the vestibular end-organs.

The Number of Myelinated Fibres in Man, their Calibres and Age-Related Changes

The total number of myelinated vestibular nerve fibres in the young and healthy human averages 18 400 (Rasmussen, 1940; Bergström, 1972, 1973b). About two-thirds of the neurons belong to the superior division; the lateral and anterior ampullary branches together contain 5900 fibres, the utricular nerve 6000. In the inferior division the saccular nerve has 4000 and the posterior ampullary nerve 2500 myelinated fibres, about 10 per cent of the latter being found in the accessory branch.

The diameters of the myelinated fibres vary between 1 and 15 μm, the majority being 6–9 μm thick. The ampullary nerves contain more thick fibres than the macular nerves; in the 'normal' adult human, 16 per cent of the superior ampullary neurons are thicker than 9 μm, and in the posterior branch 12 per cent of the fibres have diameters greater than 9 μm. Corresponding figures for the utricular and saccular nerves are 4 per cent and 7 per cent. Although the otic capsule and the inner-ear structures are fully developed at the 25th fetal week, the vestibular nerve fibres in the newborn child are considerably thinner than in the adult, the majority of fibres being 4–6 μm thick (Bergström, 1973c). Myelinization of the nerve fibres begins at the 16th fetal week (Rosenhall, 1974); the process is not completed until puberty (Langworthy, 1933) and there is a gradual increase in the thickness of the myelin sheath with increasing age (Westphal, 1897).

With increasing age, probably beginning around the age of 40, the number of sensory cells, nerve fibres, and vestibular ganglion cells gradually fall, so that an 80-year-old person may have lost as many as 40 per cent of these elements (Rosenhall, 1973; Bergström, 1972, 1973b; Richter, 1980). Not only does the number of nerve fibres become reduced, but the neurons that remain also become thinner. This is most clearly seen in the ampullary nerves: in the superior branches only 8 per cent are thicker than 9 μm, and in the posterior nerve no more than 43 per cent of the fibres are thicker than 6 μm. These morphological age-dependent changes are coupled with functional alterations, as confirmed by various vestibular tests (Mulch and Peterman, 1979).

The Intraepithelial Course of the Vestibular Neurons. Nerve Endings and Synapses

Prior to their entry into the vestibular sensory epithelia, the nerve fibres become unmyelinated; the organization of the myelin layers, with their complicated folds around the axon, is best seen where the myelin sheath terminates. After passing through the basement membrane, many neurons run almost 'horizontally' before turning upwards to the sensory cells and most of them divide into two or more branches.

The vestibular sensory cells are of two types, type I and type II, as first described by Wersäll (1956). The type II cells are found in all vertebrates and are thought to be phylogenetically older than the type I cells, which occur only in higher vertebrates.

The type I hair cell is enclosed in a nerve chalice which is the terminal portion of a thick neuron: some neurons terminate in a number of chalices, each surrounding a separate type I cell; others may terminate in a single large chalice enclosing several cells, especially in the striola (Lindeman, 1969). There is a very wide area of contact between the hair cell and the nerve chalice but the chemical synapses are very few. In some areas, mainly at the base of the hair cell, there are invaginations where the plasma membranes lie very close to each other. There are also other areas where the membranes are closely apposed: Engström, Bergström and Ades (1972) suggested that there might sometimes be a real fusion of membranes but Gulley and Bagger-Sjöbäck (1979) found no such fusions. However, in some areas, the membranes came as close as 6–7 nm, although elsewhere the distance between them varied from 30 to 35 nm. It is assumed that synaptic transmissions other than chemical ones might occur in these areas of close apposition.

The type II cell receives many club-shaped nerve endings from one or more afferent neurons, which make synaptic contacts with the cell membrane mainly in the infranuclear but occasionally also in the supranuclear region (Engström, Bergström and Ades, 1972). The afferent nerve endings can be distinguished from the efferent ones by the difference in the sizes of the mitochondria, which are smaller in the efferent than in the afferent neurons. The afferent endings are sparsely granulated and in the synaptic regions various types of synaptic structures can be seen within the hair cell. They may have the shapes of bars, bodies, or balls, and there may be a variety of intermediate forms.

The afferent fibres related to the type I cells are thicker than those in contact with the type II cells. The ampullary branches contain proportionally more thick fibres than the macular nerves and it has been shown in studies on guinea-pigs that, in the maculae, the two cell types are equally represented; by contrast, on the cristae, type I cells predominate, 60 per cent of the cells being type I and 40 per cent type II (Lindeman, 1969; Lindeman, Reith and Winther, 1981).

Occasionally, the same neuron can innervate both a type I and a type II cell, either by a collateral from the thick branch ending in a chalice or by synaptic contact between the chalice and an adjacent type II cell (Engström, Ades and Hawkins, 1965; Engström, Bergström and Ades, 1972; Bagger-Sjöbäck and Gulley, 1979).

The efferent nerve endings are richly granulated. They have no direct contact with the type I cells but terminate on the afferent neurons, usually near the stem of the nerve chalices. The type II cells receive their efferent innervation directly on the hair cell; as a rule only one ending is found contacting the cell, somewhere in the area from the nucleus downwards. These nerve endings are filled with round synaptic vesicles and inside the type II cell a sub-synaptic cistern is often present.

In the lower part of the sensory epithelium, the afferent fibres show a richly branching and complicated pattern with 'en passant' synapses both to sensory cells and to afferent neurons.

Ratio of Hair Cells to Nerve Fibres in the Cristae and Maculae. Irregularly and Regularly Firing Units

The number of sensory cells in the different vestibular regions in man has been determined by Rosenhall (1972a, b). A correlation of his results with the number of nerve fibres to these regions shows that the ratio of hair cells to nerve fibres is 2.6:1 for the superior cristae, 3.1:1 for the posterior crista, 5.6:1 for the macula utriculi and 4.6:1 for the macula sacculi. A similar correlation of Lindeman's (1969) figures for sensory cells and of Gacek and Rasmussen's (1961) for vestibular nerve fibres in the guinea-pig shows approximately the same relationships as in man.

These figures cannot show exactly how many hair cells are innervated by each nerve fibre since the variations are great – one fibre might contact only one hair cell while another innervates several cells – but they give a rough estimation of the pattern of nerve supply in the different sensory regions. The hair cells on the cristae are evidently relatively more richly innervated than those on the maculae: there are also proportionately more thick fibres to the cristae. The conduction velocity in a nerve fibre is directly proportional to its external diameter (Rushton, 1951) which might mean that we react faster to angular than to linear accelerations.

These morphological differences, with the neurons to the macula utriculi innervating on average twice as many hair cells as those to the cristae, support the theories of Walsh et al. (1972) that fibres innervating many hair cells show a regular spontaneous firing activity, and fibres innervating few hair cells show an irregular activity. These authors demonstrated that most of the fibres to the macula utriculi had regular discharge patterns, while they found both regular and irregular units when recording from the ampullary nerves.

Fernandez and his co-workers (Goldberg and Fernandez, 1971; Fernandez, Goldberg and Abend, 1972) found more irregularly discharging units among the neurons to the cristae than to the maculae, and they suggested that the irregular units corresponded to the thick fibres innervating the type I cells, the regular units to the thin fibres of the type II cells. They also recorded units that were not distinctly regular or irregular; these might correspond to neurons innervating both type I and type II cells.

Acknowledgements

I am very grateful to Ms Berit Engström, Uppsala, Sweden, for her generous assistance with the electron microscopy. I also thank the staff of the Department of Photography, Central Hospital, Karlstad, Sweden, who did the photographic laboratory work.

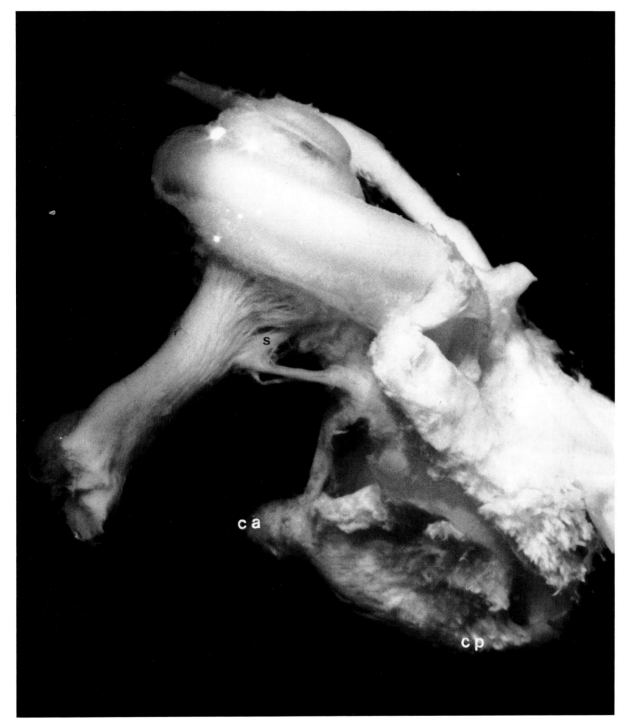

Figure 11.1 *Human inner ear and vestibulo-cochlear nerve viewed from below. The bone has been removed from the cochlea except in the round-window area. Behind the round window is the stapes and the horizontal segment of the facial nerve. The geniculate ganglion is obscured by the apical turns of the cochlea, to the left of which the greater superficial petrosal nerve emerges. Directly below the round window is the membraneous posterior ampulla; the posterior canal is partially opened. The cochlear nerve with its spirally coursing fibres enters the cochlea. The inferior division of the vestibular nerve consists of the saccular and posterior ampullary nerve; the latter nearly always has a thin accessory branch, as can be seen here. ap = Posterior ampulla; ca = anterior and cp = posterior semicircular canal; s = saccular nerve. ×10*

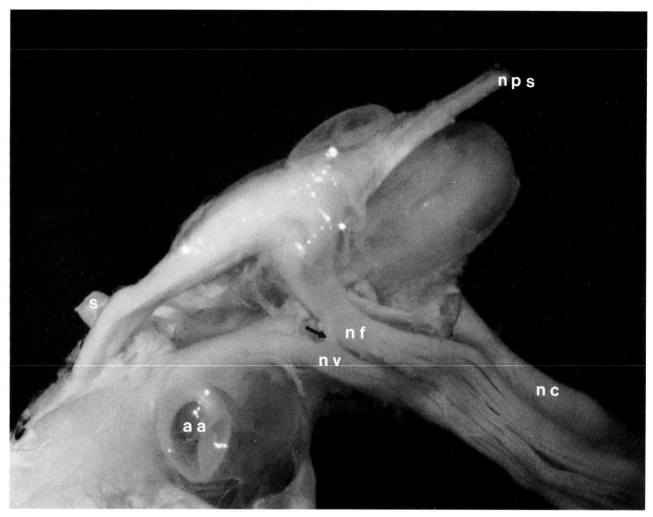

Figure 11.2 *The same ear from a superoposterior view. The superior division of the vestibular nerve runs with the facial nerve. The vestibulo-facial anastomosis is seen where the two nerves separate (arrow) and the vestibular nerve can be followed to the anterior ampulla (aa). nc = Cochlear nerve; nf = facial nerve; nv = superior division of vestibular nerve; s = head of stapes; nps = greater superficial petrosal nerve. ×10*

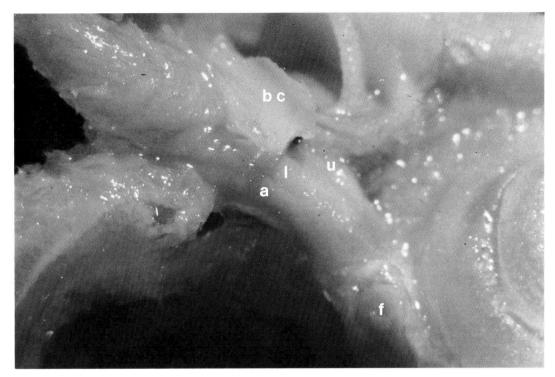

Figure 11.3 *The superior division of the vestibular nerve in man. The ampullary branches divide just prior to their entering the lateral and anterior ampulla. The utricular nerve is seen to the right between the ampullary branches and the stapes. The facial nerve has been resected at the point where it leaves the internal auditory meatus, and*

a small portion of the bony facial canal is seen between the lateral ampulla and the stapes. a = Anterior and l = lateral ampullary nerve; u = utricular nerve; f = cut end of facial nerve; bc = bony facial canal. Original magnification ×12 reduced to 96%

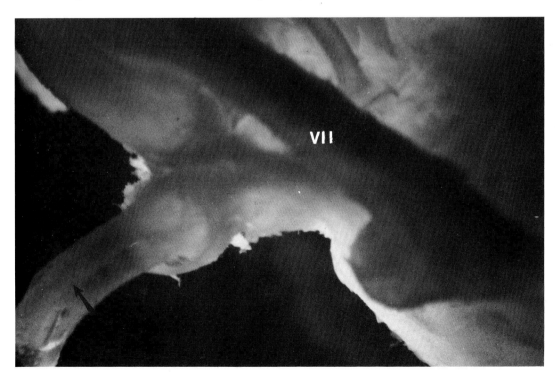

Figure 11.4 *In this case all bone has been removed, the superior ampullary nerves are seen as dark shadows below the facial nerve (VII). They run as a unit and do not divide until very close to the ampullae, where the cristae also appear as dark shadows. The*

anterior membranous semicircular canal (arrow) can be seen surrounded by the perilymphatic space. The anterior crus of the stapes stands just behind the facial nerve. Original magnification ×12 reduced to 96%

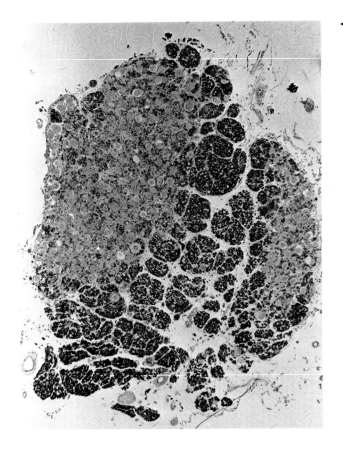

Figure 11.5 *Cross-section through the distal part of the superior portion of the vestibular ganglion in a 22-year-old woman, the anterior border of the nerve to the left. The ganglion cells belong to the ampullary nerves; those of the utricular branch will be found at a more proximal level. ×50*

Figure 11.6 *(below) Horizontal section through the superior division of the vestibular nerve in a 75-year-old man. The anterior margin at top of figure, distal end to the left. The vestibular ganglion cells are rather closely packed; their diameters vary between 15 and 50 μm. In man the superior portion of the ganglion describes an oblique line from the superoanterolateral side to the inferoposteromedial side; this line is not straight, however, but resembles more a 'Y'. At the anterosuperior margin of the nerve a small bulging of the outer contour is seen but otherwise there is no protuberance in the manner so often presented in drawings of the vestibular ganglion. The nerve is richly vascularized, especially in the ganglionic area. ×50*

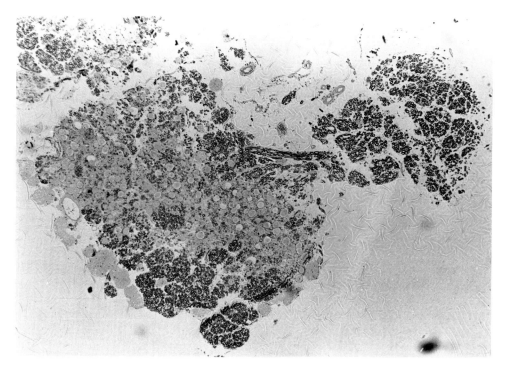

Figure 11.7 Cross-section through the inferior division of the nerve and ganglion, the posterior ampullary nerve to the right of the saccular nerve and its ganglion cells. Same case as in Figure 11.5; this section is 2 mm proximal to the section in that figure. In the upper left corner the isthmus ganglionaris and the superior portion of the ganglion can be seen. × 50

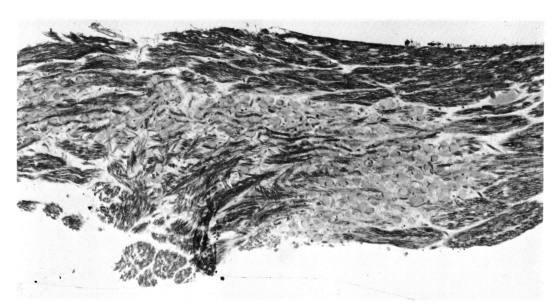

Figure 11.8 Horizontal section through the inferior division of the vestibular nerve and ganglion, same case as in Figure 11.6. The inferior portion of the ganglion has a total length of 4–4.5 mm; the cells belonging to the saccular nerve are situated anterior and distal to those of the posterior ampullary nerve, which leaves the main stem at the bottom of the picture. × 50

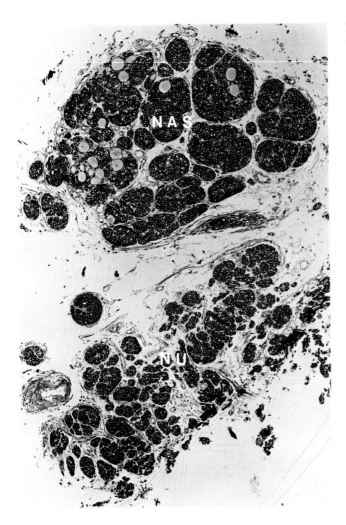

Figure 11.9 *Distal section through the superior vestibular nerve division in a 6-week-old boy. A few scattered ganglion cells are seen in the ampullary nerves, most of them in the anterior part of the nerve. Even at this distal level there is no distinct separation of the anterior and lateral branches. In the utricular nerve no ganglion cells occur at this level. Note the rich vascularization of the nerve branches; the large blood vessel anterior to the utricular nerve is a constant finding (see Figure 11.5). NAS = Superior ampullary nerves; NU = utricular nerve. Original magnification ×50 reduced to 91%*

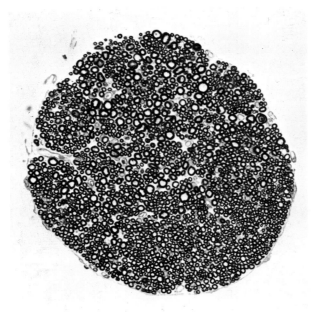

Figure 11.10 *Posterior ampullary nerve in an adult, normal guinea-pig. Most of the myelinated fibres are thin; on average 72 per cent have diameters between 3 and 5 μm, but nearly 5 per cent of the neurons are 8–10 μm thick. The guinea-pig has an average number of 1500 myelinated fibres in this nerve branch; variations occur and the specimen shown here has as many as 1850 fibres. Original magnification ×300 reduced to 91%*

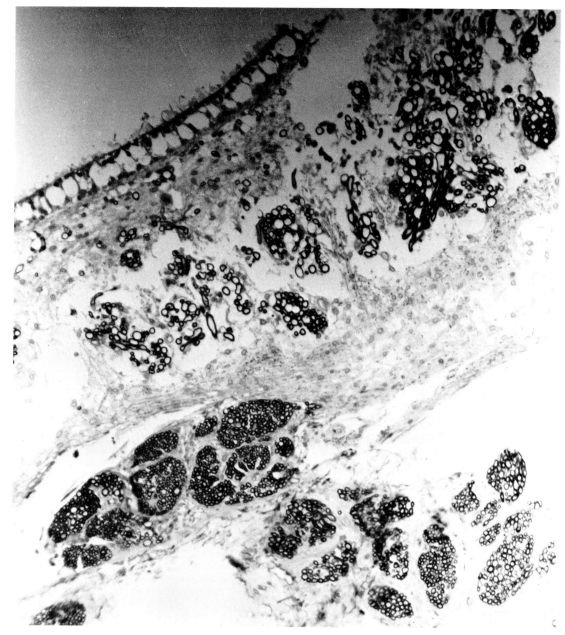

Figure 11.11 *Longitudinal section through the central area of the anterior crista ampullaris in a 22-year-old woman (same case as in Figures 11.5 and 11.7). This part of the crista is innervated by very thick fibres and the thinner fibres below run to the peripheral areas whose marginal zones contain type II cells only. ×210*

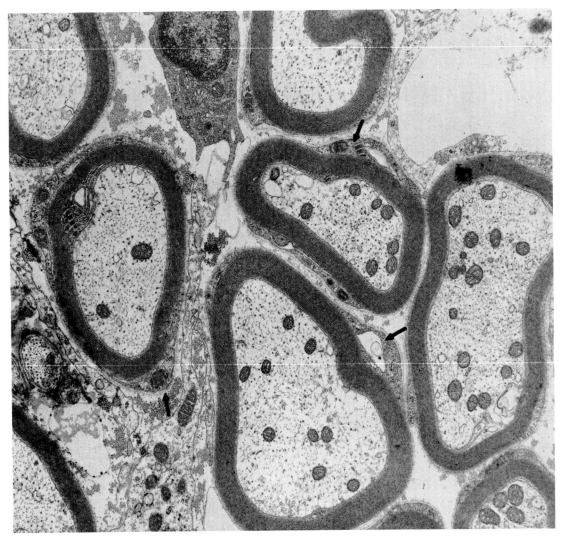

Figure 11.12 *Myelinated saccular nerve fibres surrounded by connective tissue. The contours of the fibres are affected by their close relationship to each other. In the axons a great number of mitochon-dria and neurofilaments are seen. Schwann cell cytoplasm indicated by arrows. Squirrel monkey. ×24 000*

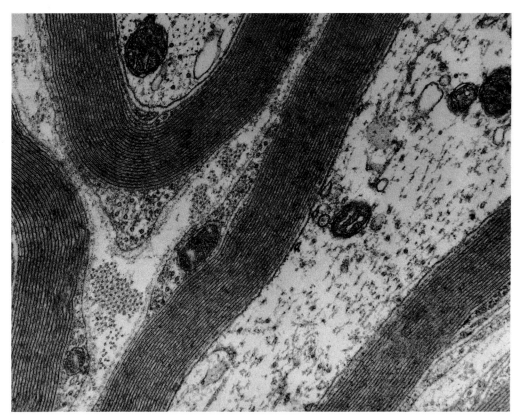

Figure 11.13 *Longitudinally and cross-cut myelinated fibres to the macula sacculi. In all three neurons the regular arrangement of the myelin layers is evident. Schwann cells with tubular structures in the* cytoplasm can be seen on the outside of the myelin sheaths. *Mitochondria, neurofilaments and vesicles occur in the axons. Squirrel monkey. Original magnification ×60 000 reduced to 90%*

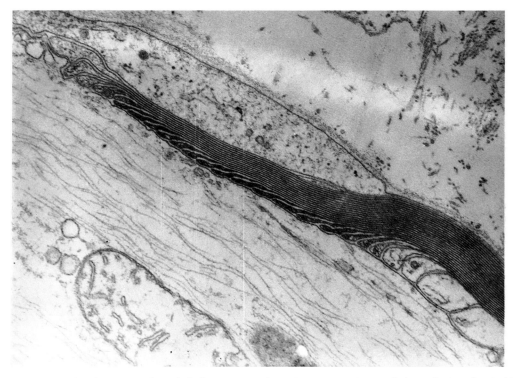

Figure 11.14 *Before passing through the basement membrane into the sensory epithelium, the neurons become unmyelinated. This* micrograph shows the 'unrolling' of the myelin layers from the axon. *Squirrel monkey. Original magnification ×39 000 reduced to 90%*

Figure 11.15 *Section through the entire sensory epithelium and the structures beneath. Four type I and three type II cells are seen; between the hair cells to the left is a supporting cell with a kinocilium (arrow). The nuclei of the supporting cells are found just above the basement membrane; between them and the hair cells, a great number of neurons with different calibres run in various directions. A fibre that has shed its myelin penetrates the basement membrane to the left, a myelinated neuron approaches the sensory epithelium in the centre and another fibre has entered the epithelium to the right. Macula utriculi. Monkey. ×2600*

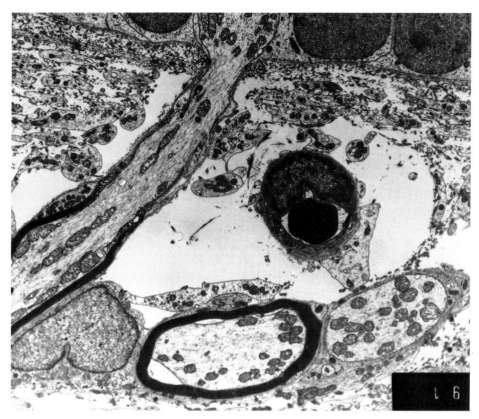

Figure 11.16 *As a rule the myelinization of the vestibular nerve fibres ceases well before passage through the basement membrane of the sensory epithelium, as shown here. One myelinated and one unmyelinated cross-cut neuron at bottom of figure. The nuclei at the upper end belong to supporting cells. Macula utriculi. Monkey. Original magnification ×6700 reduced to 83%*

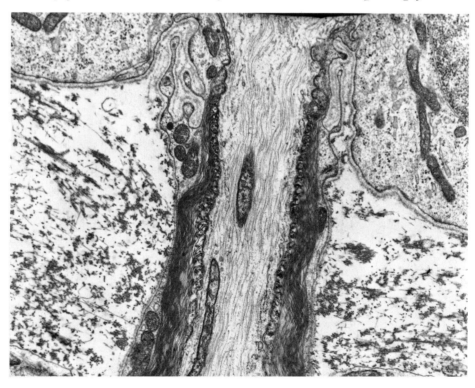

Figure 11.17 *Occasionally a neuron can be seen to retain its myelin sheath inside the sensory epithelium. Macula sacculi. Squirrel monkey. Original magnification ×22 000 reduced to 83%*

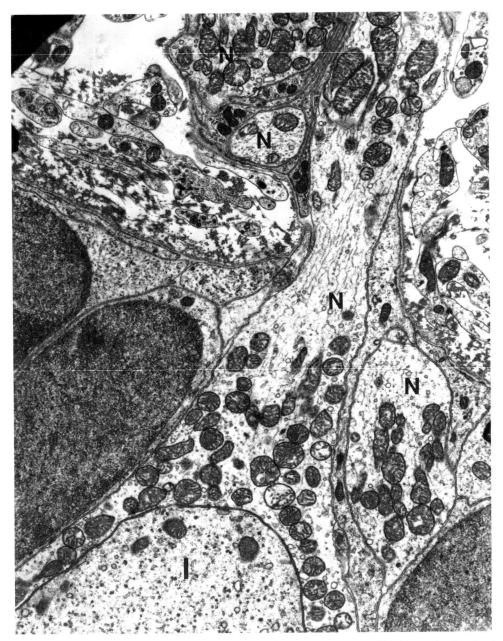

Figure 11.18 *A rather thin afferent nerve fibre, 2 μm in diameter at this level where the myelin has been shed, runs straight to a type I cell apparently without branching inside the sensory epithelium. This nerve fibre is very closely related to other afferent fibres on both sides of the basal membrane. N = Afferent nerve fibres. Macula utriculi. Monkey. ×15 000*

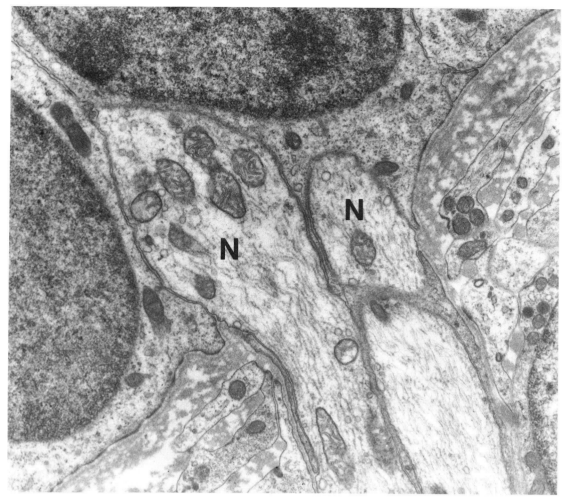

Figure 11.19 *It is unusual to see two neurons (N) that enter the epithelial layer through the same opening in the basement membrane. Macula utriculi. Monkey. ×20 000*

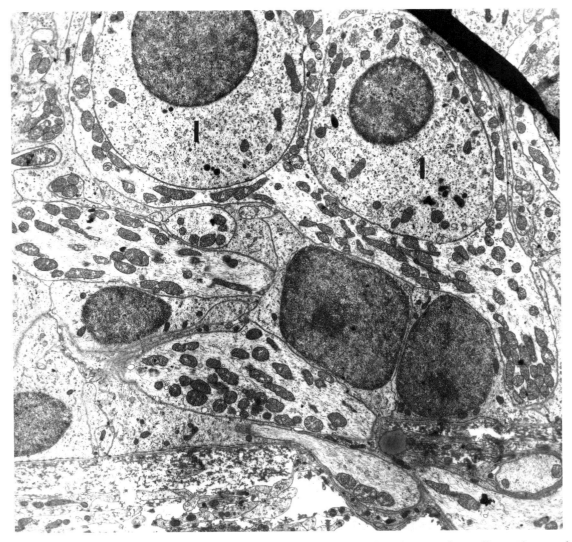

Figure 11.20 *There are nearly six times more hair cells on the macula utriculi than there are neurons in the utricular branch of the vestibular nerve. To innervate so many hair cells the neurons divide into several branches inside the sensory epithelium, which gives a complicated pattern of nerve fibres with many of these running almost parallel to the basement membrane before turning upwards to the hair cells. It is also seen how one nerve chalice encloses two type I cells. Monkey. ×7000*

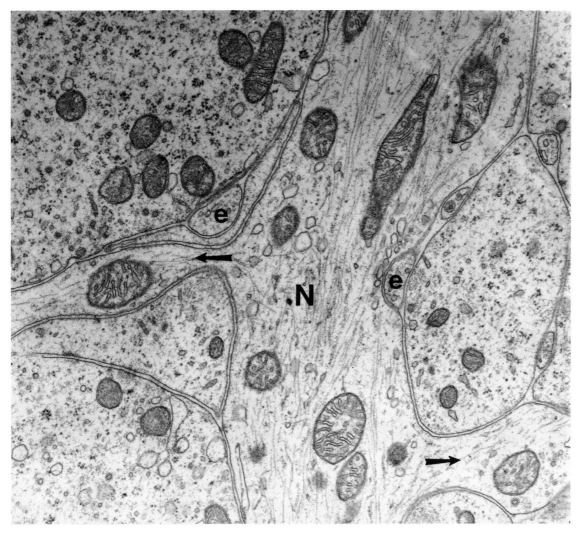

Figure 11.21 *An afferent nerve fibre (N) sending out collaterals (arrows) in different directions during its course through the sensory epithelium. There are also two efferent nerves (e); the one to the right is in close contact with the axon in which many vesicles are gathered near the border to the efferent nerve. ×29 000*

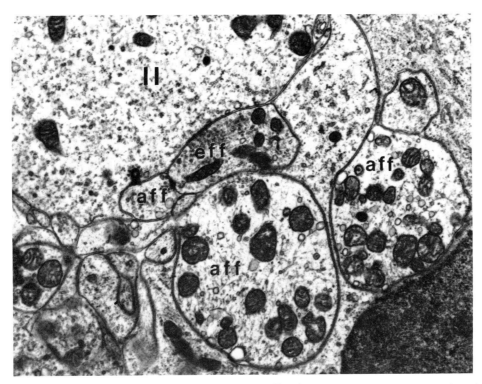

Figure 11.22 *Lower part of a type II cell with afferent (aff) and efferent (eff) nerve endings. A synaptic body is seen in the hair cell at the afferent synapse. The efferent nerve ending appears to make synaptic contact not only with the hair cell but also with the afferent nerve ending and with the thick afferent nerve below. Macula utriculi. Monkey. Original magnification ×21 500 reduced to 84%*

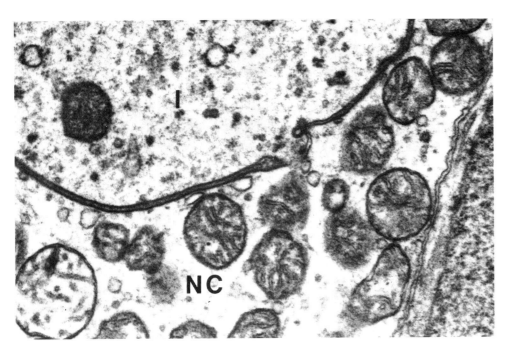

Figure 11.23 *Lower portion of a type I cell where the plasma membranes of the hair cell and the nerve chalice (NC) are disrupted and granular structures can be seen instead of the membranes. Similar and more extensive changes of the membranes have earlier been described by Engström, Bergström and Ades (1972). Macula utriculi. Squirrel monkey. Original magnification ×88 000 reduced to 84%*

References

Alexander, G. (1899) Zur Anatomie des Ganglion Vestibulare. *Sitzungsberichte der kaiserliche Akademie der Wissenschaft, math-naturw Classe,* **108,** 449.

Bagger-Sjöbäck, D. and Gulley, R. L. (1979) Synaptic structures in the type II hair cell in the vestibular system of the guinea pig. A freeze fracture and TEM Study. *Acta Otolaryngologica (Stockholm),* **88,** 401.

Bergström, B. (1972) Numerical analysis of the vestibular nerve in man. A preliminary report. *Uppsala Journal of Medical Sciences,* **77,** 205.

Bergström, B. (1973a) Morphology of the vestibular nerve. I. Anatomical studies of the vestibular nerve in man. *Acta Otolaryngologica (Stockholm),* **76,** 162.

Bergström, B. (1973b) Morphology of the vestibular nerve. II. The number of myelinated vestibular nerve fibers in man at various ages. *Acta Otolaryngologica (Stockholm),* **76,** 173.

Bergström, B. (1973c) Morphology of the vestibular nerve. III. Analysis of the calibers of the myelinated vestibular nerve fibers in man at various ages. *Acta Otolaryngologica (Stockholm),* **76,** 331.

Engström, H., Ades, H. W. and Hawkins, J. E. (1965) The vestibular sensory cells and their innervation. *Symposia Biologica Hungarica,* **5,** 21.

Engström, H., Bergström, B. and Ades, H. W. (1972) Macula utriculi and macula sacculi in the squirrel monkey. *Acta Otolaryngologica (Stockholm),* Supplement **301,** 75.

Fernandez, C., Goldberg, J. M. and Abend, W. K. (1972) Response to static tilts of peripheral neurons innervating otolith organs of the squirrel monkey. *Journal of Neurophysiology,* **35,** 978.

Gacek, R. R. (1961) The macula neglecta in the feline species. *Journal of Comparative Neurology,* **116,** 317.

Gacek, R. R. (1969) The course and central termination of first order neurons supplying vestibular end organs in the cat. *Acta Otolaryngologica (Stockholm),* Supplement **254,** 1.

Gacek, R. R. and Rasmussen, G. L. (1961) Fiber analysis of the statoacoustic nerve of guinea pig, cat, and monkey. *Anatomical Records,* **139,** 455.

Goldberg, J. M. and Fernandez, C. (1971) Physiology of peripheral neurons innervating semicircular canals of the squirrel monkey. III. Variations among units in their discharge properties. *Journal of Neurophysiology,* **34,** 676.

Gulley, R. L. and Bagger-Sjöbäck, D. (1979) Freeze-fracture studies on the synapse between the type I hair cell and the calyceal terminal in the guinea-pig vestibular system. *Journal of Neurocytology,* **8,** 591.

Langworthy, O. R. (1933) Development of behavior patterns and myelinization of the nervous system in the human fetus and infant. *Contributions to Embryology,* **24,** 1.

Lindeman, H. H. (1969) Studies on the morphology of the sensory regions of the vestibular apparatus. *Ergebnisse der Anatomie,* **42,** 1.

Lindeman, H. H., Reith, A. and Winther, F. Ö. (1981) The distribution of type I and type II cells in the cristae ampullares of the guinea pig. A morphometric investigation. *Acta Otolaryngologica (Stockholm),* **92,** 315.

Montandon, P., Gacek, R. R. and Kimura, R. S. (1970) Crista neglecta in the cat and human. *Annals of Otology, Rhinology and Laryngology,* **79,** 105.

Mulch, G. and Peterman, W. (1979) Influence of age on results of vestibular function tests. Review of literature and presentation of caloric test results. *Annals of Otology, Rhinology and Laryngology,* **88,** Supplement **56,** 1.

Okano, Y., Sando, I. and Meyers, E. N. (1980) Branch of the singular nerve (posterior ampullary nerve) in the otic capsule. *Annals of Otology, Rhinology and Laryngology,* **89,** 13.

Oort, H. (1918–19) Ueber die Verästelung des Nervus Octavus bei Säugetieren. *Anatomische Anzeige,* **51,** 272.

Rasmussen, A. T. (1940) Studies of the VIIIth cranial nerve of man. *Laryngoscope,* **50,** 67.

Rasmussen, G. L. (1946) The olivary peduncle and other fiber projections of the superior olivary complex. *Journal of Comparative Neurology,* **84,** 141.

Rasmussen, G. L. (1953) Further observations of the efferent cochlear bundle. *Journal of Comparative Neurology,* **99,** 61.

Retzius, G. (1881) Das Gehörorgan der Wirbelthiere. I *Das Gehörorgan der Fische und Amphibien.* Der Centraldruckerei: Stockholm.

Richter, E. (1980) Quantitative study of human Scarpa's ganglion and vestibular sensory epithelia. *Acta Otolaryngologica (Stockholm),* **90,** 199.

Rosenhall, U. (1972a) Vestibular macular mapping in man. *Annals of Otology, Rhinology and Laryngology,* **81,** 339.

Rosenhall, U. (1972b) Mapping of the cristae ampullares in man. *Annals of Otology, Rhinology and Laryngology,* **81,** 882.

Rosenhall, U. (1973) Degenerative patterns in the aging human vestibular neuroepithelia. *Acta Otolaryngologica (Stockholm),* **76,** 208.

Rosenhall, U. (1974) The innervation of the human vestibular sensory regions. *Equilibrium Research,* **4,** 1.

Rushton, W. A. H. (1951) A theory of the effects of fibre size in medullated nerve. *Journal of Physiology (London),* **115,** 101.

Shute, C. C. D. (1951) The anatomy of the eighth cranial nerve in man. *Proceedings of the Royal Society of Medicine,* **44,** 1013.

Voit, M. (1907) Zur Frage der Verästelung des Nervus Acusticus bei den Säugetieren. *Anatomische Anzeige,* **31,** 635.

Walsh, B. T., Miller, J. B., Gacek, R. R. and Kiang, N. Y. S. (1972) Spontaneous activity in the eighth cranial nerve of the cat. *International Journal of Neuroscience,* **3,** 221.

Wersäll, J. (1956) Studies on the structure and innervation of the sensory epithelium of the cristae ampullares in the guinea pig as revealed by the electron microscope. *Acta Otolaryngologica (Stockholm),* Supplement **126,** 1.

Westphal, A. (1897) Ueber die Markscheidenbildung der Gehirnnerven des Menschen. *Archiv für Psychiatrie und Nervenkrankheiten,* **29,** 474.

12
The Vestibular Ganglion

Jukka Ylikoski   Frank R. Galey

The afferent neurons innervating the vestibular areas are generally bipolar cells having their neuronal cell bodies within the internal auditory canal. There the vestibular ganglion or Scarpa's ganglion consists of two parts: one is the upper, corresponding to the superior vestibular nerve; the other is the lower, corresponding to the inferior vestibular nerve. The structure of the vestibular ganglion corresponds to that of other cerebrospinal nerves. It is surrounded by a loose connective-tissue sheath of arachnoid tissue and contains – in addition to the neurons and their sheaths – blood vessels, collagen fibrils and fibroblasts. A feature that makes the neuronal cell bodies of the VIIIth nerve ganglion unique is that, in most of the vertebrate species (excepting man), the cell bodies are enclosed by a myelin sheath. The following description of the vestibular ganglion cell is based on material from the Department of Otolaryngology, University of Helsinki, Finland, and from the House Ear Institute, Los Angeles, California. Most of the material consisted of surgical eighth nerve specimens from patients with varying otoneurological entities. The vestibular neurons of the rat and monkey (baboon) were also examined. All the material was fixed initially with 2–3% buffered glutaraldehyde or Karnovsky's solution, post-fixed in 1% OsO_4, dehydrated in alcohol and embedded in epoxy resin or Araldite. Thin sections were stained with lead citrate and uranyl acetate for electron microscopy.

The Neuronal Cell Body

The vestibular ganglion cells are generally round and contain cytoplasm (perikaryon) and a nucleus. They show great variation in size (*Figure 12.1*).

The Perikaryon

'Perikaryon' refers only to that part of the cytoplasm surrounding the nucleus and excluding that part contained in the processes. In vestibular ganglion cells the perikaryon is provided with a large amount of normal cytoplasmic organelles: rough and smooth endoplasmic reticulum, Golgi apparatus, mitochondria, microtubules, neurofilaments, multivesicular bodies, lysosomes and lipofuscin granules (*Figures 12.1* and *12.2*).

Granular (rough) endoplasmic reticulum (RER)

In all species studied the RER is well developed and dispersed randomly through the perikaryon. RER consists of ordered arrays of broad cisternae, stacked one on top of another with regular intervals, and of shorter cisternae of less ordered arrangement. The outer surfaces of the membranes limiting these cisternae are studded with ribosomes arranged in rows and clusters. Some ribosomes also lie free in the cytoplasmic matrix between the cisternae (*Figures 12.3* and *12.5*). Those lying free in the cytoplasm tend to be arranged in small rosettes of 5–7 surrounding a central one (*see Figure 12.13*). These free polysomes are characteristic of neuronal cell bodies. Although there are great variations in the amount of the RER among individual neurons, it has not been considered justified to classify the neuronal cell bodies into granular or filamentous types according to their content of RER, as has been suggested in some earlier studies (Rosenbluth, 1962; Ballantyne and

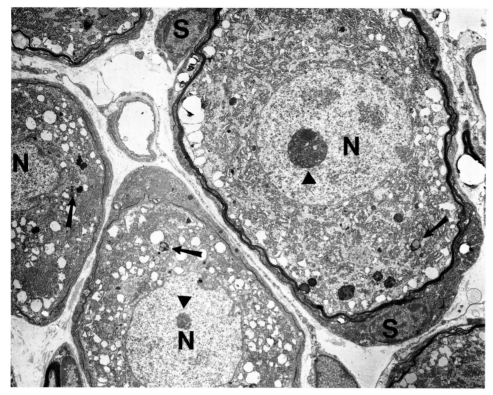

Figure 12.1 *In the rat vestibular ganglion all the neuronal cell bodies are of variable size and are surrounded by a myelin sheath of varying thickness. Perikarya show richly developed rough endoplasmic reticulum and many mitochondria, some of which are artificially swollen in the two cells on the left. All the cells have some lipofuscin granules (arrows). The nuclei (N) appear relatively light, without the usual accumulation of chromatin at the nuclear periphery. The nucleoli (arrowheads) appear strikingly dense. The cell on the right is bordered by at least two Schwann cells (S). Original magnification ×3600 reduced to 85%*

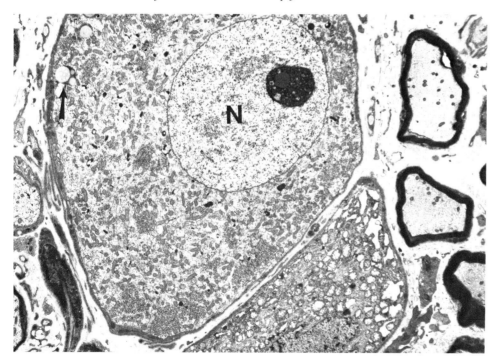

Figure 12.2 *Human vestibular ganglion cells have no myelin coat but are surrounded by a thin rim of Schwann cell cytoplasm which usually appears darker than the perikaryon. The neuron cell body has abundant perikaryonal organelles such as rough endoplasmic reticulum, mitochondria and lipofuscin granules (arrow). The nucleus (N) has a very prominent nucleolus. Original magnification ×4200 reduced to 85%*

Engström, 1969). Normally, RER is reduced at the axon hillock where it is replaced by filamentous material (*Figure 12.6*). The amount of RER, which is the Nissl substance described by light microscopy, is generally considered to represent the metabolic activity of the cell.

The margination of the Nissl bodies or RER at the periphery of the cell body in cells undergoing chromatolysis has been used for almost a century as a tool to evaluate so-called 'retrograde degeneration' (Nissl, 1892). Chromatolytic neurons with loss of RER from the central zones of the perikaryon have been recently demonstrated in human vestibular ganglion cells from patients who had earlier undergone labyrinthectomy (Ylikoski and Belal, 1981). The rearrangement of RER reflects a restorative process aimed at reconstituting the neuron to its original state. Thus, the term 'retrograde reaction' should replace the misleading term 'retrograde degeneration'.

Agranular reticulum

The smooth or agranular reticulum consists of irregularly arranged, relatively short cisternal and tubular structures without attached ribosomes. The agranular reticulum is not a prominent element in the vestibular ganglion cells.

Golgi apparatus

Vestibular ganglion cells from the middle zone of perikaryon possess several Golgi complexes (*Figure 12.3*), presenting as a stack of 5–7 long parallel cisternae separated by small spaces. Each cisterna is usually shallow, but some are widely dilated at the edges of the complex (*Figure 12.4*). Cisternae can ramify and have scattered fenestrations. In most cells the Golgi apparatus plays a role in the condensation and packaging of the secretory product, for export from the cell. However, as in other neurons, its role in the vestibular neuron is not certain. It may play a role in completing the assembly of glycoproteins, or in the formation of lysosomes.

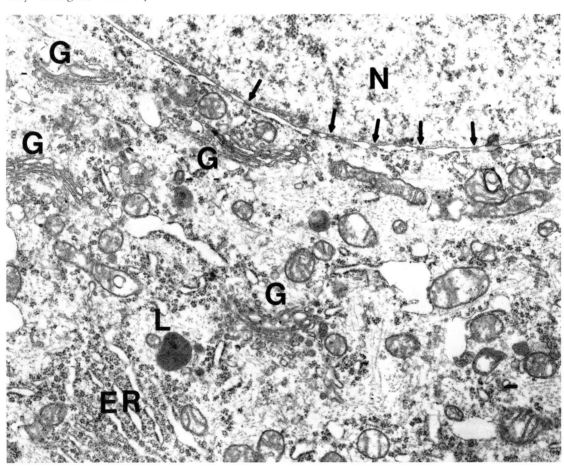

Figure 12.3 *Rat vestibular ganglion cell. The perikaryonal cytoplasm with its organelles (G = Golgi apparatus; ER = rough endoplasmic reticulum; L = lysosome) is separated from the nucleus (N) by a double membrane with a clear interspace about 30 nm wide. The two membranes, however, appear to be adherent at many sites (arrows) and the boundary appears to be formed by a chain of specialized cisternae of ER. Note rosettes of ribosomes at the cisternae and also free in the cytoplasm. ×20 900*

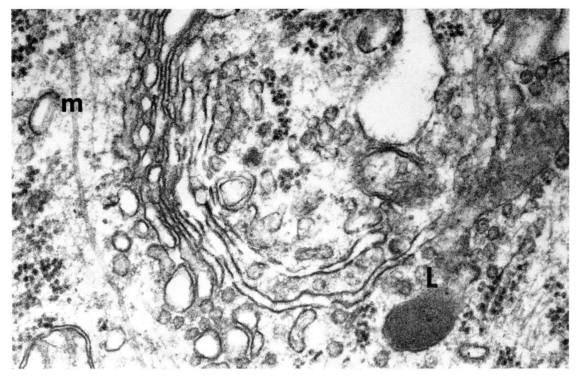

Figure 12.4 *Rat vestibular ganglion cells. The apparatus of Golgi appears as a stack of curved parallel cisternae with dilated edges. m = Microtubule; L = lysosome. ×65 000*

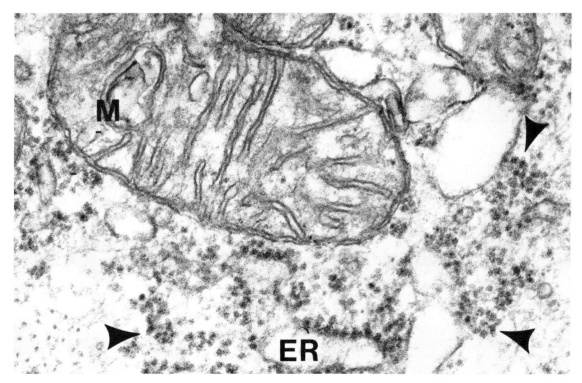

Figure 12.5 *Rat vestibular ganglion cell. The ribosomes are arranged in rows (ER) or rosettes of 5–7 granules surrounding a central one (arrowheads). The mitochondrion (M) is bounded by a smooth unit membrane enclosing another membrane from which the transverse or longitudinal cristae arise. ER = Cisterna of endoplasmic reticulum. ×65 000*

Mitochondria

The perikaryon of vestibular ganglion cells normally contains numerous mitochondria which often form elongated rodlets of about 0.1 µm diameter, 0.5 µm or more in length (*see Figures 12.1–12.3*). Each organelle is bounded by a smooth unit membrane enclosing another folded membrane which envelops the inner compartment of the mitochondrion which is filled with a dense matrix (*see Figure 12.5*). The mitochondrial cristae are often arranged in a longitudinal fashion.

Microtubules and Neurofilaments

In the perikaryon of the vestibular ganglion cells microtubules and neurofilaments occupy most of the space not occupied by other organelles. The microtubules appear as long tubular structures 20–25 nm in diameter. The neurofilaments are thinner, about 10 nm in diameter. They appear at high magnification to be tubules rather than filaments. The neurofilaments are joined together by cross-bars which are formed by globular subunits with small spike-like processes (*Figures 12.6* and *12.7*). The orientation of microtubules and neurofilaments in the perikaryon appears to be disordered. However, they tend to run parallel to one another in loose bundles that funnel into the bases of axons (*Figure 12.6*).

Lipofuscin Granules

One of the conspicuous constituents of the perikaryon of the vestibular neuron is lipofuscin granules (*see Figures 12.1* and *12.2*). Lipofuscin is generally considered to be a wear-and-tear pigment occurring in neurons and other non-renewing cells in increasing amounts with increasing age (Samorajski, Keefe and Ordy, 1964; Toth, 1968). The granules are believed to be derived from lysosomes. Lipofuscin granules are irregularly rounded structures with a diameter of 1.5–2.5 µm. They have one or more peripherally located vacuoles, the rest of the granule being filled with an electron-dense heterogeneous matrix material (*see Figures 12.6* and *12.8*). It has been shown that lipofuscin accumulates steadily throughout life and may, after accumulating to a certain point, have a detrimental effect on the neuron (Mann and Yates, 1974). We have found lipofuscin granules in all species studied. Their amount has been most striking in the human vestibular neurons from elderly patients, but some lipofuscin granules appear to be present also in the vestibular ganglion cells of relatively young rats and monkeys (*see Figures 12.1* and *12.18*).

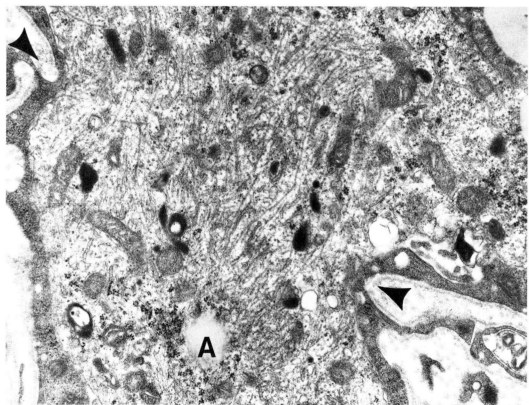

Figure 12.6 *Axon hillock area of a human vestibular ganglion cell, showing reduction of the rough endoplasmic reticulum in the area where the microtubules and neurofilaments funnel into the narrowing initial segment of the axon (A). Arrowheads indicate transition zone between the perikaryon and the axon. Original magnification ×24 000 reduced to 94%*

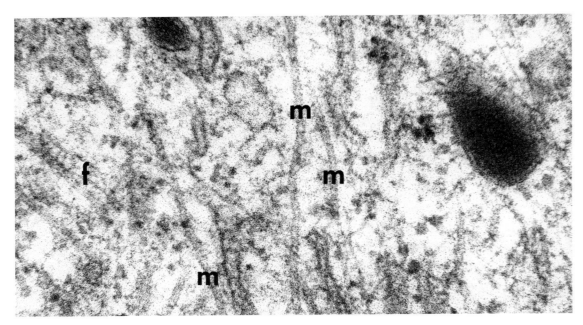

Figure 12.7 *Microtubules predominate in this zone of the perikaryon of a human vestibular ganglion cell. The microtubules (m) are long tubular structures, 20–25 nm in diameter, made up of finer* *filaments. The neurofilaments (f; diameter 10 nm), which are fewer than microtubules in this particular area, are linked by cross-bars. ×108 000*

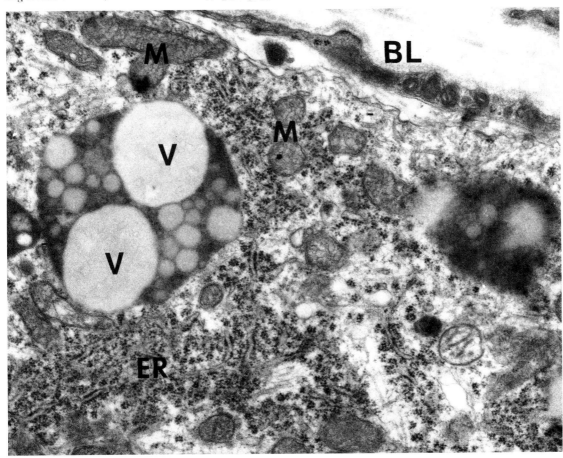

Figure 12.8 *The peripheral zones of the perikaryon of a human vestibular ganglion cell, showing dense rough endoplasmic reticulum (ER), abundant mitochondria (M) and lipofuscin granules. The largest lipofuscin granule has two large and several small vacuoles* *(V) embedded in the dense matrix material. Below the basal lamina (BL) is the dark cytoplasm of the Schwann cell and the perikaryonal cytoplasm. ×28 000*

295

Lysosomes

Lysosomes appear to be present to some extent in all vestibular neurons. They are spherical bodies, with a diameter of between 0.2 and 0.6 μm, bounded by a single unit membrane and containing a homogeneously dark matrix (*see Figures 12.3* and *12.4*).

Multivesicular Bodies, Cytoplasmic Inclusions

Multivesicular bodies are spherical structures bounded by a unit membrane with a diameter of about 0.5 μm. They contain varying numbers of small spherical or ovoid vesicles in a dark or lucent matrix. Laminated bodies – prominent structures in the neuronal perikarya of certain neurons (Morales, Duncan and Rehmet, 1964) – as well as other specialized inclusion bodies are uncommon in the normal vestibular neurons.

Nucleus

In the vestibular neurons the round nucleus is usually located in the centre of the cell body. As a rule it has one strikingly dark nucleolus but the karyoplasm appears to be lighter than the perikaryonic cytoplasm because of the lack of chromatin particles. Fine chromatin filaments are evenly dispersed without any accumulation near the nuclear envelope. This is typical for most other cell types as well as for the Schwann cell (*see Figures 12.1* and *12.2*). The chromatin consists of fine granular strands or dots about 20 nm in diameter (*Figure 12.9*). Each strand consists of one or more finer filaments, many of which appear in cross-section to be hollow. The nuclear envelope is usually unfolded and consists of a double membrane. Each membrane is about 7 nm thick. The inner membrane appears smooth, the outer more irregular, the latter being joined at several points to the inner membrane (*see Figures 12.3, 12.10* and *12.11*). Thus, the nuclear envelope consists in fact of a single membrane surrounding a space, the perinuclear cisterna.

Because several studies have demonstrated continuities between the perinuclear cisterna and the endoplasmic reticulum (ER), the nuclear envelope has been regarded as a specialized portion of the ER forming the boundary between the cytoplasm and karyoplasm (Palay, 1960). The nuclear envelope of the vestibular ganglion cells, similar to that of other cells, is fenestrated by numerous circular pores which are arranged in orderly rows (*Figures 12.12* and *12.13*). The pores are located at those points of the nuclear envelope where the two membranes become adherent (*see Figure 12.11*). Each pore has a diameter of about 65 nm, the distance between the pores and between the rows being 80–90 nm. In tangential sections through the nuclear envelope the pores appear as circular structures with thick dense walls about 20 nm thick encircling the pores (*Figure 12.13*).

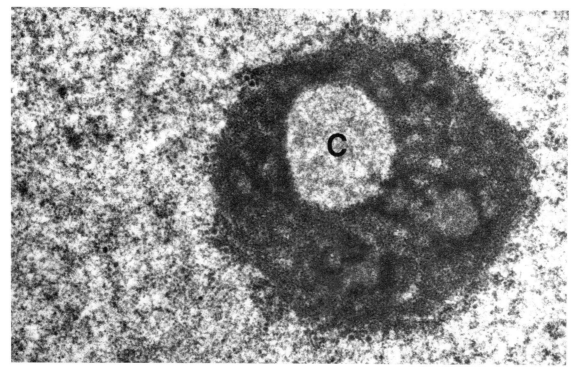

Figure 12.9 *The nuclear chromatin of a human vestibular ganglion cell consists of finely granular matter dispersed evenly throughout the nucleus. The nucleolus is composed also of a finely granular matrix of extremely high density. In the centre of the nucleolus is a spherical zone of trapped nuclear chromatin (C). ×36 000*

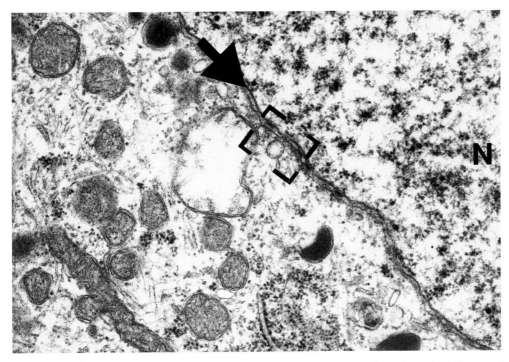

Figure 12.10 *The boundary between the perikaryon and the karyoplasm in a human vestibular ganglion cell is formed by a double* *membrane (arrow) which at certain regular points becomes confluent, forming a chain of cisternae. N = Nucleus. ×36 000*

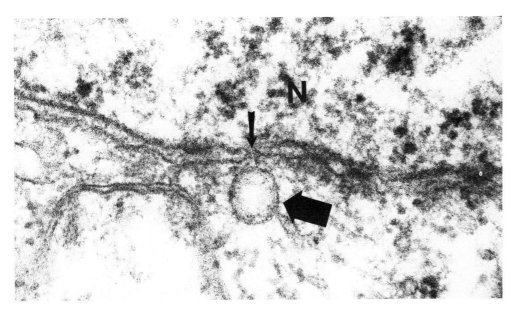

Figure 12.11 *Detail of the enclosed area in Figure 12.10 shows a point where the membranes of the nuclear (N) envelope join together (thin arrow) for a distance of 65 nm. A vesicle (thick arrow) on the* *perikaryonal side may be in the process of transport to or from the nucleus. ×108 000*

Nucleolus

Under the electron microscope the nucleolus appears as an extremely dense spherical structure. It consists of two main components: dense ribosome-like granules about 20 nm in diameter; and closely packed very fine filaments of low density (*see Figure 12.9*). These two compartments are incompletely separated from each other into a pars granulosa and a pars fibrosa.

Nuclear Inclusions

Varying nuclear inclusion bodies in the nuclei of different types of cells have been described as possible evidence of viral infection. In the vestibular ganglion cells of patients with balance disorders three types of nuclear inclusions have been demonstrated (Palva *et al.*, 1978):

1. Structures resembling interwoven yarn, in small round aggregates, each aggregate being 20–25 nm in thickness.
2. Coarse aggregates of chromatin with a diameter of 20–60 nm.
3. Light nuclear bodies with hazy fibrillar structures arranged around a few interchromatin granules. The diameter of these bodies was 0.8–1.1 μm.

However, viral cultures of such vestibular ganglia were negative.

Other types of inclusions, such as intranuclear rods and sheets, not infrequently encountered in other neurons – and also in the cochlear and vestibular nuclei (Sotelo and Palay, 1968; Feldmann and Peters, 1972) – appear not to be present in the vestibular ganglion cells.

The Schwann Cell

All the neurons of the vestibular ganglion are surrounded by a capsule of variable thickness which is continuous with the sheath around the nerve fibres. The axons are encircled by Schwann cells which are responsible for the formation of myelin in the peripheral nervous system and are believed to be of neural crest origin. When a Schwann cell ensheathes a cell body, it is sometimes called a 'satellite cell'. This division is artificial because these cells are identical, and in the present chapter only the term 'Schwann cell' is used. In most spinal and sensory ganglia the axons have a compact myelin coat and the neuron cell bodies are ensheathed by one or more layers of Schwann cell cytoplasm; however, the Schwann cells in the eighth nerve ganglia of many species form a myelin coat around the cell body. The goldfish, rat (Rosenbluth and Palay, 1961; Rosenbluth, 1962), guinea-pig and monkey (Ballantyne and Engström, 1969) have myelinated vestibular neurons. Each neuron cell body is enveloped by 1–3 Schwann cells (*Figures 12.1* and *12.14*). The number of Schwann cells ensheathing a neuron is proportional to the surface area of the neuron, each cell being estimated to cover approximately 400 μm² of perikaryal surface (Pannese, 1960).

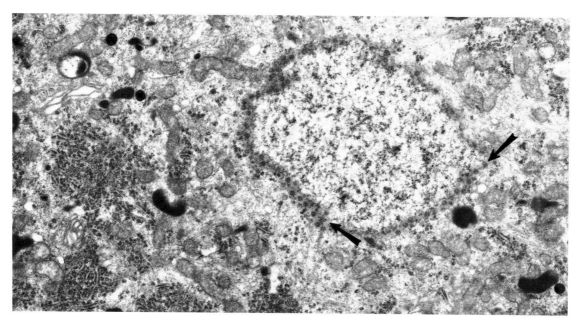

Figure 12.12 *Tangential section through part of the nucleus in a human vestibular ganglion cell, showing small circular profiles along the blurred margin of the nucleus indicating rows of pores of the nuclear envelope (arrows). ×3400*

The nucleus of the Schwann cell usually sits on the flat surface of the neuron cell body (*Figure 12.14*). The nuclear chromatin of the Schwann cell is clumped at the periphery of the nucleus. The nucleolus is much less conspicuous than that of the neurons. In the perikaryal cytoplasm the ER is often less richly developed and the mitochondria appear shorter and more rounded than in the neurons (*Figure 12.15*). Further away from the perikaryal zone relatively few organelles occur and the cytoplasmic processes form sheet-like extensions which may overlap each other and interdigitate.

Each Schwann cell as well as its processes is surrounded by a fluffy basal lamina (*Figure 12.15*). In fact, the basal lamina makes it possible to distinguish the Schwann cell processes – which sometimes surround bundles of collagen fibres forming so-called collagen pockets – from the processes of fibroblasts (*Figures 12.15 and 12.21*). The human vestibular ganglion has several 'extra' Schwann cells, a feature not present to the same extent in non-human specimens. Such Schwann cells are round, and have a relatively large and dark nucleus and scanty cytoplasm. They never have lipofuscin granules in their cytoplasm, which is common in the Schwann cells ensheathing neurons (*see Figures 12.14 and 12.15*). The extra Schwann cells have been proposed as the source of vestibular schwannomas (Henschen, 1915).

Sheaths of the Vestibular Ganglion Cells

The ensheathment of the vestibular ganglion cells varies from a capsule formed by a single layer of Schwann cell cytoplasm to a coat of compact myelin with up to 90 lamellae. We have observed no relationship between the different types of myelin sheath and other morphological features of the neurons. The main differences appear to be species related. The three species studied each represent one main type of the three observed ensheathments of vestibular ganglion cells. The following classification is arbitrary, owing to the different varieties of myelin sheaths, all of which can occur in different areas within the same sheath.

1. *Neurons encapsulated by a simple Schwann cell sheath* – Almost all neurons of the human vestibular ganglia and very few, if any, in the rat or monkey belong to this category. The cell bodies are completely invested by the cytoplasm of a Schwann cell (*see Figures 12.2 and 12.14*). There appears to be only a clear, structureless region about 20 nm wide between the neuronal plasma membrane and the Schwann cell (*Figure 12.16*). Because the Schwann cell cytoplasm is usually darker in low-power electron micrographs than that of the neuronal cell body, the neurons may appear to have a myelin coat (*see Figures 12.2 and 12.14*). However, in almost all human vestibular ganglion cells there is only a single layer of Schwann cell cytoplasm which at places attenuates to a thickness of 20 nm, completely encircling the cell body.

However, frequently the Schwann cells may overlap and display one to several (up to 15) short segments of lamellae nearest to the neuron plasmalemma (*Figure 12.17*). These 'abortive' lamellae end at one point as blind loops of cytoplasm in the middle of the sheath

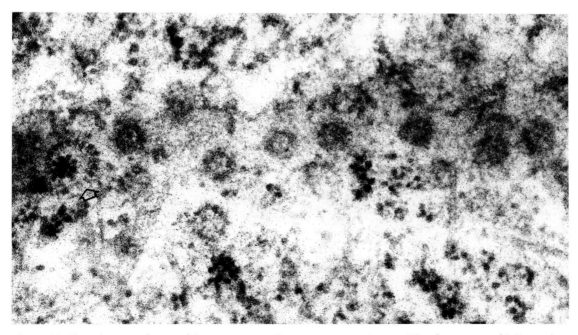

Figure 12.13 *Higher magnification of the section seen in Figure 12.12, showing the pores arranged in orderly rows, each pore having a diameter of about 65 nm. They appear to have a dense wall, about 20 nm thick. Note also the arrangement of the ribosomes in the perikaryon. They form rosettes of 5–7 granules surrounding a central one. One structure shows a peculiar helical alignment of ribosome-like particles (arrow). ×86 000*

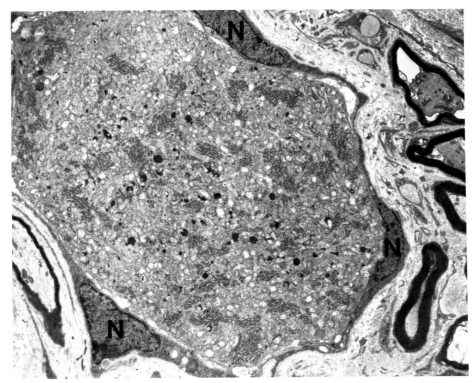

Figure 12.14 *This human vestibular ganglion cell is bordered by at least three Schwann cells containing nuclei of varying shapes (N). The neuronal perikaryon has many primary lysosomes and the* *Schwann cell cytoplasm contains some lipofuscin granules. Original magnification ×4000 reduced to 86%*

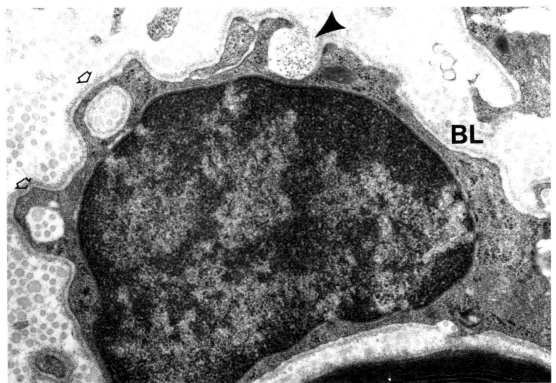

Figure 12.15 *An 'extra' Schwann cell in the human vestibular ganglion has a dense nucleus with peripheral accumulation of chromatin at the nuclear envelope and sparse perinuclear cytoplasm with the usual organelles. The cytoplasmic processes are encircling* *bundles of collagen fibrils forming collagen pockets (arrows). These processes also partially surround a bundle of thinner filaments (arrowhead). The Schwann cell is completely invested by a fluffy basal lamina (BL). Original magnification ×36 000 reduced to 86%*

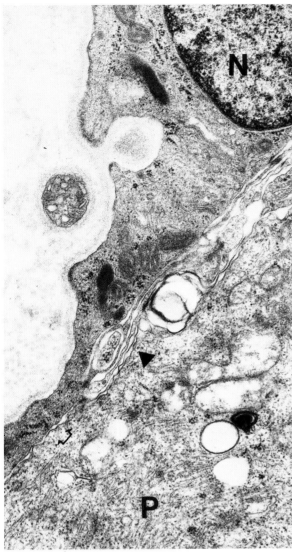

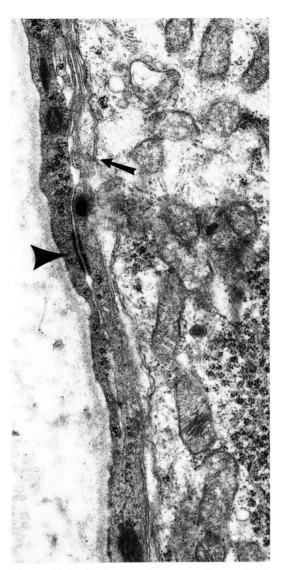

Figure 12.16 *This human vestibular ganglion cell is completely covered by a single layer of Schwann cell cytoplasm but at many points there are many lamellae between the Schwann cell and the perikaryonal cytoplasm resembling loose myelin. These lamellae seem to end as blind loops at certain points. In the lower part of the picture there is only a single cytoplasmic layer (arrow) which becomes multiplied (arrowhead). N = Schwann cell nucleus; P = perikaryon. ×34 000*

Figure 12.17 *Another area of the same cell as seen in Figure 12.16, showing several layers of Schwann cell, some of which end as blind loops (arrows). Note an intercellular junction complex between overlapping processes of two Schwann cells (arrowhead). ×3400*

rather than at an external or internal mesaxon (Brattgård, 1952).

2. Neurons encapsulated by a sheath consisting of loose myelin – In these sheaths the Schwann cell cytoplasm between the intertwining lamellae has not completely disappeared and consequently the major dense lines of compact myelin are absent. The thickness of the cytoplasmic layers varies, even within the same sheath. Usually it is in the range of 10–100 nm wide but it can be wider, particularly in the outermost parts of the sheath (*Figures 12.18* and *12.19*). The number of layers can vary from 2 to 30. In the monkey (baboon), in

which all the sheaths appear to be composed mainly of loose myelin, the number of lamellae ranges from 5 to 30. In the region of the sheath showing the closest packing of lamellae these show major dense lines of compact myelin but they lack intermediate lines and the period remains as wide as approximately 25 nm (*Figure 12.19*).

3. Neurons encapsulated by a sheath of compact myelin – This type of sheath is comparable with axonal myelin although it is thinner and irregular. It consists of multiple dense layers, with a period of approximately 15 nm, around the rat vestibular ganglion cells, almost

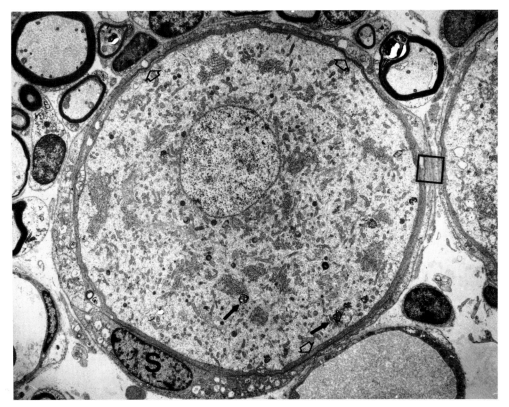

Figure 12.18 *The vestibular ganglion cell of the monkey, showing a round, relatively light perikaryon with its normal organelles. Note small lipofuscin granules (thin arrows). The spherical nucleus shows no clumping of chromatin. The Schwann cell sheath appears thick.*

Compact myelin formation can be seen at certain points (thick arrows). S = Schwann cell nucleus. Original magnification ×3600 reduced to 88%

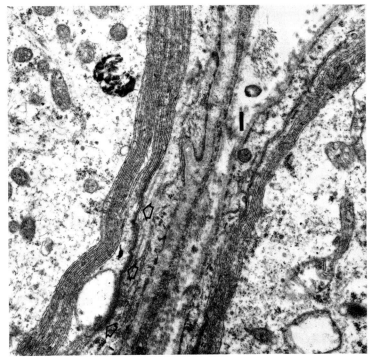

Figure 12.19 *Higher magnification of the boxed area in Figure 12.18, showing that the cell on the left has a sheath composed of 11 lamellae of loose myelin with only intermediate lines. The outermost*

lamellae become locally confluent, forming major dense lines for a short distance (arrows). I = Interstitial space. Original magnification ×21 000 reduced to 88%

302

all of which are enveloped by compact myelin of variable thickness (approximately 10 lamellae) (*Figure 12.20*). A characteristic feature of the myelin sheath is irregularity of the compact myelin pattern at almost regular intervals. The compact layers are separated and have cytoplasm between them for a short distance (*Figure 12.20*). Compact myelin is also present in most of the ganglion cells in the monkey, but it is rare in the sheath of human vestibular ganglion cells.

Fibroblasts

Fibroblasts occur both in the outer sheath of the vestibular ganglion (formed by loose arachnoid and pial tissue) and within the endoneurium. In the endoneurium the number of fibroblasts as well as the amount of collagen fibres appears to be considerably greater in the human than in animal specimens. The nuclei of fibroblasts are either oblong or quadrangular and resemble those of

Schwann cells. The perinuclear cytoplasm is usually scanty and is located mainly in the long slender processes. The cytoplasm has relatively scanty RER, numerous rounded mitochondria and different types of vesicles (*Figure 12.21*). In human specimens the cytoplasm often shows large clear vacuoles which give a fenestrated appearance to the cell (*Figure 12.22*). Similar fenestrated large fibroblasts have been described in operative specimens, from patients with probable compression neuropathy (Asbury, Cox and Baringer, 1971; Ylikoski and House, 1980).

Acknowledgements

We are grateful to Docent Dan Bagger-Sjöbäck, Karolinska Institute, Stockholm, Sweden, for providing the monkey vestibular nerve specimens; and to Marian Shiba and Hiroko Shiroishi of the House Ear Institute Los Angeles, for skilful technical assistance.

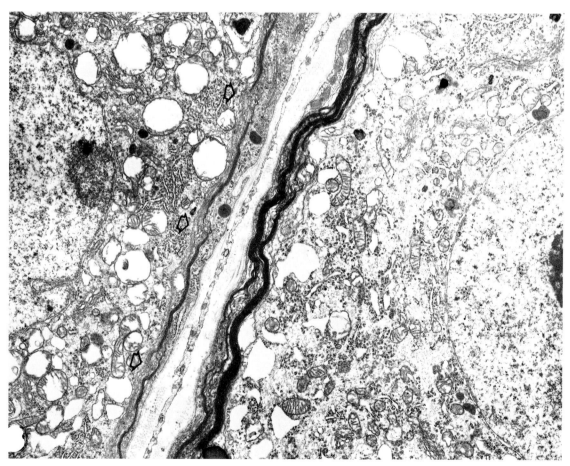

Figure 12.20 *Two neuronal cell bodies of the rat vestibular ganglion are bordered by a sheath composed of compact myelin. The cell on the left has five lamellae, the cell on the right has 20 lamellae. As seen in the left sheath, the compact myelin appears to become interrupted (arrows). At these points the compact myelin changes to loose myelin, and Schwann cell cytoplasm is incorporated between the lamellae. ×11 000*

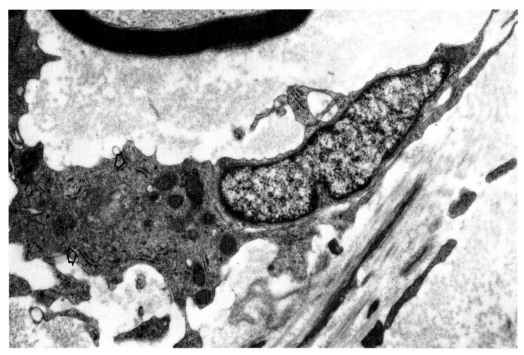

Figure 12.21 *This endoneurial fibroblast in the human vestibular ganglion shows an oblong nucleus with relatively sparse perinuclear cytoplasm which extends in different directions as thin processes. The* *ribosomes of the rough endoplasmic reticulum stain distinctly dark (arrows). The rounded mitochondria appear darker than in the neurons. Original magnification ×15 000 reduced to 92%*

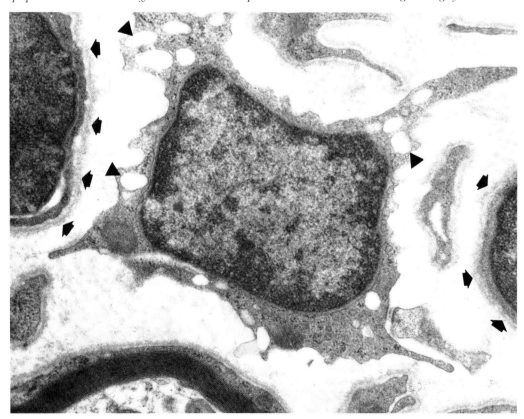

Figure 12.22 *The same specimen as seen in Figure 12.21, showing fibroblasts with a quadrangular dark nucleus and very sparse perinuclear cytoplasm which has small vesicles and larger vacuoles (arrowheads), thus producing a fenestrated appearance. Note that the* *fibroblast has no basal lamina, in contrast to that seen on the outer surface of the adjacent Schwann cell (arrows). Original magnification ×28 000 reduced to 92%*

References

Asbury, A. K., Cox, S. C. and Baringer, J. (1972) The significance of giant vacuolation of endoneurial fibroblasts. *Acta Neuropathologica (Berlin)*, **18**, 123–131.

Ballantyne, J. and Engström, H. (1969) Morphology of the vestibular ganglion cells. *Journal of Laryngology and Otology*, **83**, 19–42.

Brattgård, S. -D. (1952) The importance of adequate stimulation for the chemical composition of retinal ganglion cells during early post-natal development. *Acta Radiologica (Supplement)*, **96**, 11–80.

Feldman, M. L. and Peters, A. (1972) Intranuclear rods and sheets in rat cochlear nucleus. *Journal of Neurocytology*, **1**, 109–127.

Henschen, F. (1915) Zur histologie und pathogenese der kleinhirn-brückenwinkeltumoren. *Archiv für Psychiatrie und Nervenkrankheiten*, **56**, 20–122.

Mann, D. M. and Yates, P. O. (1974) Lipoprotein pigments – their relationship to ageing in the human nervous system. I. The lipofuscin content of nerve cells. *Brain*, **97**, 481–488.

Morales, C., Duncan, D. and Rehmet, R. (1964) A distinctive laminated cytoplasmic body in the lateral geniculate body neurons of the cat. *Journal of Ultrastructural Research*, **10**, 116–123.

Nissl, F. (1892) Über experimentell erzeugte Veränderungen an den Vorderhornzellen des Rückenmarks bei Kaninchen mit Demonstration mikroskopischer Präparate. *Allgemeine Zeitschrift für Psychiatrie und Psychisch-gerichtliche Medicin*, **48**, 675–682.

Palay, S. L. (1960) On the appearance of absorbed fat droplets in the nuclear envelope. *Journal of Biophysical and Biochemical Cytology*, **7**, 391–392.

Palva, T., Hortling, L., Ylikoski, J. and Collan, Y. (1978) Viral culture and electron microscopy of ganglion cells in Meniere's disease and Bell's palsy. *Acta Otolaryngologica (Stockholm)*, **86**, 269–275.

Pannese, E. (1960) Observations of the morphology, submicroscopic structure and biological properties of satellite cells (s.c.) in sensory ganglia of mammals. *Zeitschrift für Zellforschung und Mikroskopische Anatomie*, **52**, 567–597.

Rosenbluth, J. (1962) The fine structure of acoustic ganglia in the rat. *Journal of Cellular Biology*, **12**, 329–359.

Rosenbluth, J. and Palay, S. L. (1961) The fine structure of nerve cell bodies and their myelin sheaths in the eighth nerve ganglion of the goldfish. *Journal of Biophysical and Biochemical Cytology*, **9**, 853–857.

Samorajski, T., Keefe, J. R. and Ordy, J. M. (1964) Intracellular localization of lipofuscin age pigments in the nervous system. *Journal of Experimental Gerontology*, **19**, 262–276.

Sotelo, C. and Palay, S. L. (1968) The fine structure of the lateral vestibular nucleus in the rat. I. Neurons and neuroglial cells. *Journal of Cellular Biology*, **36**, 151–179.

Toth, S. E. (1968) The origin of lipofuscin age pigment. *Journal of Experimental Gerontology*, **3**, 19–30.

Ylikoski, J. and House, W. F. (1980) Morphologic findings in the vestibular nerve from patients with vestibular neuritis. *American Journal of Otology*, **2**, 28–35.

Ylikoski, J. and Belal, A., Jr. (1981) Human vestibular nerve morphology after labyrinthectomy. *American Journal of Otolaryngology*, **2**, 81–93.

The Endolymphatic System

13
Ultrastructural Morphology of the Endolymphatic Duct and Sac

Per -G. Lundquist   Helge Rask-Andersen   Frank R. Galey   Dan Bagger-Sjöbäck

Introduction

In 1789 Antonio Scarpa first gave an account of the human inner ear as comprising two sacs and membranous semicircular ducts and he described their position in the petrous part of the temporal bone. However, one structure, the endolymphatic sac, protruded partially from the petrous bone in close relation to the sigmoid sinus at the external orifice of the vestibular aqueduct. This labyrinthine extension had already been discovered in 1761 by Domenico Cotugno, who described it as an intradural cavity.

It was Hasse, in 1873, who named these structures the endolymphatic duct and sac, but the first detailed description of their morphology and possible function was made by Guild in 1927. His so-called 'longitudinal flow theory' (Guild, 1927b), in which the endolymphatic sac is assumed to be active in phagocytosis and fluid absorption, has remained essentially valid.

The first detailed description of the ultrastructure of the endolymphatic sac was given by Lundquist in 1965. In the following presentation, the nomenclature is based on the findings of Guild and Lundquist.

Gross Anatomy

The endolymphatic duct extends from the junction of the saccular and utricular ducts, at the medial wall of the vestibule, and it ends halfway down the vestibular aqueduct, where it dilates to form the flat funnel-shaped endolymphatic sac. This can be described as being divided into three parts. The *proximal portion*, which is the first dilatation of the sac, is located entirely inside the vestibular aqueduct. The *intermediate* or *rugose portion*, which in most animals lies partly inside and partly outside the vestibular aqueduct, is characterized by extensive epithelial folds and crypts and the lumen is often filled with cells and cellular debris. In man, the rugose portion lies entirely within the bony vestibular aqueduct. The *distal portion* is flat and lies in close contact with the sigmoid sinus in the dural fold discovered by Cotugno (*Figure 13.1*).

Normal Ultrastructure of the Endolymphatic Duct and Sac (Guinea-pig)

Endolymphatic Duct

The epithelium is of simple squamous or low cuboidal type resting on a smooth basal lamina. The cytoplasm appears dense, with a few scattered ribosomes and occasional lipid granules. The sparse mitochondria appear evenly distributed throughout the cytoplasm. The cell contacts are loose, with few tight junctions, thus classifying the epithelium as 'leaky' (Bagger-Sjöbäck, Lundquist and Rask-Andersen, 1982). The luminal surface of the duct bulges slightly into it (owing to the nucleus) and possesses a few short microvilli projecting into the endolymph (*Figure 13.2*).

The basal lamina is continuous beneath the cells and the underlying areolar, connective-tissue layer contains some scanty collagen fibres. There are small thin-walled capillaries close to the epithelium. A detailed description

(continued on page 312)

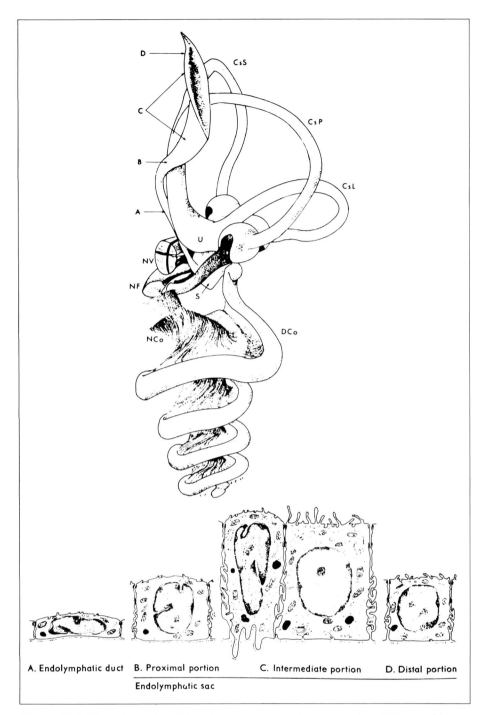

A. Endolymphatic duct   B. Proximal portion   C. Intermediate portion   D. Distal portion

Endolymphatic sac

Figure 13.1 *Schematic drawing of the endolymphatic duct and sac and the labyrinth from a posteriomedial view. The cells at the bottom of the picture illustrate the epithelial cell types found in the various parts of the endolymphatic duct and sac. CsL, CsP and CsS = Lateral, posterior and superior semicircular canals; DCo = cochlear duct; NCo = cochlear nerve; NF = facial nerve; NV = vestibular nerve; S = saccule; U = utricle. Reconstructed from an adult guinea-pig (right side) (Reproduced from Lundquist 1965, by kind permission of publishers.)*

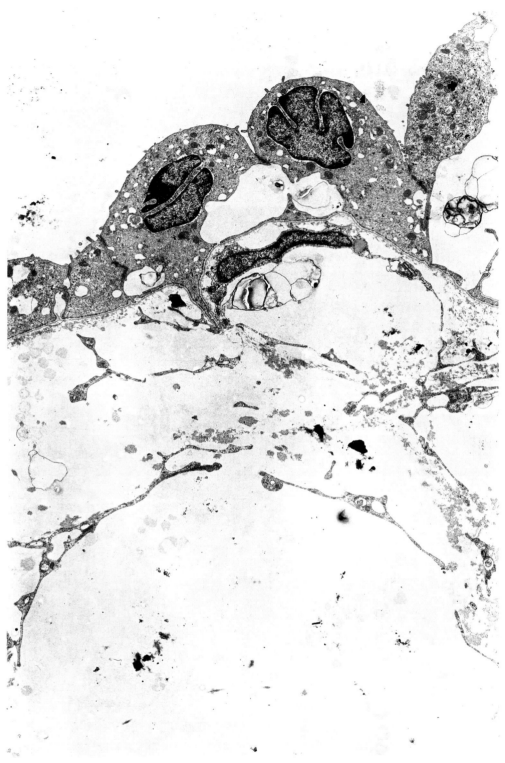

Figure 13.2 *Endolymphatic duct with low, almost squamous cells. The nuclei are irregular and the cells bulge slightly into the lumen. Surface activity is low. The epithelium is 'leaky', with very few junctional complexes. The connective tissue is of a very loose areolar type.* ×5500

of the vascular anatomy is given elsewhere (Rask-Andersen, 1979).

Endolymphatic Sac

Proximal Part

The gradually widening part of the sac exhibits a change in its epithelial structure compared with that of the duct, with the appearance of a more cuboidal type of cell. The cytoplasm is relatively dense and the nucleus is more rounded and regular than in the duct. More inclusion bodies, particularly lipid granules, are found in the cytoplasm (*Figure 13.3*).

The apical part of the cell membrane exhibits more microvilli in the sac than in the duct (*Figure 13.4*). Occasional small pinocytotic vesicles are present. The side walls of the cells are higher than in the duct, with many sealing strands of tight junctions.

The subepithelial tissue is more highly vascularized than it is in the duct. The capillaries lying close to the epithelial lining have, as in the intermediate part, pores in the endothelial walls. Collagen fibrils are more abundant than in the duct. In the lumen of the sac a population of so-called 'free-floating cells' (Iwata, 1924) is present, consisting of large vacuolated 'signet-ring'-like cells and leucocytes.

The Intermediate Part

The intermediate part constitutes most of the wide and irregular part of the sac, named *rugose* by Guild (1927a) owing to its heavily folded appearance.

The epithelial lining consists of cylindrical cells irregularly arranged in protruding papillae and deep crypts. With the electron microscope, two types of cells can be distinguished, the light cell and the dark cell, so called because of their different electron densities.

The light cells are more regular, with a rounded nucleus and many microvilli on their luminal surface. The dark cells have a more condensed cytoplasm and are often wedge shaped, with a narrow basal part and an irregular elongated nucleus (*Figure 13.5*).

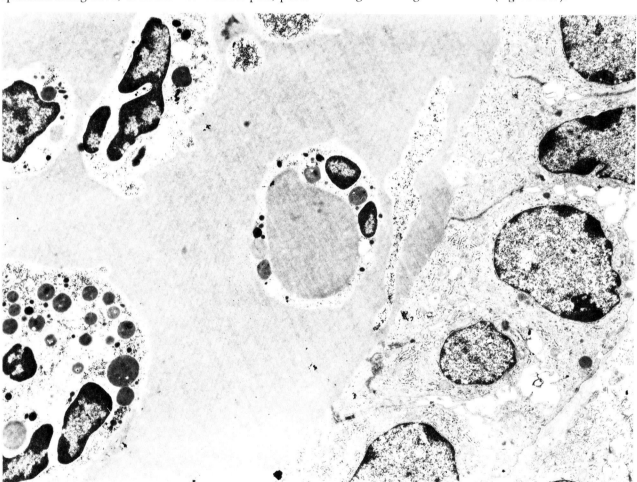

Figure 13.3 *In the proximal part of the endolymphatic sac the epithelium grows in height and becomes cuboidal. The nuclei are smooth and rounded. The pale cytoplasm contains scattered ribosomes and mitochondria. The basement membrane (basal lamina) is intact.*

There are a few junctional complexes between the cells. The lumen of the sac contains 'free-floating' macrophages, some of signet-ring pattern. Original magnification × 5500 reduced to 95%

Figure 13.4 *Freeze-fracture of the border-line between two cells of the proximal part of the endolymphatic sac. At the top of the picture impressions of short microvilli are seen. The thin lines illustrate the junctional strands. At the most, three parallel bands are present.* ×90 000

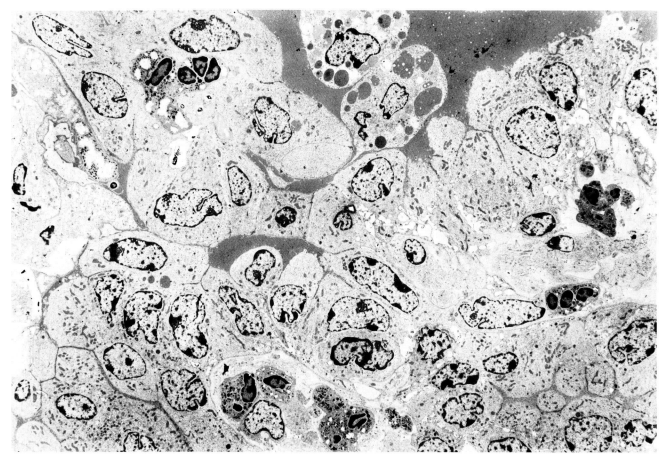

Figure 13.5 *In the intermediate part of the sac the surface area of the epithelial lining is greatly increased by numerous folds with protruding papillae and interconnecting crypts. The lumen is thus divided into channels, sometimes filled with debris. (Main lumen at top of picture, with enclosed fluid cavities centre and left.) The epithelium is cuboidal or cylindrical and constituted of light and dark cells with a dominance of light cells. The light cells exhibit a smooth, rounded nucleus and many microvilli at the fluid surface. The dark cell is denser with an irregular nucleus and does not exhibit many microvilli. In the subepithelial connective tissue leucocytes and fibroblasts are present. Original magnification ×2300 reduced to 88%*

The light-cell cytoplasm is pale, with abundant ribosomes and scattered mitochondria. The dark-cell cytoplasm is dense and has a fibrillary appearance. Both cell types are rich in mitochondria, particularly the light cells.

The apical surfaces differ in that the light cells have more microvilli than the dark cells at the fluid surface, and many pinocytotic vesicles. The neighbouring cell membranes are connected by several strands of tight junctions, thus defining the epithelium as 'tight' (*Figure 13.6*) (Bagger-Sjöbäck, Lundquist and Rask-Andersen, 1982). The basal parts of the interconnecting cell membranes often interdigitate with one another. The dark cells may appear wrinkled in their basal portion, so that small cytoplasmic papillae protrude into the subepithelial tissue; the basal portion of the light cell is more regular, although invaginations and small vesicles may often be present.

The complex folds of the epithelial lining may occasionally give the impression that there is no lumen. On closer scrutiny, however, leucocytes, cellular debris and free-floating macrophages may be found in the narrow fluid channels between the epithelial cells (*see Figure 13.5*).

The areolar connective tissue has a rich capillary network in close proximity to the epithelial lining. The capillaries may possess pores (Lundquist, 1965). Irregular tight junctions may be present between the endothelial cells, indicating a profound leakiness. Signs of high micropinocytotic activity are usually present in the endothelial cells.

There are thin-walled lymph vessels surrounded by fibroblasts and connective tissue (*Figure 13.7*) close to the side of the sac.

Distal Part

This is the flat, terminal part of the sac; in the animal, it often lies partly on the surface of the sigmoid sinus and always outside the vestibular aqueduct. The epithelial lining is cuboidal, except at its extremity, where it is squamous (*Figure 13.8*).

The cytoplasm resembles that of the cells of the intermediate portion, but the light cells predominate

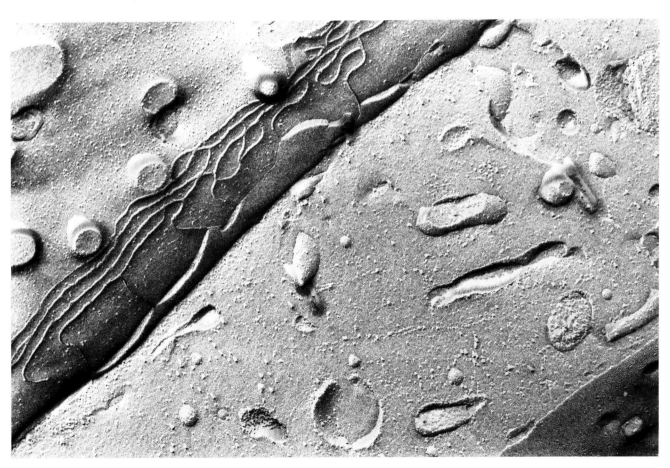

Figure 13.6 *Freeze-fracture of cell borders in the intermediate part of the endolymphatic sac; this shows several interconnecting strands, indicating a tighter, non-leaky, epithelium. About five strands are usually present. Original magnification × 70 000 reduced to 88%*

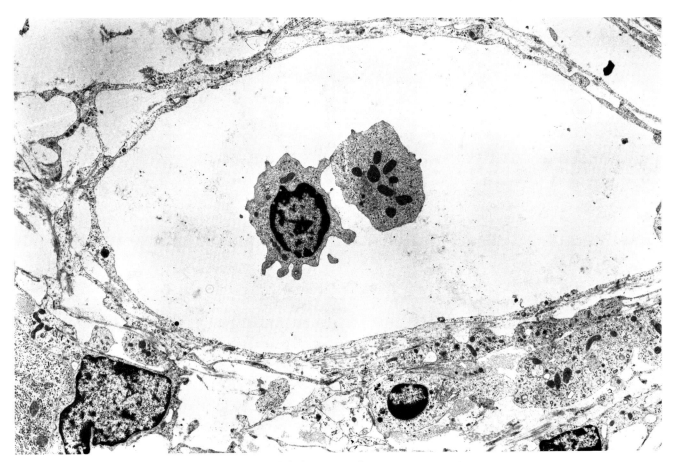

Figure 13.7 *The areolar connective tissue beneath the epithelial lining of the endolymphatic sac contains many thin-walled capillaries, as illustrated, with the features of lymphatic vessels with lymphoid cells present in the lumen. Original magnification × 5200 reduced to 88%*

here. The cell surface exhibits fewer microvilli and pinocytotic vesicles. The neighbouring cell borders display several rows of junctional complexes, usually with about six sealing strands, forming an apparently tight non-leaky epithelium (*Figure 13.9*) (Bagger-Sjöbäck, Lundquist and Rask-Andersen, 1982).

The connective tissue of the distal part is richly vascularized by a network of capillaries. In the extreme distal part, the connective tissue surrounding the sac is denser and it blends completely with that around the sigmoid sinus.

The Ultrastructure of the Endolymphatic Duct and Sac in Man

The ultrastructure of the endolymphatic sac in man was first described by Schindler in 1980, and then by Lundquist and co-workers in 1983. Essentially the epithelial lining resembles that of other animals. However, even though the light-cell type is dominant in the rugose part of the sac, there is much more variation in shape and cytoplasmic density than in the lower animals. Thus, it is sometimes difficult to classify the cells into the basic light and dark types. Furthermore,

the rugose part is so profusely folded that the lumen may appear to consist of interconnecting canaliculi, as suggested by Antunez *et al.* (1980) and by Linthicum and Galey (1981), based on computer-aided reconstruction of the endolymphatic sac in human temporal bones. For comparison, some typical features of the human sac are presented in *Figures 13.10–13.14*.

The endolymphatic duct exhibits the same features as in the lower animals, with squamous-to-cuboidal cells, mostly of the light type. In man there is little morphological distinction between the end of the endolymphatic duct and the proximal part of the sac (*Figure 13.10*).

Intermediate (Rugose) Part

The most striking features of this part are the projecting papillae and rugose folds of the surface epithelium. The single sheet of lining cells varies from cuboidal to columnar but, owing to the heavy folds, a pseudostratified appearance sometimes exists. With scanning electron microscopy, the differences between light and dark cells are demonstrated with respect to surface morphology, with a distinctive content of microvilliform processes on the former type (*Figure 13.11*). In this part, the height distribution of the epithelial cells varies but

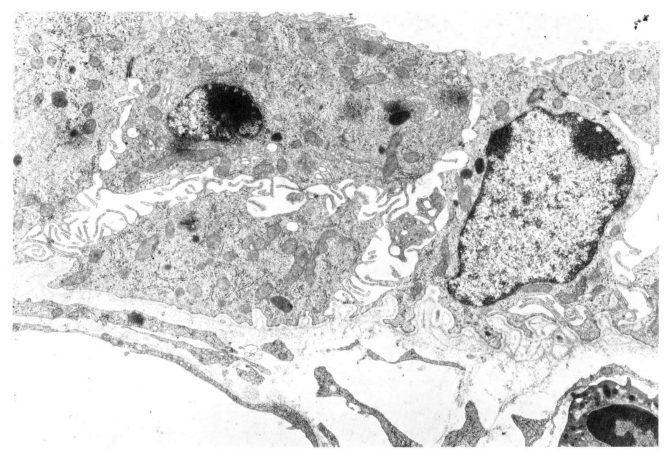

Figure 13.8 *In the distal part of the endolymphatic sac the epithelium reverts to the cuboidal type. The nuclei become spherical. The interdigitating cell borders are interconnected by several tight junctions. The surface activity appears to be low. Original magnification ×7500 reduced to 88%*

generally the columnar cell predominates. The cellular configuration can be seen to vary from broad cylindrical to narrow wedge-shaped types. Lipid granules are often found in the cytoplasm, as well as apical pinocytotic vesicles and inclusion vacuoles (*Figure 13.12*).

The cellular junctions are similar to those of the guinea-pig and sometimes the typical infoldings of neighbouring cell membranes (as described in the guinea-pig) are also present, as well as the distinctive appearances of the light- and dark-cell cytoplasm (*Figure 13.13*).

Distal Part

The distal part, extending from the orifice of the vestibular aqueduct, is usually very thin and flat. The cells are cuboidal-to-squamous in type and similar to those of the guinea-pig. The lumen is often very thin, so that the two epithelial layers are in close contact.

The subepithelial connective tissue is much denser than in the animal, reflecting a firmer dural attachment (*Figure 13.14*).

(*continued on page 324*)

Figure 13.9 *Typical freeze-etched appearance of cell borders of the distal part of the endolymphatic sac. Here the cell borders are very tight, sometimes with up to ten or more junctional strands.* ×90 000

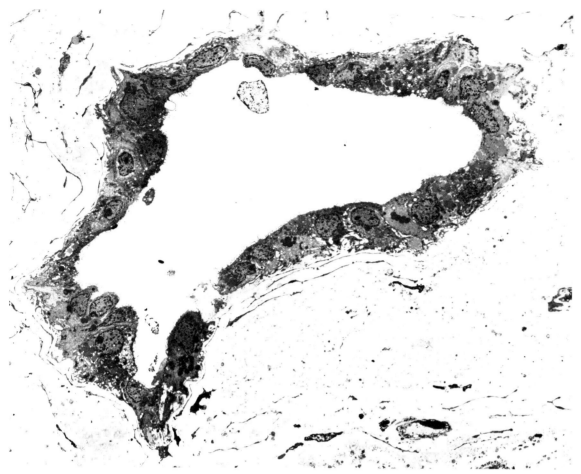

Figure 13.10 *The human endolymphatic duct exhibits an epithelium similar to that described in the lower animals. However, the transition to the proximal portion of the endolymphatic sac is not as clearly marked as in other animals. Even the distal part of the duct contains the characteristic cell types, i.e. wedge-shaped darker and cylindrical lighter cells, the latter with many microvilli, the former with a smoother surface. ×1600*

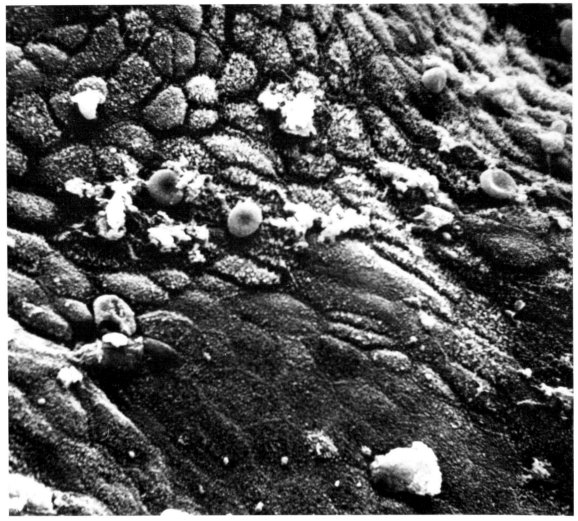

Figure 13.11 *Scanning electron microscopy clearly illustrates how one type of cell exhibits a distinct, slightly bulging pinocytotic fluid surface (top), while the other cell type is smoother, with a fluid surface devoid of any sign of activity (bottom). Cellular debris and blood corpuscles are often found attached to the cell surfaces.* ×1600

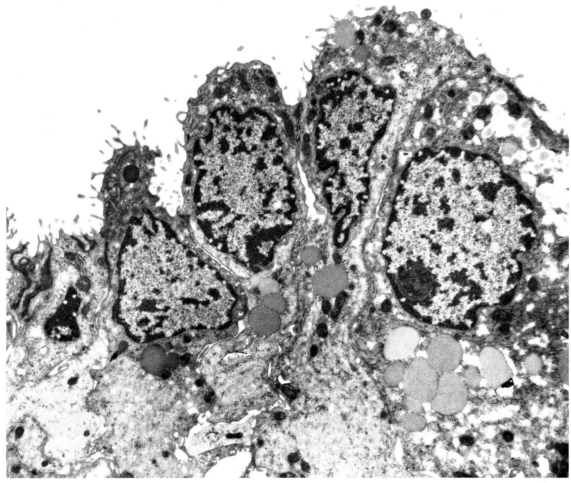

Figure 13.12 *Typical appearance of human endolymphatic sac (intermediate, rugose portion). In contrast to the lower animals, the cytoplasmic differences between the light and dark cells are not so clear. As illustrated, neighbouring cells can vary from wedge shaped to cylindrical. In the cytoplasm many vacuoles and lipofuscin granules are found.* ×9900

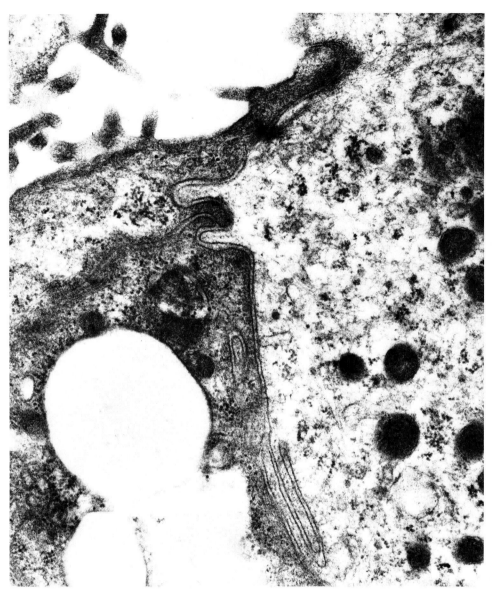

Figure 13.13 *Sometimes the conspicuous difference between light and dark cells, as described in the animal, can be recognized in man. Here a typical dark-cell cytoplasm, with lots of fibrils and a dense network of microtubules, is seen (left); and a typical light cell, with clusters of ribosomes, few fibrils and scattered microchondria, is also seen (right). Note the interdigitating cell borders at the bottom of the picture and the tight junctions at the top.* ×42 000

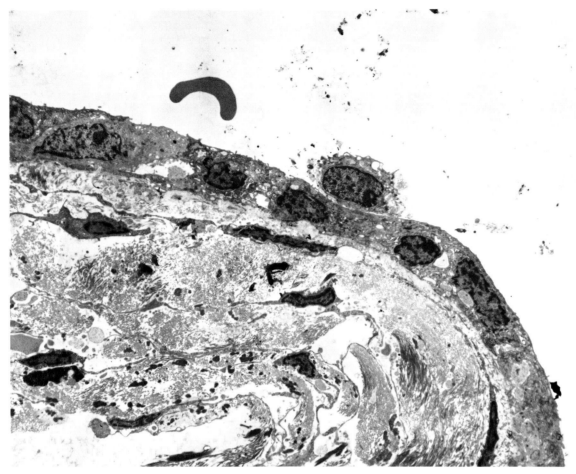

Figure 13.14 *The distal part of the endolymphatic sac protrudes from the orifice of the vestibular aqueduct and is enclosed in a dural sheath. Here the epithelium changes into a squamous type, with low cells with very few microvilli. Signs of surface activity are absent. In man the connective tissue is denser in character and changes directly into a dural type with abundant collagen fibrils (bottom).* ×3400

Discussion

The endolymphatic sac is differentiated from the superior part of the developing labyrinth and contains highly specialized epithelial cells of non-sensory origin. It has been suggested by many authors that it is of great importance in the metabolism and defence mechanisms of the inner ear (Iwata, 1924; Guild, 1927; Lundquist, 1965). The ultrastructural features are consistent with the assumption that the epithelium of the sac may be active in fluid transport and phagocytosis (Lundquist *et al.*, 1983).

Two principal theories have been suggested with regard to the microcirculation of the inner-ear fluids: the first was proposed by Corti, and subsequently by Guild. They suggested that endolymph was formed by the secretory epithelia in the cochlear and vestibular partitions and that the endolymph was reabsorbed in the endolymphatic sac; studies on ionic gradients by Naftalin and Harrison (1958) and by Lawrence, Wolsky and Litton (1961) proposed instead that the endolymph was formed and reabsorbed locally, so that a radial flow might exist.

The two conflicting concepts of endolymph circulation have been linked, however, by Lundquist (1976) in his 'dynamic flow theory' in which it is proposed that energy and ionic metabolism are maintained in a normal state by a radial mechanism, while a slow longitudinal flow towards the endolymphatic sac may help to remove high-molecular-weight waste products and cellular debris.

With the help of freeze-fracture techniques, Bagger-Sjöbäck and co-workers (1982) described the intercellular junctions in the epithelium of the endolymphatic sac. They found that the duct and the proximal part of the sac exhibit relatively few continuous parallel sealing strands of tight junctions around the cells, whereas in the intermediate and distal parts of the sac several rows of junctional strands are present.

These findings also suggest that, with regard to fluid absorption, the endolymphatic duct and sac exhibit a capacity for diffusion of fluid and ions in the duct and proximal part of the sac, owing to the fact that the epithelium here is 'leaky'. In the intermediate (rugose) part, on the other hand, the epithelium is linked by several parallel sealing strands of tight junctions, giving the impression of a 'tight' epithelium. This suggests the presence of a barrier against the passive flow of ions and macromolecules due to diffusion. On the other hand, the conspicuous signs of surface activity, with microvilli and pinocytotic vesicles and vacuoles, indicate that active reabsorption occurs here. Schindler, Lundquist and Morrison (1974) destroyed the vestibular labyrinth unilaterally in guinea-pigs by cryosurgical trauma and 2–4 days later they were able to demonstrate markedly increased activity in the endolymphatic sac on the operated side; signs of increased fluid transport, with widened intercellular spaces and loss of junctional complexes, were also observed.

Thus the normal reabsorption of endolymph may occur mainly in the leaky duct and in the proximal part of the endolymphatic sac, where lymph channels are also present; and labyrinthine stress-induced activity may occur in the intermediate part of the sac, in which either pinocytotic or macrophagic activity can be induced by exogenous or endogenous trauma. Digestion of cellular debris is probably performed by the free-floating macrophages, which could explain the high protein content demonstrated in the sac. Failure of the intermediate part of the sac to react to changes in the endolymph content could cause an accumulation of high-molecular-weight debris, thus inducing a hyperosmotic state in the endolymph and giving rise to a secondary 'hydrops', as has been demonstrated in animal experiments by Kimura and co-workers (Kimura, 1967; Kimura and Schuknecht, 1965).

The probable role of the endolymphatic sac in the aetiology of Ménière's disease cannot be ascertained from morphological data; and, so far, ultrastructural analysis of sac specimens from patients with Ménière's disease has not demonstrated any marked differences from the findings usually noted in human biopsy samples. This may be due largely to the fact that the specimens have been taken from the intradural distal part of the sac, and not from the rugose part. In man, this is located entirely within the vestibular aqueduct, a most important difference compared with the findings in lower animals. When the intra-osseous part of the human endolymphatic sac is also removed, high cylindrical cells (indicative of metabolic activity) are always present. The greatly enlarged surface area, as demonstrated by computer reconstruction, is also indicative of a special and important role of the endolymphatic sac in endolymph circulation.

From a morphological point of view, the suggestion (which has sometimes been made) that the endolymphatic sac is only vestigial in nature must be discarded. The highly differentiated epithelial cells present in the intermediate (rugose) part of the sac, together with the marked differences in junctional characteristics in its different parts, must be important for maintaining the inner-ear fluid in a normal state.

References

Antunez, J. -C. M., Galey, F. R., Linthicum, F. H. and McCann, G. D. (1980) Computer-aided and graphic reconstruction of the human endolymphatic duct and sac. *Annals of Otology, Rhinology and Laryngology*, Supplement **76**, 23–32.

Bagger-Sjöbäck, D., Lundquist, P-G. and Rask-Andersen, H. (1982) Intercellular junctions in the epithelium of the endolymphatic sac: a freeze-fracture and TEM study on the guinea pig's labyrinth. In *Ménière's Disease*, eds. Vosteen, Morgenstern, Schuknecht and Wersäll, pp. 127–140. Stuttgart: Georg Thieme Verlag.

Cotugno, D. (1761) *De Aquaeductibus Auris Humanae Internae*. Simoniana: Neapoli.

Guild, S. R. (1927a) Observations upon the structure and normal contents of the ductus and saccus endolymphaticus in the guinea pig (*Cavia cobaya*). *American Journal of Anatomy*, **39**, 1–56.

Guild, S. R. (1927b) The circulation of the endolymph. *American Journal of Anatomy*, **39**, 57–81.

Hasse, S. (1873) Die Lymphbahnen des inneren Ohres. *Anat. Studien. Bd.*, **1**, 765.

Iwata, N. (1924) Über das Labyrinth der Fledermaus mit besonder Berucksichtigung des statischen Apparates. *Aichi Journal of Experimental Medicine*, **1**, 41–173.

Kimura, R. S. (1967) Experimental blockage of the endolymphatic sac and duct and its effect in the inner ear of the guinea pig. A study of endolymphatic hydrops. *Annals of Otology, Rhinology and Laryngology*, **76**, 664–688.

Kimura, R. S. and Schuknecht, H. F. (1965) Membranous hydrops in the inner ear of the guinea pig after obliteration of the endolymphatic sac. *Practica Oto-Rhino-Laryngologica*, **27**, 343–354.

Lawrence, M., Wolsky, D. and Litton, W. B. (1961) Circulation of the inner ear fluids. *Annals of Otology, Rhinology and Laryngology*, **70**, 753–756.

Linthicum, F. H., Jr. and Galey, F. R. (1981) Computer-aided reconstruction of the endolymphatic sac. *Acta Otolaryngologica*, **91**, 423–429.

Lundquist, P. -G. (1965) The endolymphatic duct and sac in the guinea pig. An electron microscopic and experimental investigation. Thesis. *Acta Otolaryngologica*, Supplement **201**, 1–108.

Lundquist, P-G. (1976) Aspects on endolymphatic sac morphology and function. *Archives of Oto-Rhino-Laryngology*, **212**, 231–240.

Lundquist, P-G., Bagger-Sjöbäck, D., Galey, F. and Ylikoski, J. (1983) Aspects of the ultrastructure of the human endolymphatic sac – with discussion on comparative anatomy. *American Journal of Otolaryngology* (In press).

Naftalin, L. and Harrison, M. S. (1958) Circulation of labyrinthine fluids. *Journal of Laryngology and Otology*, **72**, 118.

Rask-Andersen, H. (1979) The vascular supply of the endolymphatic sac. *Acta Otolaryngologica*, **88**, 315–327.

Rask-Andersen, H. and Stahle, J. (1979) Lymphocyte–macrophage activity in the endolymphatic sac. *Oto-rhino-laryngology*, **41**, 177–192.

Scarpa, A. (1789) *Anatomicae Disquisitiones de Auditu et Olfactu*. Ticini: P. Galeatius.

Schindler, R. A. (1980) The ultrastructure of the endolymphatic sac in man. *Laryngoscope*, Supplement, **21**.

Schindler, R. A., Lundquist, P-G. and Morrison, M. D. (1974) Endolymphatic sac response to cryosurgery of the lateral ampulla. *Annals of Otology, Rhinology and Laryngology*, **83**, 202–216.

Index